Lecture Notes in Mechanical Engineering

Lecture Notes in Mechanical Engineering (LNME) publishes the latest developments in Mechanical Engineering - quickly, informally and with high quality. Original research reported in proceedings and post-proceedings represents the core of LNME. Volumes published in LNME embrace all aspects, subfields and new challenges of mechanical engineering. Topics in the series include:

- Engineering Design
- Machinery and Machine Elements
- Mechanical Structures and Stress Analysis
- Automotive Engineering
- Engine Technology
- Aerospace Technology and Astronautics
- Nanotechnology and Microengineering
- Control, Robotics, Mechatronics
- MEMS
- Theoretical and Applied Mechanics
- Dynamical Systems, Control
- Fluid Mechanics
- Engineering Thermodynamics, Heat and Mass Transfer
- Manufacturing
- Precision Engineering, Instrumentation, Measurement
- Materials Engineering
- Tribology and Surface Technology

To submit a proposal or request further information, please contact the Springer Editor in your location:

China: Dr. Mengchu Huang at mengchu.huang@springer.com
India: Priya Vyas at priya.vyas@springer.com
Rest of Asia, Australia, New Zealand: Swati Meherishi at swati.meherishi@springer.com
All other countries: Dr. Leontina Di Cecco at Leontina.dicecco@springer.com

To submit a proposal for a monograph, please check our Springer Tracts in Mechanical Engineering at http://www.springer.com/series/11693 or contact Leontina.dicecco@springer.com

Indexed by SCOPUS. The books of the series are submitted for indexing to Web of Science.

More information about this series at http://www.springer.com/series/11236

Lihui Wang · Vidosav D. Majstorovic ·
Dimitris Mourtzis · Emanuele Carpanzano ·
Govanni Moroni · Luigi Maria Galantucci
Editors

Proceedings of 5th International Conference on the Industry 4.0 Model for Advanced Manufacturing

AMP 2020

Springer

Editors
Lihui Wang
Department of Production Engineering
Royal Institute of Technology
Stockholm, Sweden

Dimitris Mourtzis
Department of Mechanical Engineering and Aeronautics
University of Patras
Patras, Greece

Govanni Moroni
Department of Mechanical Engineering
Politecnico di Milano
Milan, Italy

Vidosav D. Majstorovic
Faculty of Mechanical Engineering
University of Belgrade
Belgrade, Serbia

Emanuele Carpanzano
Department of Innovative Technologies
University of Applied Sciences and Arts
Manno, Switzerland

Luigi Maria Galantucci
DMMM
Politecnico di Bari
Bari, Italy

ISSN 2195-4356 ISSN 2195-4364 (electronic)
Lecture Notes in Mechanical Engineering
ISBN 978-3-030-46211-6 ISBN 978-3-030-46212-3 (eBook)
https://doi.org/10.1007/978-3-030-46212-3

This Springer imprint is published by the registered company Springer Nature Switzerland AG
The registered company address is: Gewerbestrasse 11, 6330 Cham, Switzerland

Preface

The 5th International Conference on the Industry 4.0 model for Advanced Manufacturing, INDUSTRY 4.0 AND CLOUD MANUFACTURING—AMP 2020, will be held at University of Belgrade, Faculty of Mechanical Engineering, Belgrade, Serbia, from June 1 to 4, 2020. It is organized by the University of Belgrade, Faculty of Mechanical Engineering, and Belgrade Chamber of Commerce and Industry. This year's conference attracted more than 200 participants, including academics, practitioners and scientists from 23 countries, who contributed 34 papers (26 keynotes) on plenary and workshop sessions.

The previous conferences on the Industry 4.0 model for Advanced Manufacturing—AMP were as follows:

1. First International Conference USA-EU-Japan-Serbia Manufacturing Summit, Belgrade, May 31–June 2, 2016, Serbia (AMP Conference 2016), with main topic: Advanced Manufacturing Program—INDUSTRY 4.0 model for Serbia (AMP Conference 2017).
2. Second International Conference USA-EU-Japan-Serbia Manufacturing Summit, Belgrade, June 7–9, 2017, Serbia—Smart And Intelligent Products (AMP Conference 2017).
3. Third International Conference USA-EU-Japan-Serbia Manufacturing Summit, Belgrade, June 5–7, 2018, Serbia—INDUSTRY 4.0 for SMEs (AMP Conference 2018).
4. Four International Conference USA-EU-Japan-Serbia Manufacturing Summit, Belgrade, June 4–7, 2019, Serbia—INDUSTRY 4.0 and Internet of things for manufacturing (AMP Conference 2019).

The main objective of these conferences is to bring together leading world experts to discuss the challenges and opportunities of the new Industry 4.0 model of manufacturing. Our hope is that such an event will assist in the development and growth of new innovative manufacturing industries in Serbia, producing smart products with intelligent characteristics, and relying on modern, new manufacturing processes and systems.

Conference is hosted by the Faculty of Mechanical Engineering of the University of Belgrade. Belgrade is the capital city of Serbia, located at the scenic confluence of two major European rivers, with a uniquely remarkable and turbulent history and a vibrant cultural and entertainment scene. University of Belgrade has a long tradition of academic excellence, where great minds from Nikola Tesla and Mihajlo (Michael) Pupin to Milutin Milankovic held lectures or were faculties. Its engineering still remains exceptionally respected in Europe, with its alumni scattered in top universities around the globe. Faculty of mechanical engineering in Belgrade is the largest such school in south-eastern Europe and one of the largest in Europe.

Main topics of interest for this conference include:

- Industry 4.0 model framework
- Cloud manufacturing models
- Design of smart and Intelligent products
- Innovative design and development of intelligent products
- Internet of things for Manufacturing
- Big data challenges, data integrity, accuracy and authenticity
- Cloud computing, cloud-based products, cloud manufacturing
- Cyber-physical manufacturing
- Manufacturing automation in the Industry 4.0 model
- Manufacturing systems and enterprise models for Industry 4.0
- Advanced manufacturing
- Engineering education for Industry 4.0
- What we can to do?
- Roadmap for AM based on I4.0 model in Serbia.

We acknowledge the outstanding contributions of the following colleagues and friends for conference establishing and development as following:

Conference Founder and Honorary Chair

Prof. Dr. Jun Ni, Department of Mechanical Engineering, University of Michigan, Ann Arbor, MI, USA.

Conference Chair 2020

Prof. Dr. Lihui Wang, Royal Institute of Technology, Dept. Production Engineering Stockholm, Sweden.

Conference Co-chairs 2020

1. Prof. Dr. Jozsef Vancza, (Chair of the CIRP STC-O), SZTAKI, TU Budapest, Budapest, Hungary—for EU region;
2. Prof. Dr. Emanuele Carpanzano (CIRP Fellow), SUPSI–ISTePS, Manno, Switzerland—for EU region.
3. Prof. Dr. Luigi Maria Galantucci, (CIRP Fellow), Politecnico di Bari, Milano, Bari– for EU region.
4. Prof. Dr. Giovanni Moroni, (CIRP Fellow), Politecnico di Milano, Milano, Italy—for EU region.
5. Prof. Dr. Dimitris Mourtzis, (CIRP Fellow), University of Patras, Patras, Greece—for EU region.
6. Prof. Dr. Yashiro Takaya, Osaka University, Osaka, Japan—for Far East region.
7. Dr. Ömer S. Ganiyusufoglu, CIRP President CMAG, Chairman of Shenyang Machine Tool (Group) Co., Ltd., China—for Middle East region.
8. Prof. Dr. Marco Taisch, Politecnico di Milano, Milano, Italy—for EU region.
9. Prof. Dr. Dragan Djurdjanovic, Department of Mechanical Engineering, University of Texas, Austin, TX, USA—Conference co-founder.
10. Prof. Dr. Vidosav Majstorovic, Belgrade University, Faculty of Mechanical Engineering, Belgrade, Serbia – Conference co-founder.
11. Prof. Dr. Slavenko Stojadinovic, Belgrade University, Faculty of Mechanical Engineering, Belgrade, Serbia.

International Program Committee 2020

Prof. Dr. D. Aleksandric, MEF, Belgrade, Serbia; Dr. A. Archenti, KTH Royal Institute of Technology, Stockholm, Sweden; Prof. Dr. E. Budak, Sabanci University, Istanbul, Turkey; Dr. J. Caldeira, INESC TEC, Porto, Portugal; Prof. Dr. E. Carpanzano, Institute of Systems and Technologies for Sustainable Production, Switzerland; Prof. Dr. E. Chlebus, TU Wroclaw, Wroclaw, Poland; Prof. Marcello Colledani, Politechico de Milano, Milano, Italy; Prof. Dr. G. Constatntin, TU Bucharest, Romania; Prof. Dr. N. Durakbasa, TU Vienna, Austria; Prof. Dr. L. M. Galantucci, Politecnico di Bari, Bari, Italy; Prof. H. Hohonoki, Japan; Prof. Dr. S. J. Hu, The University of Michigan, Ann Arbor, Michigan, USA; Dr. L. Jalba, Microelectronics S. A., Romania; Prof. Dr. J. Jedrzejewski, Wroclaw University of Science and Technology, Poland; Prof. Dr. F. Jovane, Politechnico de Milano, Italy; Prof. Dr. A. Jović, University of Belgrade, Belgrade; Dr. Ö. S. Ganiyusufoglu, Chairman of Shenyang Machine Tool (Group) Co., Ltd., China; Prof. Dr. D. Kiritsis, Ecole Polytechnique Fédérale de Lausanne STI-IGM-LICP, Switzerland; Prof. Dr. S. Krile, University of Dubrovnik, Maritime Department, Dubrovnik, Croatia; Prof. Dr. Y. Koren, The University of Michigan, Ann Arbor, Michigan, USA; Prof. Dr. D. Kramar, University of Ljubljana, Slovenia; Prof. Dr. B. Lalić, FTS, Novi Sad, Serbia; Prof. Dr. B. Iung,

Lorraine University, Nancy, France; Prof. Dr. G. Lanza, Karlsruhe Institute of Technology, Germany; Prof. J. Lee, U of Cincinnati, Cincinati, USA; Prof. Dr. L. Zhongqin, President of Shanghai Jiao Tong University, Shanghai, China; Prof. Dr. Xi Lifeng, Vice president of Shanghai Jiao Tong University, Shanghai, China; Prof. Dr. M. Macchi, Politecnico di Milano, Italy; Prof. Dr. V. Majstorovic, BU–FME, Belgrade, Serbia; Prof. Dr. M. Milošević, FTS, Novi Sad, Serbia; Prof. Dr. R. Mitrović, University Belgrade; Prof. Dr. P. Monka, TU Presov, Slovakia; Prof. Dr. L. Monostori, TU Budapest, Hungary; Prof. Dr. G. Moroni, Politecnico di Milano, Milano, Italy; Prof. Dr. D. Mourtzis, University of Patras, Rio-Patras, Greece; Prof. Dr. J. Ni, MEF–Ann Arborn, Minchigan, USA; Dr Augusta Paci, CNR, Rome, Italy; Prof. Dr. M. Piska, TU Brno, Czech Republic; Prof. Dr. V. Patoglu, Faculty of Engineering and Natural Sciences, Sabanci University, Istanbul, Turkey; Prof. Dr. C. Pupaza, TU Bucharest, Romania; Prof. Dr. Y. Takaya, University of Osaka, Japan; Prof. Dr. T. Tolio, Politechico de Milano, Milano, Italy; Prof. Dr. D. Đuričin, BU—Faculty of Economics, Belgrade, Serbia; Prof. Dr. D. Đurđanović, The University of Texas at Austin, Austin, USA; Prof. Dr. A. Shih, University of Michigan's Mechanical Engineering, Ann Arbor, USA; Prof. Dr. G. Seliger, Production Center Berlin Institute of Machine Tools, Berlin, Germany; Prof. Dr. S. V. Sreenivasan, UT Austin, USA; Prof. Dr. M. Zimmermann, Technische Universität München, Germany; Prof. Dr. J. Vancza, TU Budapest, Hungary; Prof. Dr. L. Wang, KTH Stockholm, Sweden.

Organizing Committee 2020

Assist. Prof. Dr. Slavenko Stojadinovic, FME—University of Belgrade, Chair; Andrijana Sajko, Sava Sajko, Sofija Popovic, Sajko team, Belgrade.

AMP 2020 conference can be regarded as a leading global conference in the area of modern manufacturing to several of its special dimensions: (i) It presented a spectrum of scientific and practical advancements in the field of advanced manufacturing (Cyber-physical manufacturing, Industry 4.0), and (ii) it offered practical applications and solutions for various problems in the world of modern manufacturing.

Conference planning, preparation and realization required engagement of a number of persons and organizations. We express our gratitude to all of them, and especially to

- Founder, Chair, co-chairs and Conference International Programme Committee Members,
- All authors, and especially to the authors that prepared keynote papers thus contributing to the high scientific and professional level of the conference,
- All members of the International Programme Committee for the review of the papers and chairing the conference sessions,
- Springer and Mr. Pierpaolo Riva for publishing AMP conference proceedings within the edition Lecture Notes in mechanical engineering,

- Ministry of Education, Science and Technological Development of the Republic of Serbia for the support of the conference and
- Chamber of Commerce and Industry, Serbia, Belgrade, conference co-organizer.

We wish to express a special gratitude to all colleagues at the University of Belgrade, Faculty of Mechanical Engineering, for their invested efforts that enabled preparation and realization of the AMP conference in the possible best manner, especially to Sajko team for conference technical organizer.

March 2020

Lihui Wang
Conference Chair

Vidosav D. Majstorovic
Dragan Djurdjanovic
Conference Co-chairs

Contents

An Adaptive Scheduling Method Based on Cloud Technology: A Structural Steelwork Industry Case Study

Dimitris Mourtzis(✉), Antonis Gargallis, John Angelopoulos, and Nikos Panopoulos

Laboratory for Manufacturing Systems and Automation, Department of Mechanical Engineering and Aeronautics, University of Patras, 26504 Patras, Greece
mourtzis@lms.mech.upatras.gr

Abstract. Decision making at the shop floor level has become more complex than ever before due to the massive growth in available data. The increasing market demands, concerning product quality and delivery times, make critical judgement and decision-making crucial requirements of the modern manufacturing problems. Human decision making has become insufficient and struggles to achieve manufacturing goals. Cutting edge technologies like the Internet of Things (IoT) and Cyber Physical Systems (CPS), that are the cornerstones of the Industry 4.0 smart factories, can contribute to efficient decision making. Therefore, more accurate and improved critical decisions can be achieved for the current as well as for the future status of a manufacturing system. Furthermore, production scheduling is one of the main issues that engineers have to address. The decision support tools of the Industry 4.0 era contribute to effective production scheduling, while considering a larger amount of data and constraints than ever before. This research work proposes a production scheduling method, that uses past and near real-time data to check resource and task status, providing insight to production engineers and enabling enhanced decision making. The results are validated in a structural steelwork industry shop case study.

Keywords: Production scheduling · Production monitoring · Industry 4.0 · Decision support tools

1 Introduction

Since the advent of the fourth industrial revolution, the traditional manufacturing approach is entering the digitalization and Cloud-based connection era. The modernization of work towards Digital Manufacturing is considered as a top priority among enterprises in the industry. First of all, advanced connectivity has to be achieved in order to enable simple collaboration between different agents like personnel, machines and IT tools within a factory. This means that information can be distributed correctly, and decision making can be more efficient. Moreover, intelligence that is provided by tools such as data analytics, machine learning and digital twins has to be exploited in order to make use of the immense amount of data found in that environment. Finally,

L. Wang et al. (Eds.): AMP 2020, LNME, pp. 1–14, 2020.
https://doi.org/10.1007/978-3-030-46212-3_1

flexible automation tools have to be used to augment the workforce resulting in improved operational efficiency. As a result, manufacturing companies are able to keep up with the advances in the modern-day market and meet demands with greater ease. With all the benefits of Digital Manufacturing, the adoption of these new technologies is becoming more widespread every day across all the industrial sectors. The creation of pilot cases within the industry reaches 64% in the communication domain, 70% in the intelligence domain and 61% in flexible automation. Nevertheless, these pilot cases do not always evolve into full working systems, meaning that the actual numbers of digitalization in the industry are much lower [1]. Although the general concept of Digital Manufacturing is ubiquitous between the manufacturing industry, the implementation methods vary considerably depending on the nature of each manufacturing system. This means that there is no certain roadmap for industry digitalization. Every sector has its differences and every factory setting has different Digital Manufacturing capabilities.

Consequently, each case has to be analyzed and examined independently in order to meet the different needs of the organization [2]. The digitalization of the factory environment can be achieved with the merge of the shop floor systems with Information Technology (IT) solutions in order to constitute possible shop floor awareness [3]. A key factor to enable smart decision-making is the remote near real-time monitoring of the whole factory. This can not only support adaptive control within the factory, but it can also help to bridge the gap in information distribution. The structure of the Interconnected Factory is presented in Fig. 1.

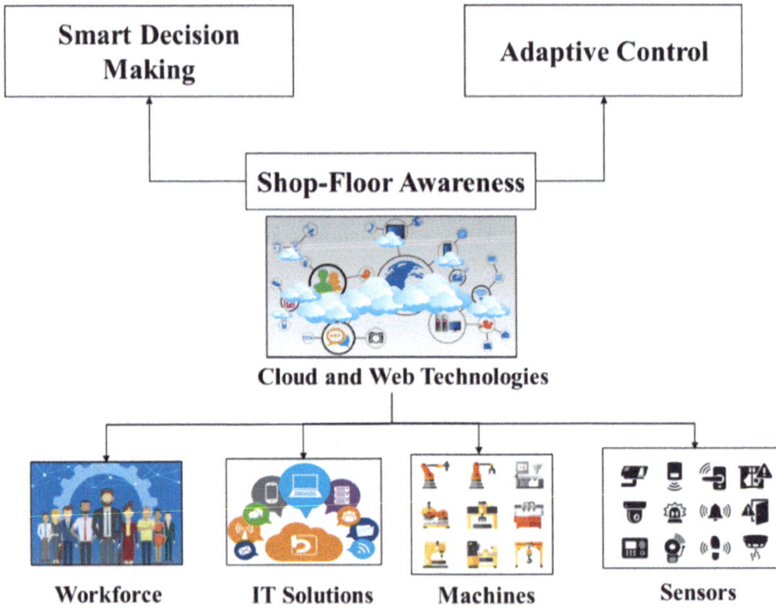

Fig. 1. Interconnected factory

2 Literature Review

Scheduling is a decision-making process that is used on a regular basis in many manufacturing industries. Concretely, scheduling deals with the allocation of resources to tasks over given time periods aiming at the optimization of one or more manufacturing objectives [4]. Production scheduling is one of the main decision-making functions that takes place in most of the production environments where certain tasks have to be assigned and processed by specified resources [5, 6]. Moreover, the significance of the enterprise systems and simulation integration in improving shop floor's short-term production planning capability have been addressed in [7]. Next, an approach to support the management of a ship repair yard by integrating into an open and flexible system a number of critical business functions with production planning, scheduling, and control is described in [8]. Last but not least, a networked Virtual manufacturing system, which was composed of 3D dimensional CAD/CAM modelling, VR, production planning and control and simulation, to address knowledge management issues has been proposed in [9]. The flow of information in manufacturing systems can be shown in Fig. 2.

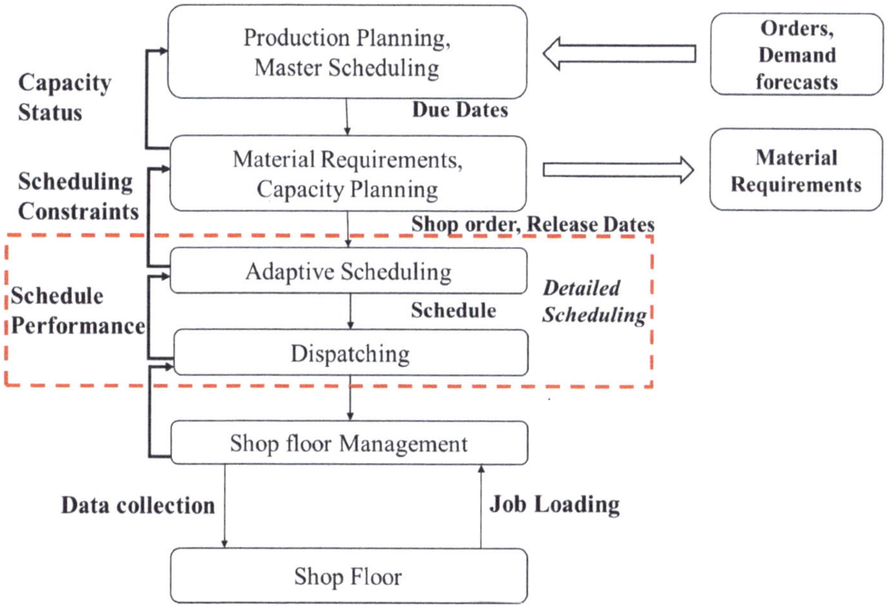

Fig. 2. Information flow diagram in a manufacturing system (Adopted from [4])

A characteristic definition for Manufacturing Systems was coined by Chryssolouris (2006), describing MFG systems as "the combination of humans, machinery and equipment that are bound by a common material and information flow". Indeed, scheduling is of great importance for the design, control and operation of MFG systems [10]. What is more, the main scheduling issues that engineers have to deal with, can be summarized to: i) the requirement generation, ii) the processing complexity, iii) the scheduling criteria, and iv) the scheduling environment. A common technique in scheduling modeling, is the provision of energy-efficient production plans by incorporating machine reliability [10]. As such, Chen et al. (2019) based on the ant-colony optimization method, have developed a mathematical programming model aiming at minimizing tardiness cost as well as energy cost [11]. Moreover, a comprehensive review of the state-of-the-art and new trends, as well as the major historical milestones in the development of simulation engineering for manufacturing systems and recent approaches to business and study in key industrial areas, is presented in [13]. Furthermore, an adaptive decision-making system that supports distributed execution control and emergent scheduling is analyzed in [12].

Supervisory Control and Data Acquisition (SCADA) is the technology that enables data collection from one or more distant facilities of the production line, while sending limited control instructions to those facilities [14]. A methodology for deriving main scheduling decisions is introduced for the exploitation of machine information in near real-time machine condition monitoring [15]. Moreover, a model was implemented in the form of an Internet-enabled software framework, offering a set of characteristics including virtual organization, scheduling, and monitoring, in order to support cooperation, flexible planning and monitoring across extended manufacturing enterprises [16]. In addition to that, the application of the simulation approach to a supplier selection example is described in [17]. Wang (2013) developed a web-based service-oriented CAPP system based on function blocks for machine availability monitoring and process planning, supporting Cloud manufacturing [18]. In the last few years, digital twin has arisen as a technology that combines data-driven real-time monitoring of real entities with scenarios simulation, aiming to improve its functionality and maximize its efficiency [19]. Digital twin is not solely focused on one entity but may be expanded to whole systems, thus creating a holistic approach for the digitalized factories of the future.

The preferred approach in tackling scheduling problems is mainly approximating good solutions that work and provide production environments within known error margins. Towards, the globalization paradigm, an investigation of the performance of decentralized manufacturing networks through the Tabu Search and Simulated meta-heuristic methods in conjunction with the Artificial Intelligence method is analyzed in [20]. Following the shift towards high demand and highly customized products, an intelligent method that uses three variable control parameters and can be used to define secure, globalized production network configurations capable of manufacturing mass-customized goods are described in [15].

The growing need for higher product customization in combination with uncertainty about demand needs efficient ways to design manufacturing network configurations. A tool to help decision-making in realistic manufacturing network design issues that examine the reliability and viability of centralized and decentralized production networks under intense product customization is presented in [21]. The product complexity, especially in highly personalized markets affects the overall performance of the manufacturing systems. To address the challenge of high flexibility, production scheduling is a vital part of a decision-making process.

3 Proposed Scheduling Method

This research work aims to create an approach to adaptive scheduling based on the production scheduling problem. The main issues of this approach concern the type of production sequence, the products and the resources [22]. The proposed method takes into account specific problems encountered in this type of industrial scenario and uses tools provided by digital manufacturing methods. The proposed scheduling method presented, is used in order to lever these advantages with the scope of creating high performance schedules within the concept of digital manufacturing, and has already been validated in previous publications [3, 10, 23]. In the core of the scheduling method, lies an Intelligent Search Algorithm (ISA). The three control parameters used, are the Maximum Number of Alternatives (MNA), the Decision Horizon (DH) and the Sampling Rate (SR). These parameters can be adjusted according to the needs of the manufacturing system and when selected correctly can help the ISA to identify good solutions quicker and more effectively. More analytically, MNA Controls the number of alternatives that are going to be evaluated, thus setting the width of the search. The DH constitutes the depth of the exhaustive search of the algorithm, defining the depth of time in which there are jobs assigned to resources. Finally, at the end of the search, there can be still unassigned jobs that are left out of resources due to the MNA – DH combination. Then, SR controls the number of alternatives that is created by randomly attaching the rest of the available jobs to each alternative previously created. For each alternative created by the MNA – DH combination, a utility score is derived. This utility score is based on the weights of the different criteria assigned and is created as a mean of all further alternatives created by the SR parameter for each initial alternative. Then, a decision matrix is formed, as illustrated in Fig. 3, with all the utility scores of the alternatives, thus finally using the schedule alternative with the highest utility.

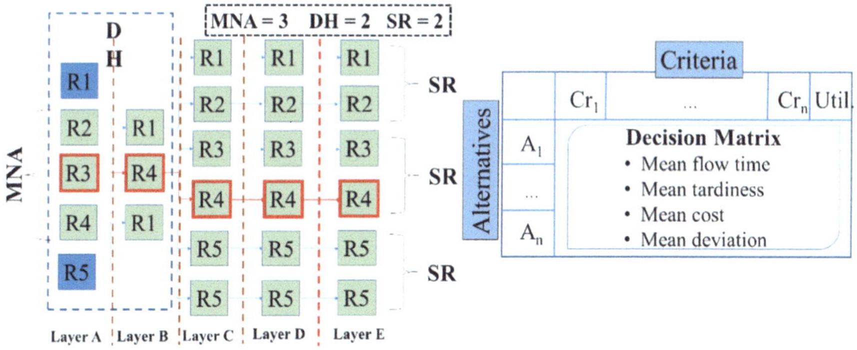

Fig. 3. Multivariable scheduling method (Adopted from [6, 23])

A significant benefit of the proposed algorithm is the fast and accurate schedule adaptation according to the job shop availability and new order arrivals. When the creation of a new schedule is needed due to the arrival of new jobs to be performed, the jobs that are being executed currently stay fixed, while the new tasks are mixed in a pool with the rest of the tasks that are pending and exist in the previous schedule. Then, rescheduling occurs taking into account this new sum of pending tasks. With this functionality, the shop floor can process jobs more efficiently by adapting to uncertainty factors such as unforeseen demand, rush orders and disruptions in the factory setting such as sudden changes in resource availability. Lastly, the flowchart of the proposed method is presented in Fig. 4.

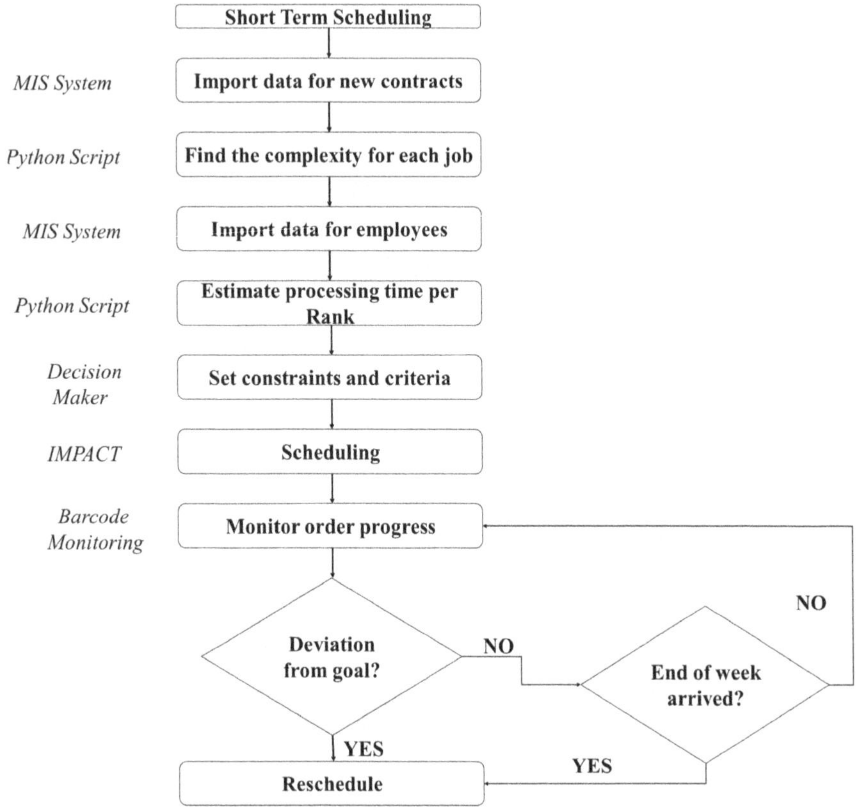

Fig. 4. Flowchart of proposed method

4 A Structural Steelwork Industry Case Study

The structural steelwork industries provide a huge application area for Industry 4.0 challenges. Current practices in structural steelwork companies have bottlenecks, and therefore there is a need for the integration of new technologies. Moreover, there is a need for unique and personalized products. This fact strongly complicates the production automation. The shop floor is divided into multiple sectors that are based on different workshops and job shops. The raw material is usually steel based parts such as beams and sheet metal. These parts are processed by cutting and holing machines. Afterwards, the parts are processed by blasting machines to provide clean surfaces, in order to begin the main fabrication. The main fabrication processes that are used for the production of steel structures are fit-up and welding. First, the fitters assemble the parts to form the metal structures. Then, the parts are welded together using GMAW welding and stud welding techniques. Finally, the structures are quality tested to ensure minimal structural flaws exist in the final products. The general workflow diagram of the steel shop floor is depicted in Fig. 5.

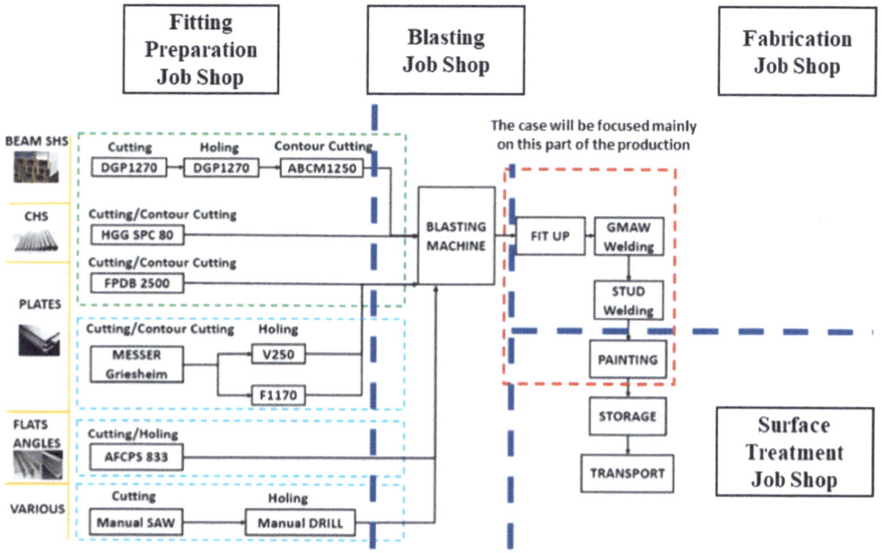

Fig. 5. Workflow diagram of the structural steelwork shop floor

4.1 Complexity Calculation and Task Duration Estimation

The complexity index (CI) of an assembly is calculated using three parameters namely the assembly length, the assembly weight, and the number of fittings. Moreover, a base value must be added to account for factors outside assembly features. The different weights in the complexity calculation equation need to be set after the application of the proposed model and be tuned empirically to the needs of each shop floor. The equation for calculating the complexity index for each assembly is presented below.

$$Cl = \frac{W_{\alpha l} \cdot A_L + W_{\alpha w} \cdot A_w + W_{nf} \cdot NF}{W_{\alpha l} + W_{\alpha w} + W_{nf}} + BC \quad (1)$$

Where, $W_{\alpha l}$, W_{aw}, W_{nf}, are the weights for the assembly length, the assembly weight, and the number of fittings respectively. The A_L, A_W, NF are the assembly length, the assembly weight, and the number of fittings respectively. The Base Complexity (BC) is the complexity that depends on the assembly type. The n denotes the total number of parts that the operator manufactured. Hence, by dividing the total working hours of the operator to each assembly based on complexity, the Assembly Production Time (APT_i) can be more accurately estimated. This is justified by the fact that the production time is affected by various parameters, such as product preparation, and setup. The processing time of a task equals the complexity of a given task, divided by the productivity in complexity per hour of the resource that fulfills that given task (Eq. 2). The output is the time it takes for a certain resource to finish a certain task.

$$PT = \frac{Task\ Complexity}{Resource\ Productivity} \quad (2)$$

Although the complexity of each task given the weights in the complexity formula is set, productivity varies heavily between human resources and even for the same resource during a given time frame. For this reason, the productivity of human resources has to be modeled and set to certain averages, in order for the algorithm to produce reliable schedules with a good approximation of the real-life shop floor.

4.2 Monitoring and Rescheduling Through Product Complexity

For the purpose of this case study, a monitoring and estimating processing times method is introduced, through product complexity. Furthermore, with proper monitoring and processing time estimation (Fig. 6), rescheduling can be triggered in order to reshape schedules according to the needs of the shop floor in order to be able to cope with given goals or system failures and maintenance. This method aims to solve the issue of shop floor monitoring and processing time estimation by proposing a method that will retrieve information during the fabrication (welding and fit-up processes) of the assemblies.

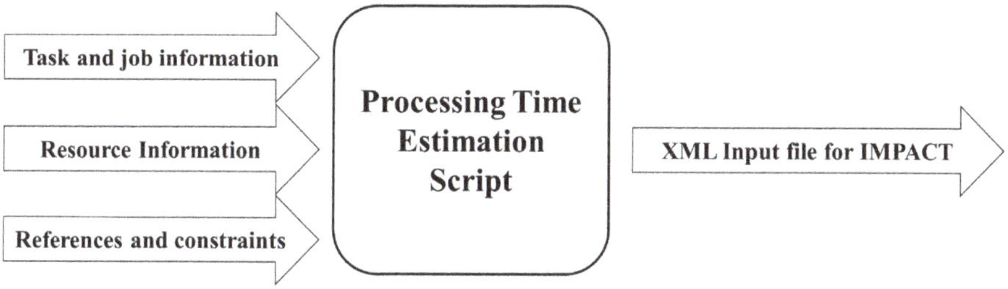

Fig. 6. Processing estimation script input/output diagram

5 Simulation Results

The parameters for the scheduling algorithm are SR = 3, DH = 2 and MNA = 1000. Moreover, the criteria used for the ISA are the mean flow time and mean tardiness. The jobs and the resources in the scheduler UI are presented in Fig. 7 and a resulted FIFO Gantt Chart is illustrated in Fig. 8 (Mean Flow Time = 40.4 h and Mean Tardiness = 4.9 h). The available scheduling methods are as follows:

- EDD (Earliest Due Date)
- FIFO (First In First Out)
- SPT (Shortest Processing Time)
- ISA – Multicriteria (Intelligent Search Algorithm)

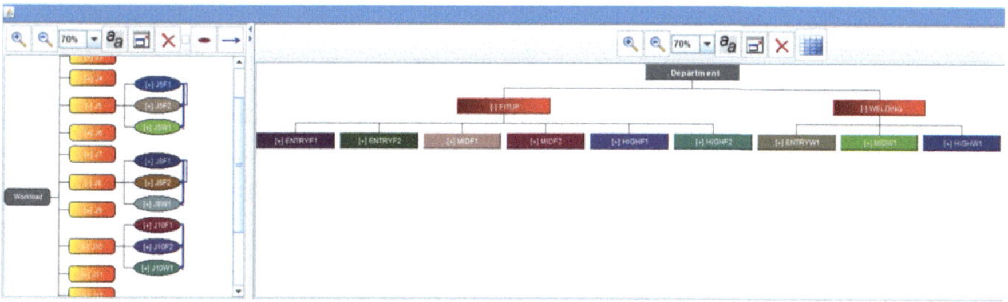

Fig. 7. Jobs and resources in scheduler UI

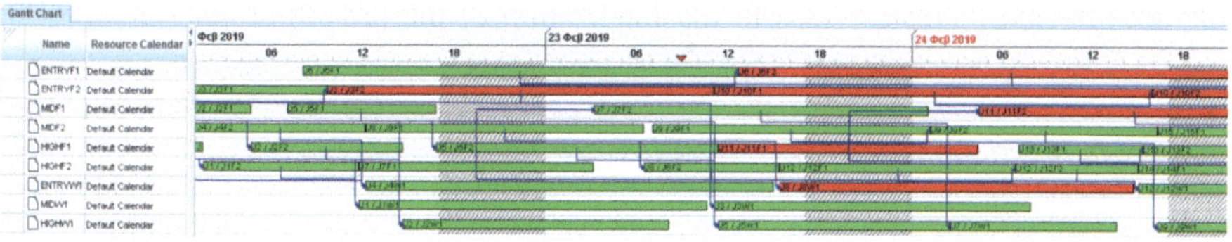

Fig. 8. FIFO Gantt chart (Mean Flow Time: 40.4 h - Mean Tardiness: 4.9 h)

Due to the randomness involved in the ISA used in the case study, repetition of the scheduling process for the same data provides slightly different, although still viable, schedules. For this reason, the scheduling process is performed 10 times and the mean flow time and tardiness for each, produce a new schedule. A multi criteria Gantt Chart is presented in Fig. 9.

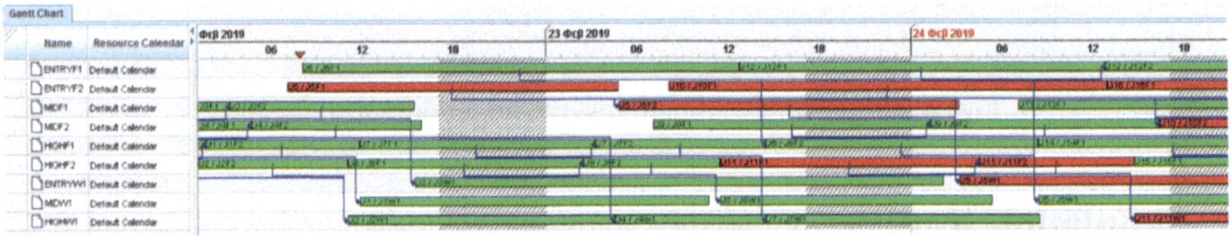

Fig. 9. Multicriteria Gantt chart (Mean Flow Time: 38.9 h - Mean Tardiness: 3.5 h)

Based on the results of these multiple schedules using the multicriteria algorithm, the mean flow time is taken and compared with the schedule produced using the FIFO dispatch rule. There is significant improvement both in mean flow time and mean tardiness for the jobs scheduled when using the multicriteria algorithm compared to the FIFO dispatch rule (Table 1).

Table 1. Multicriteria and FIFO comparison

Multicriteria	FIFO	Variables
38.9	40.4	Mean Flow Time (Hours)
3.5	4.9	Mean Tardiness (Hours)

When the scheduling process is terminated, the schedule is produced and is ready for the decision maker. The information gained from the produced schedule, in combination with the mean flow time and mean tardiness are guidelines for the decision maker in order to organize the whole production process.

5.1 Rescheduling Scenario Based on Barcode Order Tracking

In order to proceed to rescheduling, a completion delay is added to two of the tasks that are processed by the shop floor and are part of this simulated order. Therefore, task J4F2 is delayed by 10 working hours and task J1F2 is delayed by 8 working hours. This does not only create issues in the finalization of the tasks, but also changes the start and finish times for tasks that are either processed by the same resources or are constrained by these tasks. The job dispatch and end of processing hours, without the added delays, are provided by the scheduling program, in the form of an XML file. Then after the start of production, the barcode monitoring system provides the status of production and the completion times for the tasks. In this way, the delay in the production of the two delayed tasks, as well as the rest of the affected tasks can be measured. The data from the scheduler and the barcode system are used to create the order tracking graph, as shown in Fig. 10. The blue points in the scatter plot correspond to the completion timestamps for every task in the schedule. The orange points are created using the task completion data from the barcode system.

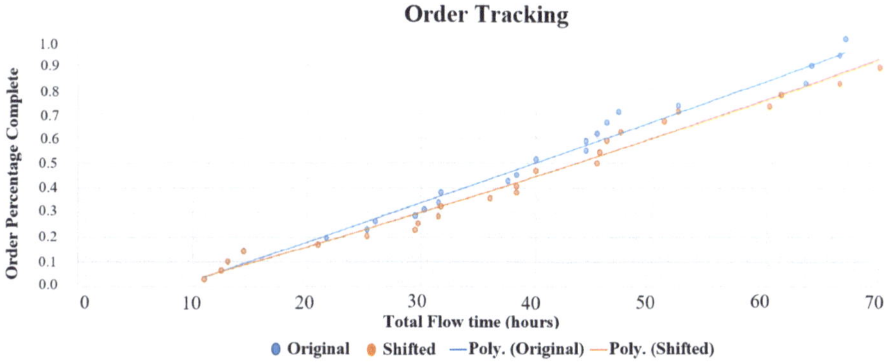

Fig. 10. Estimated order completion and order monitoring

The proposed scheduling method provides great flexibility concerning rescheduling capabilities. Whenever the decision maker decides that rescheduling is needed, the scheduling process can be performed quickly, effectively with high flexibility. When a schedule has been created, with a resource in mind, the failure of that resource can bring great variation to the outcome of the production system. Therefore, the flexibility of the proposed scheduling method can greatly help overcome such production problems. Such a scenario is shown in Fig. 11. First, a schedule is created with 9 resources active. During the production process, resource MIDF2 is not able to produce anymore (Fig. 11a). As such, rescheduling is initiated and a new viable schedule is created, excluding the MIDF2 resource from the production process (Fig. 11b).

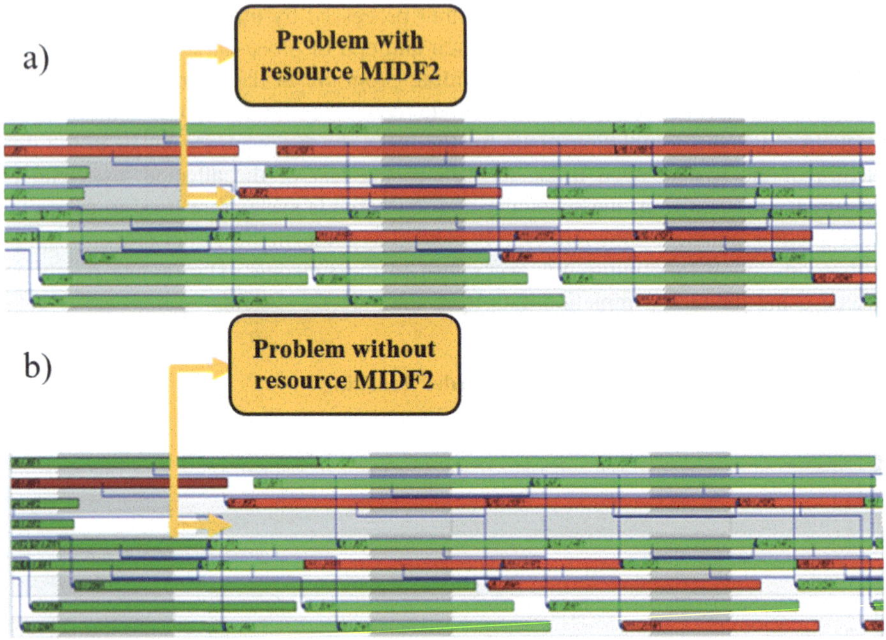

Fig. 11. Rescheduling due to resource problems

6 Conclusions and Future Work

An adaptive scheduling method proposed and validated in a structural steelwork industry has been presented in this research work. The case study acts as a guideline for decision makers to create production schedules. These schedules help finish orders in time, attain factory goals and promote stability and growth to the production enterprise. This aim is achieved based on factory stored and near real-time data. The data is provided to the system through the integration of factory IT tools and an ISA algorithm. This approach, provides flexibility to the system, making it adaptive to different shop floor types and changing shop floor status. The algorithm used for task scheduling, creates viable schedules by taking shop floor data into account. This helps create better schedules that fit the shop floor. The simulation results resulted in significant improvements in mean tardiness and mean flow time when compared to simple dispatch rules. The monitoring aspect of the scheduling system promotes advanced order tracking. Therefore, decision makers can have updated shop floor awareness, promoting enhanced decision making. As a result, unwanted fluctuations in delivery times are decreased and the extra cost is reduced.

Future work will be focused on further integration of the systems used in this case study for monitoring, scheduling and rescheduling. Furthermore, different approaches to the scheduling method like neural networks could be used for the scheduling aspect of the system. Finally, the system could be used on a full-scale real-life case study, for better determination of the different parameters in the scheduling algorithm and complexity calculation equation.

References

1. Behrendt, A., Kelly, R., Rettig, R., Stoffregen, S.: Digital Manufacturing – escaping pilot purgatory. Copyright McKinsey & Company. https://www.mckinsey.com/~/media/mckinsey/business%20functions/operations/our%20insights/how%20digital%20manufacturing%20can%20escape%20pilot%20purgatory/digital-manufacturing-escaping-pilot-purgatory.ashx. Accessed 13 Dec 2019
2. Mourtzis, D., Milas, N., Vlachou, K., Liaromatis, I.: Digital transformation of structural steel manufacturing enabled by IoT-based monitoring and knowledge reuse, pp. 295–301 (2018). https://doi.org/10.1109/CoDIT.2018.8394874
3. Mourtzis, D., Vlachou, E.: A cloud-based cyber-physical system for adaptive shop-floor scheduling and condition-based maintenance. J. Manuf. Syst. **47**, 179–198 (2018). https://doi.org/10.1016/j.jmsy.2018.05.008
4. Pinedo, M.L.: Scheduling: Theory, Algorithms, and Systems, 4th edn., pp. 1–670 (2016). https://doi.org/10.1007/978-3-319-26580-3
5. Harjunkoski, I., Maravelias, C.T., Bongers, P., Castro, P.M., Engell, S., Grossmann, I.E., Wassick, J.: Scope for industrial applications of production scheduling models and solution methods. Comput. Chem. Eng. **62**, 161–193 (2014). https://doi.org/10.1016/j.compchemeng.2013.12.001
6. Mourtzis, D., Vlachou, E., Doukas, M., Kanakis, N., Xanthopoulos, N., Koutoupes, A.: Cloud-based adaptive shop-floor scheduling considering machine tool availability. In: ASME International Mechanical Engineering Congress and Exposition, pp. 1–11 (2015)
7. Rashid, A., Tjahjono, B.: Achieving manufacturing excellence through the integration of enterprise systems and simulation. Prod. Plan. Control **27**(10), 837–852 (2016). https://doi.org/10.1080/09537287.2016.1143132
8. Mourtzis, D.: An integrated system for managing ship repair operations. Int. J. Comput. Integr. Manuf. **18**(8), 721–733 (2005). https://doi.org/10.1080/09511920500234044
9. Peng, Q., Chung, C., Yu, C., Luan, T.: A networked virtual manufacturing system for SMEs. Int. J. Comput. Integr. Manuf. **20**(1), 71–79 (2007). https://doi.org/10.1080/09511920600877494
10. Chryssolouris, G.: Manufacturing Systems: Theory and Practice, 2nd edn. Springer, New York (2006). https://doi.org/10.1007/0-387-28431-1
11. Chen, L., Wang, J., Xu, X.: An energy-efficient single machine scheduling problem with machine reliability constraints. Comput. Ind. Eng. **137** (2019). https://doi.org/10.1016/j.cie.2019.106072
12. Rolón, M., Martínez, R.: Agent-based modelling and simulation of an autonomic manufacturing execution system. Comput. Ind. **63**(1), 53–78 (2012). https://doi.org/10.1016/j.compind.2011.10.005
13. Mourtzis, Dimitris: Simulation in the design and operation of manufacturing systems: state of the art and new trends. Int. J. Prod. Res. (2019). https://doi.org/10.1080/00207543.2019.1636321
14. Boyer, S.: SCADA: Supervisory Control and Data Acquisition, 3rd edn. ISA-The Instrumentation, Systems and Automation Society, Research Triangle Park, North Carolina (2009)
15. Mourtzis, D., Doukas, M., Vlachou, K., Fragou, K., Vandera, C.: Knowledge enriched short-term scheduling for engineer-to-order products. Procedia CIRP **19**(C), 160–167 (2014). https://doi.org/10.1016/j.procir.2014.05.012
16. Mourtzis, D.: Internet-based collaboration in the manufacturing supply chain. CIRP J. Manuf. Sci. Technol. **4**(3), 296–304 (2011). https://doi.org/10.1016/j.cirpj.2011.06.005

17. Mourtzis, D., Vlachou, E., Giannoulis, C., Siganakis, E., Zogopoulos, V.: Applications for frugal product customization and design of manufacturing networks. Procedia CIRP **52**, 228–233 (2016). https://doi.org/10.1016/j.procir.2016.07.055
18. Wang, L.: Machine availability monitoring and machining process planning towards Cloud manufacturing. CIRP J. Manuf. Sci. Technol. **6**(4), 263–273 (2013). https://doi.org/10.1016/j.cirpj.2013.07.001
19. Tuegel, E.J., Ingraffea, A.R., Eason, T.G., Spottswood, S.M.: Reengineering aircraft structural life prediction using a digital twin. Int. J. Aerosp. Eng. (2011). https://doi.org/10.1155/2011/154798
20. Mourtzis, D., Doukas, M., Psarommatis, F.: A toolbox for the design, planning and operation of manufacturing networks in a mass customisation environment. J. Manuf. Syst. **36**, 274–286 (2015). https://doi.org/10.1016/j.jmsy.2014.06.004
21. Mourtzis, D., Doukas, M., Psarommatis, F.: Design of manufacturing networks for mass customisation using an intelligent search method. Int. J. Comput. Integr. Manufacturing. **28**(7), 679–700 (2015). https://doi.org/10.1080/0951192X.2014.900867
22. Mourtzis, D., Doukas, M., Bernidaki D.: Simulation in manufacturing: Review and challenges. Procedia CIRP, 213–229 (2014). https://doi.org/10.1016/j.procir.2014.10.032
23. Mourtzis, D., Doukas, M., Psarommatis, F.: A multi-criteria evaluation of centralized and decentralized production networks in a highly customer-driven environment. CIRP Ann. **61** (1), 427–430 (2012). https://doi.org/10.1016/j.cirp.2012.03.035

Overview of Human-Robot Collaboration in Manufacturing

Lihui Wang(✉), Sichao Liu, Hongyi Liu, and Xi Vincent Wang

Department of Production Engineering, KTH Royal Institute of Technology, Stockholm, Sweden
lihuiw@kth.se

Abstract. Human-robot collaboration (HRC) in the manufacturing context aims to realise a shared workspace where humans can work side by side with robots in close proximity. In human-robot collaborative manufacturing, robots are required to adapt to human behaviours by dynamically changing their pre-planned tasks. However, the robots used today controlled by rigid native codes can no longer support effective human-robot collaboration. To address such challenges, programming-free and multimodal communication and control methods have been actively explored to facilitate the robust human-robot collaborative manufacturing. They can be applied as the solutions to the needs of the increased flexibility and adaptability, as well as higher effort on the conventional (re)programing of robots. These high-level multimodal commands include gesture and posture recognition, voice processing and sensorless haptic interaction for intuitive HRC in local and remote collaboration. Within the context, this paper presents an overview of HRC in manufacturing. Future research directions are also highlighted.

Keywords: Human-robot collaboration · Multimodal control · Manufacturing

1 Introduction

Robots are capable of offering fast, precise, repetitive and heavy-duty task execution on manufacturing shop floors but they lack the flexibility and adaptability of humans. Therefore, recent robotics research has focused on the study and characteristics of human-robot collaboration (HRC) [1], in all industrial tasks and applications, that combines the accuracy and strength of robots with the cognitive ability and flexibility of humans. In the context of manufacturing, HRC aims to realise an environment where humans can work side by side with robots in close proximity [2]. Due to smaller lot sizes of customised products and complex product tasks, manufacturing is experiencing continuously rising demands of increased flexibility and adaptability to changing manufacturing demands. The main target of HRC is to integrate the repeatability and accuracy of robots with the flexibility of the humans that can be applied as the solutions to the needs of manufacturing systems. HRC used in manufacturing systems also facilitates higher overall productivity with better product quality. In addition, the multi-interface design of the HRC system has better compatibility in terms of functionality and enables multimodal control for better performances.

L. Wang et al. (Eds.): AMP 2020, LNME, pp. 15–58, 2020.
https://doi.org/10.1007/978-3-030-46212-3_2

The lack of standards and safety solutions in industrial applications (e.g., automobile production lines) results in low acceptance of the human-robot combination for users in the manufacturing systems, even though the safety solutions in robot labs are trusted and reliable. Therefore, in any HRC system, human safety is of great importance. The challenge is not only collision detection in real time but collision avoidance at runtime via active closed-loop robot control. In the last decade, many research efforts have been focused on HRC, and varying approaches to facilitating multimodal communication, adaptive robot control, and collaborative manufacturing have been reported in the literature. However, the relationship between robots and humans focusing on different scenarios: coexistence, interaction, cooperation, and collaboration can give confusions and the roles of humans in the co-workspace with robots are less clear. Therefore, a systematic overview and analysis on this subject is needed, which is the motivation and objectives of this paper.

This paper starts with current states of HRC in manufacturing and then provides detailed treatments on relevant issues with a focus on HRC in manufacturing. The remainder of this paper is organised as follows: Sect. 2 presents the definition, classification and characteristics of HRC in manufacturing; Sect. 3 reviews the industrial scenarios of HRC in manufacturing; Sect. 4 provides insights on multimodal robot control driven by the high-level commands (e.g., gesture, voice, and haptic interaction) and finally, Sect. 5 concludes this paper and highlights the remaining challenges and future research directions.

2 Definition, Classification, and Characteristics of HRC

In this section, the definition of HRC is presented after analysing the relationship between humans and robots. Then, the classification and characteristics of HRC in the manufacturing context are introduced.

2.1 Definition of HRC

Before giving the definition of HRC, the relationship between humans and robots from different perspectives is analysed [3]. In fact, the relationship between humans and robots is complex and there still exists confusions which are classified from a number of perspectives. Wang et al. [4] analysed and summarised the relationship between humans and robots from the following perspectives: (1) workspace, (2) direct contact, (3) working task, (4) simultaneous process, and (5) sequential process.

1. *Workspace*: In a human-robot cell, humans need to work in a shared workspace with robots. Here, the shared workspace indicates whether the human and the robot are working in the same working area with no physical or virtual fences for separation.
2. *Contact*: indicates whether the robot and the human have a direct physical contact.
3. *Shared working task*: represents whether there is the same operation the human and the robot work on towards the same working objective.

4. *Simultaneous process*: indicates that the human and the robot are working at the same time, but on the same or different tasks.
5. *Sequential process*: represents that there are no overlapping operations for the human and the robot and they are arranged one after another in the temporal scale.

According to the above discussions, the relationship between humans and robots can be classified and defined as human-robot coexistence, human-robot interaction, human-robot cooperation, and human-robot collaboration. These four classifications of human-robot relationship are explained as follows:

Human-Robot Coexistence: This is a basic situation where an industrial robot and a human coexist in a physical workspace without overlapping other's workspace. They occupy a separated location without any direct contact between them. The robot and the human work on the tasks in parallel without interactions, and might exchange the work object between them but the process is completed independently and simultaneously.

Human-Robot Interaction (HRI): This involves the physical interaction of humans and robots in a shared working environment without any direct contact. The operator and robot work on different tasks and the task is completed step by step, sequentially. Interaction, by definition, requires communication between robots and humans. HRI will certainly happen at the cognitive level, fundamentally concerning communication between humans and robots through many channels, and one party may guide or control the other locally or remotely where physical HRI may happen in the same workspace.

Human-Robot Cooperation: The cooperation relationship can be developed between the human and robotic agents who have their own autonomy for the common goals, objectives, utility or profits. This is called *cooperation*, an interactive relationship that makes it possible to harness the knowledge of other system components in the service of joint interests. A shared workspace of the robot and human in human-robot cooperation is required but there is no direct contact between the robot and human. They share the same resource but complete respective working tasks in a sequential order (Table 1).

Table 1. Features of different human-robot relationships (adapted from [1])

		Coexistence	Interaction	Cooperation	Collaboration
Shared	Workspace		✓	✓	✓
	Direct contact		✓		✓
	Working task		✓		✓
	Resource			✓	✓
	Simultaneous process	✓		✓	✓
	Sequential process		✓	✓	

Human-Robot Collaboration: Collaboration is the joint activity of humans and robots in a shared symbiotic working environment, with the definite objective to accomplish together a set of given working tasks. It requires typically a coordinated, synchronous activity from all parties where physical contact is also allowed. In any case, collaboration assumes a joint, focused goal-oriented activity from the parties who share their different capabilities, competences, and resources.

Based on the classification of the human-robot relationship, the definition of HRC is a *state in which a purposely designed robot system and an operator work on simultaneous tasks within a collaborative environment* [5]. The goal is to enable close collaboration between humans and robots, in all service and industrial tasks, that require the adaptability of humans to be merged with the high performance of robots. There is no temporal or spatial separation of the robotic and human's activities.

Many academic and application-oriented studies on HRC have been successfully developed in recent years. A number of factors motivate HRC being a spotlight in the field and they are summarised as the combined strength of robots and humans, collaborative and shared working conditions, promising potential of increased productivity, flexibility, and adaptability, increased robustness in the control performances and higher degree of resilience, and improved ergonomics [1].

2.2 Classification of HRC

Industrial robots had been expected to work as the assistant of human workers for a long time, comprising a fast and automatic assembly system and collaborative manufacturing environment [6]. A systematic analysis of HRC requirements and possible solutions is always needed for industrial cases and application scenarios. In fact, systematic analysis and classification of HRC was reported [7]. Due to its specific constraints, manufacturing usually occupies a subset of possibilities. The key properties highlighted in most of studies that define distinct classes of HRC instances across all applications are summarised as follows:

- *Temporal* and *spatial sharing*: robots and humans collaborate in a sharing workspace. The shared workspace means that the close spatial relationship between robots and humans exists for the local HRC collaboration [8]. The same goes for the remote HRC collaboration; even though working together in the same workspace for the given tasks, the processes or operations of robots and humans do not does not necessarily have a temporal separation [9]. For example, close collaboration with physical contact—e.g., common handling of large workpieces—does, naturally, require co-location and simultaneous operations [10].
- *Multiplicity*: industrial HRC cells can be distinguished as single, multiple, and team settings (shown in Fig. 1) and robots and humans can be considered as agents. The team setup assumes the parties in an HRC cell as a group acting by consensus or coordination, and interacting with the environment and other parties in single or multimodal control commands (e.g., gesture, voice, and haptic). Multiple parties can compete for resources and other parties' services [11]. For example, multi-robots are haptically controlled or guided by an operator for a specified collaborative manufacturing task.

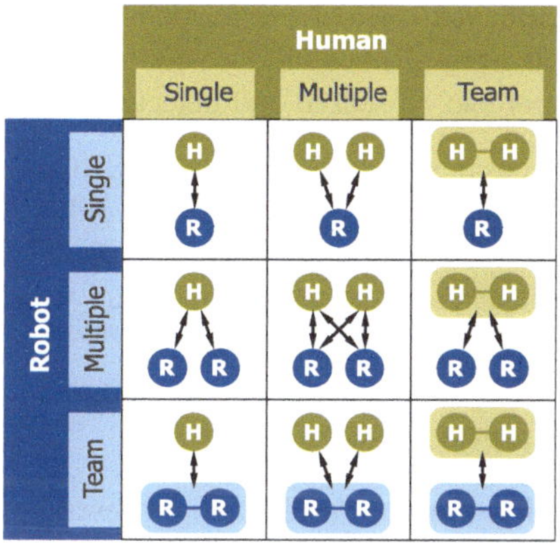

Fig. 1. Possible schemes of the human and robotic agents' multiplicity (adapted from [2]).

- *Autonomy* closely related *leader–follower* relationships: Autonomous agents (human and robotic agents) have enough knowledge and authority to operate and act on their own without direct instructions or intervention from the environment or other agents [12, 13]; This represents how much of robot action is directly determined by human agents, or which agent takes the lead in the given task. In HRC, robots and humans need to take actions for implementing their own duties and performing the leadership in any industrial instances [11]. In the manufacturing context, four types of the roles for the task execution scenarios are classified: inactive (resting), active (leading), supportive (following), and adaptive (the robot can dynamically change the role) or intuitive (role subject to human's own decision), shown in Fig. 2. Some of studies in HRC applications (e.g., in a steady-ready environment and task/process planning) give the assumption that these roles are pre-planned and assigned before task execution except the adaptive and intuitive roles [14]. Robots and humans can

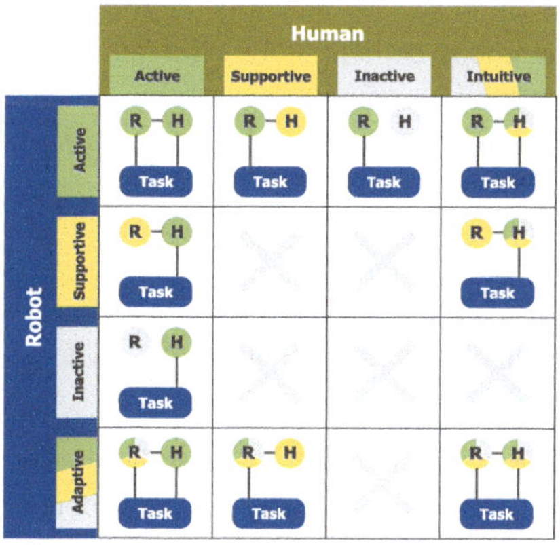

Fig. 2. Possible cases of the human and robotic agents' roles (adapted from [2]).

(re)-assign the leader/follower roles in a dynamic environment where the robot can automatically change the pre-planned tasks to collaborate with an operator and the intention of the operator can be sensed and detected in real time [15].

It needs to be pointed out that the classification of HRC in manufacturing does not cover all the properties of the HRC applications, and only focuses on the key properties. Some others covered in its full property set are typically reflected in the specified applications and considered as independent characteristics which are rarely defined as a distinct class.

2.3 Characteristics of HRC

In the last decade, HRC has emerged as a spotlight research topic in manufacturing industries which is a top priority of many research institutes, universities, and companies. Why HRC is so welcomed and attractive? The motivations or reasons behind this are that HRC is characterised by the combination of complementary strengths of human and robotic skills, the system design of increased productivity, flexibility and adaptability, increased robustness and higher degree of collaboration and coordination, improved ergonomics, and more attractive work environments [1, 16, 17]. The new topic of HRC offers many new possibilities. During collaborative manufacturing with HRC, a number of industrial applications have been a focus in recent years and research efforts on HRC in manufacturing have been numerous [18], e.g., HRC for automotive manufacturing [19], HRC in cellular manufacturing [20], HRC in hybrid manufacturing cell [21], and relevant issues with HRC in manufacturing [22]. These paradigms or cases are typically characterising HRC in manufacturing as follows, which are the requirements of HRC systems [23]:

- Flexible and adaptive configuration for system deployment with specified functionalities and user-friendly interfaces used for integration into the overall production
- Attractive potential of increased productivity and profits, better product quality, and improved robustness of collaborative manufacturing systems
- Selective and scalable approach for industrial automation, increased flexibility and responsiveness of the robot collaborating with humans in response to dynamic changes of pre-planned tasks
- Active/passive support, assistance, and collaboration for human workers in tedious or non-ergonomic tasks
- Intuitive human-robot-interface or multimodal communication channels for a higher level of compatibility and adaptability
- Safety standards and operation regulation for all parties in HRC systems as well as responsibility and legal obligation for suppliers, owners, and operators of HRC systems

The manufacturing industry is required to meet the need of the next generation of assembly, which is performed by safe and reliable robots working in the shared workplace collaborating with humans. In response to such trend/need, many research efforts have been focused on human-robot collaborative assembly in hybrid/cellular manufacturing. In fact, human-robot collaborative assembly is indeed the most common and definitively successful one of HRC in manufacturing. Therefore, detailed treatments on relevant characteristics of HRC assembly are discussed below.

During HRC assembly, objects/assembled parts are arranged in space by actions in time so that the manufacturing of products specified by design can be performed [1]. Parts of the product and the applied technological resources and humans occupy the space, and the key objectives require short response time for the execution of actions. In HRC assembly, objects and operations are closely related and also have a constraint relationship with each other in many aspects [24, 25]. In addition, the functionality of efficient and dynamic human-robot task allocation is considered in the workplace design, which makes HRC assembly safe, ergonomic, and symbiotic [26].

Recently, symbiotic human-robot collaboration attracts more attention in the research field. The symbiotic relationship with respect to related work in human-robot relationships is defined in [27], which can be considered as asynchronous group coordination between the human and the robot. Symbiotic human-robot collaboration formalises mutual '*benefit*' of humans and robots into a cyber-physical environment [28], and a symbiotic interaction between humans and robots in a shared workspace is performed to overcoming the limitation of the robots while executing complex tasks via the combination of humans' strength. In such symbiotic collaboration, humans and robots can be viewed as an agent team where symbiotic agents perform asynchronous actions for solving problems that team members could not have performed alone in a working environment with uncertainty [11]. Distinguishing with conventional HRC, symbiotic collaboration is characterised as follows [1]:

- *Autonomy*: symbiotic agents (human and robotic agents) are autonomous and they do not control or direct each other [29]. All the agents take asynchronous actions for a given goal and coordinate asynchronous communication [30]. The leadership and roles of symbiotic agents in the team are assumed and changed dynamically according to the requirements of tasks and actual situations.
- *Multimodal intuitive programming*: this offers opportunities for robot programming without requirements of expert knowledge of the system for operators. Multi-communication channels between robots and humans can be used for system control and task execution.
- *Programming-free*: the multimodal fusion of human commands (e.g., hand gesture, voice command, haptic instructions) and other forms of natural inputs without the need of coding contribute to programming-free robot control.
- *Symbiotic collaboration*: the interplay of humans and machines benefits each other with the help of different devices, e.g. screens, goggles, wearable displays, and collaborative parties in the group formulate a symbiotic relationship via team coordination and communication, and
- *Context-aware dependency*: the system should be capable of interleaving autonomous humans with robot decisions based on trustworthy inputs from onsite sensors and monitors inspecting both humans and robots.

3 Industrial Scenarios of HRC in Manufacturing

Human-robot collaboration in the area of manufacturing (i.e. collaborative manufacturing with HRC [31] and symbiotic HRC assembly [1]) is widely discussed in a number of applications. HRC is also applied as the solution to the needs of the advanced collaborative manufacturing systems of the future industry. In such a manufacturing context, topics related to HRC in manufacturing covered in this section include symbiotic HRC assembly, safety of HRC in manufacturing, and cyber-physical-system-enabled HRC.

3.1 Symbiotic HRC Assembly

Symbiotic human-robot collaboration places the interplay of humans and machines into a shared working environment where a symbiotic interaction of humans and robots for specified tasks and operations can be achieved [1]. The research of utilising robots working as collaborating partners in assembly lines in an unstructured environment has been active recently, such as an automated assembly system [32] and a collaborative manufacturing environment [9]. Compared with conventional assembly, i.e. automatic assembly and manual assembly, one potential application of HRC in the assembly industry that is actively explored is the symbiotic HRC assembly where human and robotic agents interact within a shared workspace to finish complex assembly activities which requires the combination of the strengths of both parties.

However, the systematic design of solutions for such symbiotic collaboration between humans and robots requires an overall framework from the definition and analysis of the problem to the solution design and implementation [2]. To address such challenges and relevant problems in the European manufacturing industry, the SYMBIO-TIC project (funded by EU Horizon 2020) [17] aims to offer adequate solutions of these important issues towards a safe, dynamic, intuitive and cost-effective working environment where symbiotic collaboration between human workers and robots can take place. The elaboration of the structure of a symbiotic collaboration environment can be the first step in the symbiotic HRC assembly. Within this context, a systematic approach used for creating a symbiotic HRC environment can be understood as a series of analysis and synthesis steps containing the definition of application scenarios, the accurate description of tasks and agents, proceeding towards a feasible solution. Meanwhile, the consideration of reiteration and introduction of implicit knowledge at points of manual intervention is required to be included in the elaboration process. While human workers participating in the manufacturing activities and especially working in a shared working environment in a collaborative role can make the procedure more demanding due to the diversity of actions arising from increased human autonomy.

Here, the EU project (SYMBIO-TIC) [17] developed a symbiotic system structure for HRC-enabled assembly, as shown in Fig. 3. It comprises of active collision avoidance, planning and control cockpit, adaptive robot control, and mobile worker assistance modules. In any manufacturing case, human safety is of utmost importance. Therefore, the active collision avoidance module is developed to ensure a safe and reliable working environment with/without fence where the sensor-based virtual three-dimensional (3D) models of robots with point-cloud images of human workers from depth cameras for efficient perception and monitoring is established. In the structured symbiotic robotic environment with safety guidelines, dynamic assembly tasks shared by humans and robots in a collaborative environment need to be planned for and assigned to available and capable resources. Next, adaptive robot control is facilitated by the use of IEC 61499-based function blocks [33] with the embedded smart decision-making algorithms, and within such context, diverse programming modes (i.e., programming-free robot control and multimodal robot control) are included for symbiotic intuitive programming control. In addition, mobile worker assistance modules consist of mobile worker tracking and identification and AR (Augmented Reality) driven in-situ decision support. These modules aim to use diverse intelligent and assistive devices to make human operators work easily in the HRC assembly and implement intuitive and ergonomic interaction between humans and robots. Under such a symbiotic system structure, a selection of industrial cases of local and remote HRC assembly have been successfully deployed.

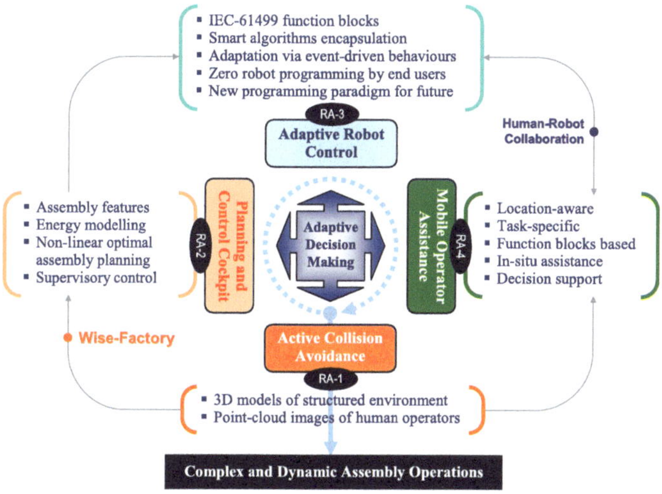

Fig. 3. A symbiotic system structure for HRC-based assembly.

An exemplified symbiotic HRC system is developed in three industrial scenarios, including food packaging, aeronautic assembly, and automotive engine assembly in which a private cloud is structured to encapsulate the capability of computing and access. Figure 4 illustrates a supercharger assembly case as a part of the automotive engine. This assembly case aims to validate and evaluate HRC system discussed in [2] where humans and robots combine their complementary strengths instead of limiting each other. In such collaborative assembly activities, humans and robots take their own roles in the process of assembly. The assembly steps within the tasks start with picking and placing a resonator on the side of the robot, while human operators take responsibility of controlling seals. Next, the robot starts to print decal and prepare a tube for the next assembly operation. Then, humans individually participate in multiple assembly steps from placing all the parts together, mounting cables to placing a throttle and screws, and a tool mounted on the end-effector of the robot is used for tightening screws controlled by the multimodal communication channels. Finally, complex assembly tasks can be finished via a symbiotic collaboration between humans and robots.

The assembly steps within the task are classified and defined where some of the steps are active HRC actions, e.g., place and hold parts and tighten screws, while some are supportive for the preparation of the next step, e.g., take and place. Additionally, the wide availability of the HRC system can get access to general mobile devices, e.g., smartphones and tablets. Meanwhile, the private cloud in the backend provides adequate support for the dynamic task planning and mobile worker assistance. The assembly case study facilitates to validate the feasibility of transforming conventional industrial robotic cells into collaborative environments based on the symbiotic system architecture. Compared with the investment in new devices that are designed as collaborative robots, the legacy industrial robots have higher speed, payload, and better stiffness.

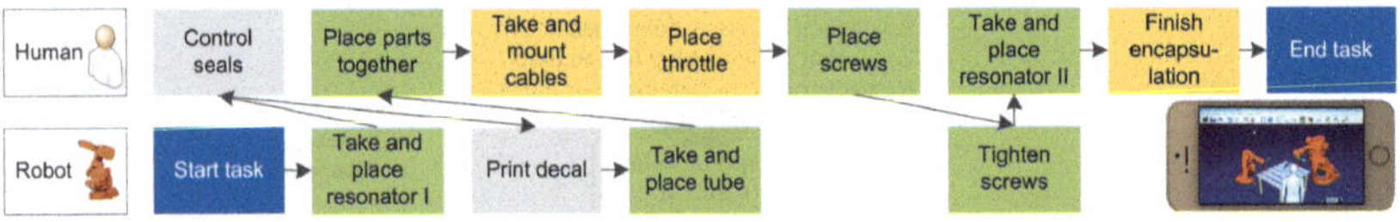

Fig. 4. A supercharger assembly case within a symbiotic HRC environment (adapted from [2]).

In HRC assembly, the workplace design that allows automatic and dynamic task allocation and trajectory planning is of importance to the assembly efficiency through decreasing the required time of a new assembly premise design [34, 35]. Given an assembly environment, complex assembly tasks need to be handled in a dexterous and flexible manner. Tasks that require repeatability, heavy load lifting, and accuracy are mainly assigned to robots. In this scenario, tasks that are allocated to robots and humans may be re-planned due to the availability and suitability of the resources against the allocated time [36]. This challenge makes human-robot task planning be widely investigated [37]. An intelligent decision-making method of task planning in HRC hybrid assembly cells was developed in which the resource and workload models of humans and robots are included [34]. A hybrid assembly cell for the assembly of a vehicle dashboard is applied to depict the proposed task planning approach. The assembly premise is composed of a dual-arm robot and a human operator where some of assembly tasks are handled by human operators, while some others by the dual-arm robot [38]. The assembly steps include traverse placement, computer placement, and cable installation (Fig. 5(a)). A Gantt chart of the best alternative is illustrated in Fig. 5 (b). This assembly case offers the potential application of single-arm robots, collaborating with humans but also in a more complex cell with multi-arm robots and humans. The focus is rather given to the human-robot coexistence for the execution of sequential tasks, in order for the automation level in manual or even hybrid assembly lines to be increased.

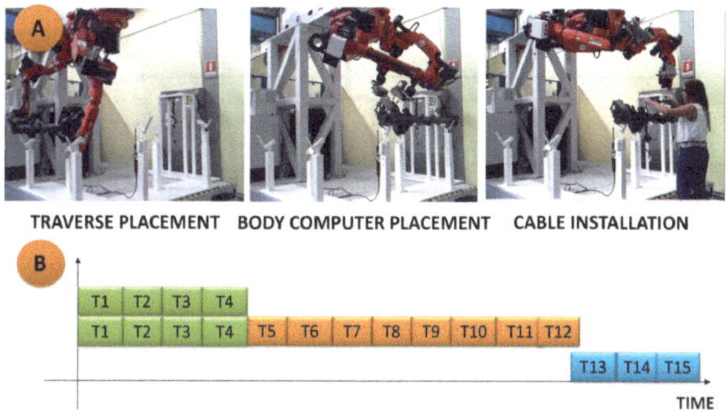

Fig. 5. a) Dashboard assembly case; b) Gantt chart of the best alternative (adapted from [38]).

Manufacturing companies nowadays need to maintain a high level of flexibility and adaptability to deal with uncertainties on dynamic shop floors caused by the part supply, urgent job insertion or delay, cutting tool shortage, and machine unavailability. Such challenges are also faced by the robotic assembly industry [39]. Within the context, an adaptive assembly system, that is based on real-time manufacturing information and can respond to the changing demands and disturbances adaptively with alternative assembly solutions, shows the potential for addressing these challenges in a dynamic environment. A function block-based assembly planning and robot control system with enhanced adaptability is developed to form the capacity of real-time decision making at the controller level [40]. This robotic assembly system can also be viewed as a remote human-robot collaborative assembly system where humans hidden behind the system take the role of developing algorithms embedded in constructed function blocks, while the robot is controlled for the task execution and assembly operation. A mini human-robot collaborative assembly cell is developed to illustrate the enhanced adaptability and dynamism enabled by incorporating the event-driven function blocks with smart decision-making algorithms (Fig. 6). This assembly case consists of three parts with three assembly operations and they are placing the top part in a fixture, placing the top part on the base component, and inserting the pin into the hole.

In this assembly cell, the tasks handled by human operators include modelling assembly operations and decision process for assembly planning and relevant algorithm development. To implement the assembly operations and the designed functionality, three types of function blocks are established: (1) assembly feature function blocks for placing and inserting, (2) material handling function blocks, and (3) management function blocks for HRC assembly cell management. The assembly sequences determined by assembly feature function blocks in the HRC assembly cell are specified as follows:

- picking the base component from the pin feeder and placing it in the fixture
- moving the top part from the pin feeder and placing it on top of the base component
- grasping the pin from the pin feeder and inserting it in the hole aligned between the top and the base component
- removing the assembled product and placing it on the conveyer

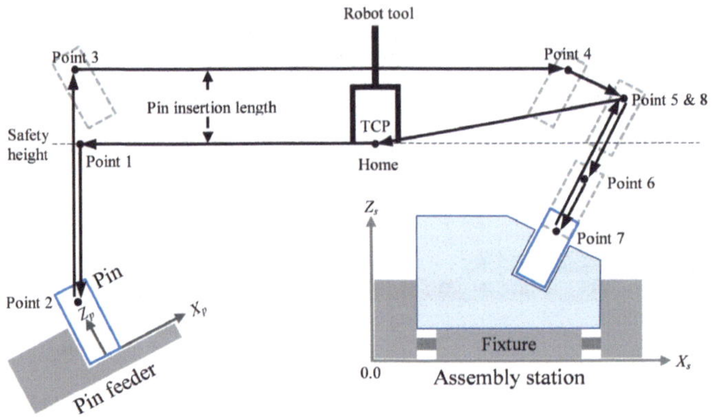

Fig. 6. Function-block-based robotic assembly with enhanced adaptability (adapted from [39]).

For robotic assembly, the *insert* tasks handled by a robot can be classified as two steps: (1) picking up the pin, and (2) inserting the pin into a hole. During this assembly process, the trajectory of the inserting operation is planned/re-planned by the algorithm embedded in the assembly feature function blocks. These two assembly steps can be viewed as a series of the ordered sequence of the robot movement and actions along the calculated trajectories of the robot's end-effector. The whole process of the robotic assembly consists of eight movement operations, one grasping movement, and one releasing operation. They are depicted as follows: (1) moving towards and approaching the pin, (2) moving over the pin, (3) grasping the pin, (4) lifting the pin vertically, (5) moving the pin over the hole, (6) approaching the hole at a given configuration, (7) preparing for insertion, (8) inserting the pin into the hole, (9) releasing the pin, and (10) moving end-effector away from the hole.

Additionally, this type of HRC assembly scenario can make human operators not necessarily work in a shared workspace together with the actual robot in a physical environment, instead a remote human-robot collaborative assembly can be achieved.

3.2 Safety of HRC in Manufacturing

Robots (industrial or collaborative) are becoming flexible and versatile robotic co-workers such that they are expected to help humans handle complex or physically demanding work in industrial settings. As a common characteristic in these foreseen applications, robots should have the capacity of operating in very dynamic, unstructured, and partially unknown environments, sharing the workspace with a human user.

As a consequence of the introduction of HRC technologies, great importance has been attributed to robot safety standards, which have been updated to address new co-working scenarios. ISO 10218-1/2 [5, 41] identify four collaborative modes, which are summarised as follows: Safety-rated Monitored Stop, Hand Guiding, Speed and Separation Monitoring, Power and Force Limiting. Lasota et al. [42] presented a survey of potential methods of ensuring safety during HRI and classified them into four major categories: safety through control, motion planning, prediction, and consideration of psychological factors. In particular, HRC in manufacturing allows human operators working side by side with robots in close proximity, and human safety is of paramount importance. Additionally, cost-efficient, fast, and reliable handling of possible collisions in the collaborative environment is needed, along with control strategies for safe robot reaction. Therefore, this section presents the detailed treatment of relevant issues with a focus on safety of HRC in manufacturing. This section starts with the safety standards for industrial robots from the perspective of the possible causes of safety issues, standard design, and relevant international standards. The seven elementary phases of the collision event pipeline, namely pre-collision, detection, isolation, identification, classification, reaction, and collision avoidance, are introduced with a focus on collision detection and collision avoidance in the HRC environment [43]. This is due to the fact that typical challenges are not only collision detection in real time but collision avoidance at runtime via active closed-loop robot control strategies [44].

The causes resulting in potential accidents in HRC can be classified into three categories, and they are engineer failures, human errors, and poor working environment and conditions, as shown in Fig. 7 [45, 46]. All of these failures can lead to incorrect response and interaction by both the robot and human operators. Engineer failures indicate the failures induced by the hardware and software of the robot system. For example, the vision-based 3D model of the human workers cannot be used to monitor the human motion in a shared environment in real time [47], as it may lead to a possible collision in a case of human locating in a distance with injury risk. Another example is that, approaching the singularity of the robot joint configuration, a tiny change in the Cartesian space can induce a huge change of joint configuration in the joint space, which may lead to a possible collision in a close working distance [48]. Human errors can be summarised from design mistakes and unintended interaction errors. Design mistakes are induced by the faults or defects during the design, planning, and any production modification to a robotic manufacturing cell. Interaction errors are caused by the violation of operating procedures and safety guidelines, and human's carelessness in the collaborative interaction. The causes from the environment can be classified into poor working conditions with sensing and lighting, extreme temperature, and these factors are common for sensor-based monitoring and tracking scenarios, i.e. camera-based collision detection and voice-based robot control.

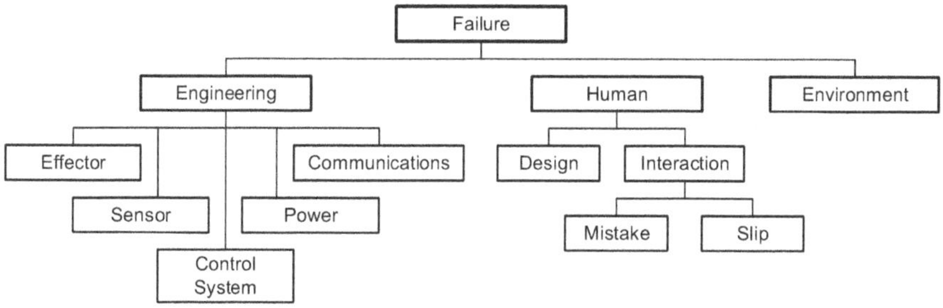

Fig. 7. Possible causes of potential accidents and classification (adapted from [45, 46]).

For the safe operation in the field of engineering, strict and comprehensive standards and directives are developed. EU directives [26], indicative general standards, and robot standards are developed to guarantee the safety of operation procedures. They aim to standardise the design and prevent engineering failures from the design phases and they are illustrated in Tables 2, 3, and 4. In the safety of machinery, ISO 13855:2010 [49] established the positioning of safeguards with respect to the approaching speeds of parts of the human body. It specifies parameters based on values of the approaching speeds of parts of the human body and provides a methodology to determine the minimum distances to a hazard zone from the detection zone or from actuating devices of safeguards. One of the significant differences between HRC and conventional industrial manipulation is that the former allows closer collaboration between a robot and a human in a shared workspace [50]. This requires the safety standards and directives for HRC to be rethought. To meet the safety requirements for HRC, a new technical specification on safety for HRC (ISO/TS 15066:2016 [51]) is under development, which specifies safety requirements for collaborative industrial robot systems and the work environment. The process parameters including the speed, force, and transferred energy are specified to limit the interaction procedures of both robots and humans.

Table 2. EU directives [26]

Title	Description
2006/42/EC	Machinery Directive (MD)
2009/104/EC	Use of Work Equipment Directive
89/654/EC	Workplace Directive
2001/95/EC	Product Safety Directive
2006/95/EC	Low Voltage Directive (LVD)
2004/108/EC	Electromagnetic Compatibility Directive (EMC)

Table 3. Indicative general standards [52]

Title	Description
EN ISO 12100	Safety of machinery – General principles for design – Risk assessment and risk reduction
EN ISO 13949-1/2	Safety of machinery – Safety-related parts of control systems – Part 1: General principles for design, Part 2: Validation
EN 60204-1	Safety of machinery – Electrical equipment of machines – Part 1: General requirements
EN 62061	Safety of machinery – Functional safety of safety-related electrical, electronic and programmable electronic control systems

Table 4. Robot standards [52]

Title	Description
EN ISO 10218-1	Robots and robotic devices – Safety requirements for industrial robots – Part 1: Robots
EN ISO 10218-2	Robots and robotic devices – Safety requirements for industrial robots – Part 2: Robot systems and integration
ISO/PDTS 15066	Robots and robotic devices – Collaborative robots

In the past years, many research efforts have been focused on tackling the relevant safety issues of human operators who work side by side with robots in a shared working environment from the perspective of design, construction, control strategies, and methodologies. Accordingly, a number of safety strategies have been introduced with a focus on three types of robot safety, and they are classified as crash safety (only 'safe'/controlled collisions), active safety (timely detecting imminent collisions), and adaptive safety (intervening in the operation of the hardware equipment and applying corrective actions) [26, 53]. Based on such classification, a metrics that relies on the relative distance between the human worker and the robot as well as on the robot itself was introduced to evaluate the safety in the HRC environment [54]. Meanwhile, a kinematic control strategy that adaptively modifies the velocity of the robot on an assigned path was proposed to enforce a safe HRC.

When coming to the control strategies of the safety of HRC, many robotic control strategies have tackled the safety issues to guarantee human safety. Heinzmann and Zelinsky [55] described the formulation and implementation of a control strategy for robot manipulators, which provides quantitative safety guarantees for the user of assistive robots. A control scheme for robot manipulators was introduced to restrict the torque commands of a position control algorithm to values that comply with pre-set safety restrictions. These safety restrictions limit the potential impact force of the robot in the case of a collision with a person. Calinon et al. [56] proposed a learning-based control strategy for a robotic manipulator operating in an unstructured environment while interacting with a human operator, and this control scheme considers the important characteristics of the task and the redundancy of the robot to determine a controller that is safe for the user. Additionally, Laffranchi et al. [57] presented an energy-based control strategy to be used in robotic systems working closely or cooperating with humans. The presented method bounds the dangerous behaviour of a robot during the first instants of the impact by limiting the energy stored in the system to a maximum imposed value. Geravand et al. [58] developed an energy-monitoring-based control strategy for safe physical human-robot collaboration where the collaborative task of manipulating a rigid object is performed by a human-robot team. Within this control scheme, energy flows among the different subsystems involved in physical HRC can be observed, and then the energy of the robots and the power transferred to the human are limited and shaped to enhance the human safety according to selected metrics. Meguenani et al. [59] proposed physically meaningful energy-related safety indicators for robots sharing their workspace with humans. The kinetic and potential energy of the robotic system based criteria are used to limit the amount of dissipated energy of a KUKA LWR4 serial robot in case of collision and modulate the contact forces, respectively.

Advancements in the safety of HRC with a focus on safety standards and relevant control strategies are discussed above, while the challenge for safe HRC is not only efficient collision detection in real time but active collision avoidance at runtime via efficient closed-loop robot control [44].

Collision detection between humans and robots is crucial for human safety in a shared working environment where humans work side by side with robots in close proximity. Many research efforts have been taken on the robotic collision detection in the past years. A rather intuitive approach is to monitor the measured currents in robot electrical drives, looking for fast transients possibly caused by a collision [60, 61]. In addition, the forces applied can be limited by a robot manipulator during contact without the use of external sensors [62–64]. Haddadin et al. [43] reviewed and expanded methods for generating a collision monitoring function and derived a new concept for the reconstruction of the contact wrench. A generalised collision monitoring signal within the force/torque control was produced to detect the possible collision between a robot and a human. Another proposed scheme compared the actual commanded motor torques (or motor currents) with the nominal model-based control law (e.g., the instantaneous motor torque expected in the absence of collision), with any difference being attributed to a collision [65]. Focusing on the torque changes at the joints, De Luca et al. [66, 67] developed a physical collision detection/reaction method based on a residual signal and a collision avoidance algorithm based on depth information of the HRC. This idea was refined by considering the use of an adaptive compliance control [68]. A torque residual signal of the robotic system between the actual input torque to the manipulator and the reference input torque is used for detecting collisions [65, 69]. The monitoring signal that shows a detected collision triggers the stop of the manipulator. However, tuning of collision detection thresholds in these schemes is difficult because of the highly varying dynamic characteristics of the control torques [43].

Additionally, two methods were widely considered: (1) using a vision system to perform 3D inspection through 3D models as well as skin colour detection to perform 3D tracking of human body in a robotic cell [70], and (2) an inertial sensor based approach [71] using geometry representation of human operators through a special suit for motion capturing. The latter requires the testers to wear a special suit with sensors which can limit motion capture of the suit wearer in unmonitored zones. Therefore, this approach is validated to be not feasible for real-world applications due to the possible safety leak, e.g., a mobile object may hit a stationary operator.

Thus, Ebert and Henrich [72] presented a sensor-based collision detection method for an industrial robotic manipulator based on images taken from several stationary cameras in the work cell. The collision test works entirely based on the images and does not construct a representation of the Cartesian space. Henrich and Gecks [73] developed a vision-based collision detection method based on several stationary and calibrated video cameras, and potential collisions of the known objects in any of their (future) configurations with a priori unknown dynamic obstacles can be detected via collected images. Vogel et al. [74] presented a projector-camera based approach for dynamic safety zone boundary projection and violation detection. Krüger et al. [70] utilised multiple 2D cameras to monitor the workspace, and then calculated a 3D model of the scene. This model is used to determine the spatial distance between the worker and the robot and therefore governs the decision whether to intervene in the control programme of the robot. A projector-camera-based approach was presented by Vogel, Walter, and Elkmann [75], which consists of defining a protected zone around the robot by projecting the boundary of the zone. The approach is able to dynamically and visually detect any safety interruption. In [76], a triple stereovision system was reported for capturing the motion of a seated operator (upper body only) by wearing colour markers. Nonetheless, relying on the colour consistency may not be suitable in uneven environmental lighting conditions. In addition, tracking markers of mobile operators may not appear clearly in the monitored area. Instead of markers, a time-of-flight (ToF) camera was adopted in [77] for collision detection, and an approach using 3D depth information was proposed by Fischer and Henrich [78] for the same purpose. Using laser scanners in these approaches offers suitable resolution but requires longer computational time, since each pixel or row of the captured scene is processed independently. On the other hand, ToF cameras provide high-performance solutions for depth images acquisition, but with insufficient level of pixel resolution (capable of reaching 200×200) and with high expense. Rybski et al. [79] acquired data from 3D imaging sensors to construct a 3D grid for locating foreign objects and identifying human operators, robots, and background.

In addition, other researchers focused on combining different sensing techniques to track humans and robots on shop floors such as the work presented in [80], which used both ultrasonic and infrared proximity sensors to establish a collision-free robotic environment. Moreover, other researchers like Cherubini et al. [31] incorporated both force/torque sensors and vision systems into a hybrid assembly environment to provide a direct contact between the human and the robot.

Among commercial systems of safety protection solutions, SafetyEYE® (Pilz GmbH & Co. KG) is a popular choice. It computes 2½D data of a monitored region using a single stereo image and detects the violation of predefined safety zones. Accessing into any of the safety zones will trigger an emergency stop of the monitored environment. However, these safety zones cannot be updated during active surveillance.

In recent research, depth sensors that can output the dynamic reflection of objects in 3D models are widely applied in the collision detection for a safe human-robot collaborative environment. Fischer and Henrich [78] used multiple 3D depth images for fast collision detection of multiple unknown objects where the depth sensors placed around the work cell are used to observe a common surveilled 3D space. The acquired depth images are used to calculate the minimum distance to any obstacle and then maximum velocity can be limited. Flacco et al. [58] developed a fast method to evaluate distances between the robot and possibly moving obstacles (including humans), based on the depth space. The distances are used to generate repulsive vectors that are used to control the robot while executing a generic motion task. Lately, Morato et al. [81] developed a multi-Kinect system to build an explicit model of the human and a roll-out strategy, which forward-simulates the robot's trajectory into the near future. Real-time replication of the human and robot movements inside a physics-based simulation of the work cell can enable the stop of the robot whenever an imminent collision between the human model and any part of the robot is detected. In parallel, Schmidt and Wang [44, 47] developed a depth camera (Kinect sensor)-based approach for collision detection between the robot and human [82, 83] and the experiment shows that the collision between depth images and 3D models in an augmented environment can be timely detected (as shown in Fig. 8) [47]. 3D models linked to real sensors are used to represent a structured manufacturing environment and monitor the robot, and then the depth images of the operator captured by Kinect sensors are used to calculate the relative distance between robots and humans based on the captured position of the robot. The procedures of depth images acquisition and processing are illustrated in Fig. 9 [47]. Finally, the potential collision between a robot and a human can be effectively detected.

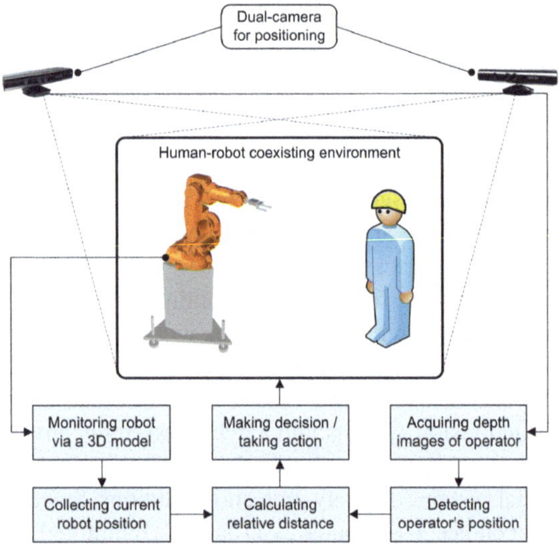

Fig. 8. Depth camera-based collision detection and avoidance in an augmented environment (adapted from [82]).

Once an undesired physical collision has been detected, the robot is switched as fast as possible from the control law associated with normal task execution to a reaction control law. Graham et al. [84] reported an avoidance system where the computer-controlled sensor system was developed to monitor the workspace of the robot, and this system sent corrective commands to the robot to avoid personnel injury or damage to the robot or equipment when dangerous situations occur. Based on the classification of the collision, some examples of successful collision reaction strategies have been specified in [85–87], including stopping the robot, slowing down the velocity, retracting, reflex control, and impedance relaxation. Flacco et al. [88] proposed a real-time collision avoidance approach for safe human-robot coexistence, and a collision avoidance algorithm was developed where different reaction behaviours are set up for the end-effector and for other control points along the robot structure. Bala and Bone

[89] studied a 3D dynamic human-robot collision avoidance problem where a collision avoidance algorithm was developed to search for collision-free paths by moving the end-effector along a set of predefined search directions. Seto et al. [90] proposed a real-time self-collision avoidance control method for the robot during human-robot cooperation. The self-collision avoidance motion could be realised based on a reaction force generated by the contacts between the elastic elements before the actual self-collision of the robot. Ratsamee et al. [91] reported a human-robot collision avoidance method using a modified social force model with body pose and face orientation. The physical and social components (i.e. human position and body pose) are integrated into the modified social force model which allows robots to predict human motion and perform smooth collision avoidance. Additionally, Polverini et al. [92] proposed a real-time collision avoidance method for redundant manipulators in a human-robot interactive environment based on kinetostatic safety field developed entirely on the kinematic level, where the kinematic redundancy is exploited for simultaneous task execution and collision avoidance. Tamura et al. [93] proposed a method to determine whether a pedestrian performs an avoiding behaviour or not, and developed a robot that smoothly avoids a collision against the pedestrian.

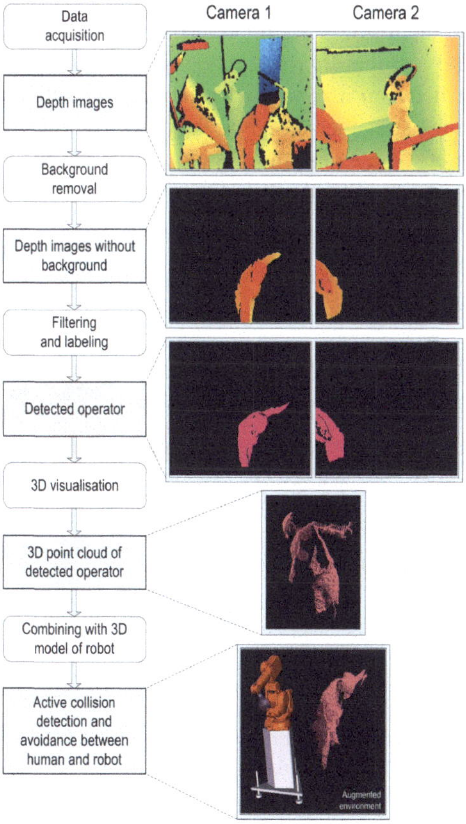

Fig. 9. Procedure of depth image processing (adapted from [47]).

Lately, some of the research attempted to deploy active collision avoidance strategies for safe human-robot collaboration. Calinon et al. [56] proposed an active control strategy based on task space control with variable stiffness, and combined it with a safety strategy for tasks requiring humans to move in the vicinity of robots. A risk indicator for human-robot collision is also defined, which modulates a repulsive force distorting the spatial and temporal characteristics of the movement according to the task constraints. Wang et al. [83] reported a novel methodology of real-time active collision avoidance in an augmented environment, where virtual 3D models of robots and real camera images of operators are used for monitoring and collision detection. The procedures of depth images acquisition and processing are illustrated in Fig. 9. The depth cameras are used to acquire and then process the depth images when the robot is in action. After the background removal via the back projection of the robot model to the images, the depth images of the detected operator by applying a noise-removal filter and a connected-component algorithm are converted to point clouds in the robot coordinate system, and merged into a single 3D point cloud after image registration. Finally, the single 3D point cloud of the operator is superimposed to the 3D model of the robot in augmented virtuality (as shown in the bottom image in Fig. 9). In [47], a depth camera-based approach for cost-effective real-time safety monitoring of a human-robot collaborative assembly cell was reported where active collision avoidance via path modification and robot control in real-time with zero robot programming was deployed. According to the result of the processed depth images, the minimum distance between the point cloud and the 3D robot model is calculated, and then the actions of avoiding the collision between a robot and a human are actively taken. In [82], the active collision avoidance strategies based on the detected collision were proposed and classified into four safety strategies: the system can alert an operator, stop a robot, move away the robot, or modify the robot's trajectory away from an approaching operator, as shown in Fig. 10.

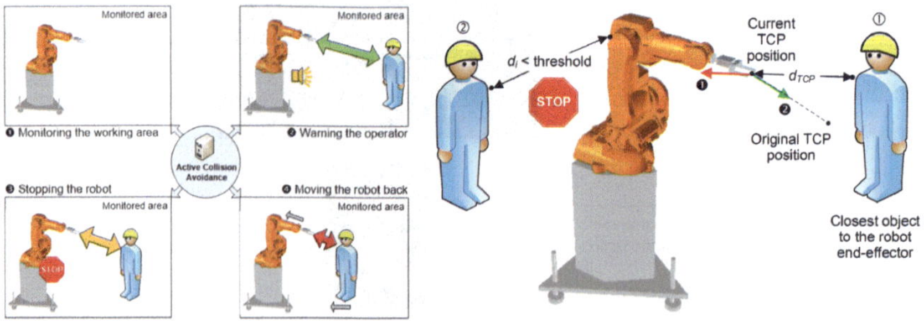

(a) (b)

Fig. 10. (a) Active collision avoidance strategies (adapted from [44]); (b) Robot behaviours in active collision avoidance (adapted from [47]).

3.3 HRC in Cyber-Physical Systems

Human-robot collaboration places the interplay of humans and machines into a cyber-physical environment where human and robotic agents interact in a shared work environment to solve some complex tasks which require the combination of their best, complementing competencies [1, 11, 28]. Cyber-physical systems (CPS) are systems of collaborating computational entities that are in intensive connection with the surrounding physical world and its ongoing processes, providing and using, at the same time, data-accessing and data-processing services available on the Internet [94–96]. The embedded computational entities and physical processes are connected through communication networks (i.e. the Internet of Things) [96] and can coordinate with each other in real time, i.e. physical processes are monitored and controlled by the embedded computers, and feedback loops are designed to allow both sides to affect each other [11]. According to the concept and characteristics of CPS, the human-robot collaborative system within a shared working environment can be viewed an advanced CPS, which is supported by the dynamic control algorithms and online monitoring devices [97]. The close and multi-directional interaction of virtual and physical entities forms a CPS where automated components and humans can be integrated into a cybernetic and collaborative environment combining their complementary strengths instead of the mutual restriction of their potentials. The physical and software components are tightly coupled to each other while the physical world (robots, monitors, products, etc.) is reflected and controlled by the cyber world (3D models, monitoring data, software, task plans, and so forth) [2]. CPS enables such human-robot collaboration in areas of dynamic task planning, active collision avoidance, and adaptive robot control [3].

In this section, representative examples of HRC in CPS are presented, and they are (1) a CPS application of HRC assembly, (2) a CPS-enabled remote HR system and ARbased CPS for remote HRC, and (3) remote HRC: a CPS application for a hazard manufacturing environment.

3.3.1 CPS Application in HRC Assembly

Compared with fully automated assembly or purely manual assembly [32, 98], HRC assembly can combine the skills of the robot and human, and offer the opportunities of improved overall performance. From the concept of CPS, HRC places the interaction and cooperation of humans and robots into a cyber-physical environment where the CPS related entities are the human operators and the robotic arms and the cyber part includes the control algorithms, communication service, and data exchange [1, 101].

The CPS-enabled collaboration scenario shown in Fig. 11 is a CPS industrial application of HRC assembly [94]. The concept of CPS is applied in HRC assembly, and the assembly scenario is that the robot is used to assemble a shaft and a washer, and insert the assembled parts in an output magazine. The operator's responsibility is to take out the assembled parts from the output magazine and fill fresh parts into an input magazine. The human is the physical part of the CPS and human instructions (voice commands, gesture instructions, and haptic interactions) are defined as the high-level control commands to control the assembly task execution. More details and procedures of the HRC assembly as a CPS application are illustrated as follows.

In the cyber part [82], the kinetic sensors above the workspace are used to monitor the whole working environment, including the robot and human and acquire the depth images. The background is removed from the depth images using the reference images captured in the calibration stage. Depth information related to the movement of the robot is subtracted as well from the captured depth information by projecting back the robot model to the depth images. When the human preforms the assembly tasks by following the assembly steps and procedures as discussed above, for example, the robot follows the operator's hand to deliver needed assistance, and the human instructs the robot to execute the specified task by voice and gesture commands (i.e. part grasping and posture change). The distance between the robot and human is calculated via a link between the human point cloud and a monitored virtual 3D model of the robot. The active collision avoidance is activated when the possible collision between the robot and human is detected. Avoiding collision is restricted only to control the robot at the time of picking and delivering parts. For seamless HRC, switching between following and avoiding behaviours of the robot can be implemented by a simple button press or through a voice command.

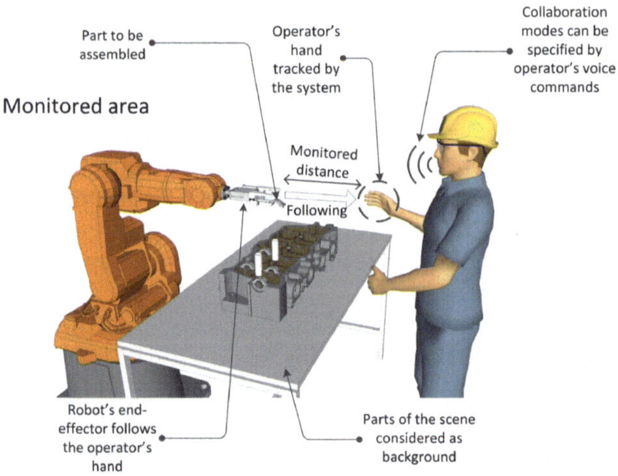

Fig. 11. HRC assembly: a CPS application (adapted from [94]).

3.3.2 AR-Based and CPS-Enabled Remote HRC

A CPS consists of a "cyber" part and a "physical" part [11]. In a CPS, sensors or other communicating tools embedded in physical entities are responsible for real-time data acquisitions. The data are then sent to the computational part of the system through a network/cloud environment (allowing decentralised control) [102]. The computational parties are responsible for monitoring and controlling the actions of the physical entity through embedded communicating equipment such as actuators.

The potential applications of CPS in manufacturing have attracted many attentions with the focus on the research and implementation of HRC production systems. In this section, a remote HRC system that applies the concept of CPS is presented as shown in

Fig. 12 [103], which achieves the features of haptic lead-through remotely with a local collaborative robot (UR5 [104]) and an industrial robot (ABB IRB 120 [105]).

In the physical part, a collaborative robot in a collaborative workstation is controlled by a human operator's haptic lead-through with respect to the display of the AR assembly parts from the cyber representations of a remote workstation, as shown in the left-bottom of Fig. 12. For the implementation of controlling the industrial robot, control parameters (joint positions of the collaborative robot) are monitored and acquired in real time for constructing the motions of the robot in the joint space and the Cartesian space via forwards kinematics, respectively. The motion of the collaborative robot can be replicated for and followed by the remote industrial robot. To implement the remote HRC, the control and communication blocks in the cyber part are designed and implemented. The joint positions of a collaborative robot can be obtained from the robot controller and transmitted into the cyber part of the system as a control input. The robot kinematics is encapsulated and integrated into the robot kinematics database module which is used for the kinematic control both for the robot in the collaborative workstation and the robot in the remote workstation. Forward kinematics allows the joint positions of the robot in the joint space to be transferred as the reference position in the Cartesian space, while the inverse kinematic function is the opposite of the forward kinematics. The robot end-effector can be controlled in real time by the embedded control algorithms where the robotic workspace position is transferred into the joint positions. Finally, the remote industrial robot can be controlled respectively in the remote workstation.

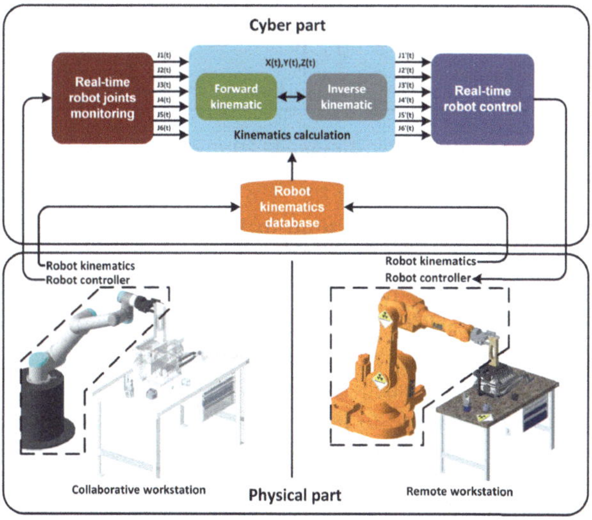

Fig. 12. CPS-enabled remote human-robot system (adapted from [103]).

The second case is an AR-based CPS for remote HRC assembly, as shown in Fig. 13. To achieve the remote HRC, the assembly parts and environment of the remote workstation should be acquired and displayed for the human operator physically located in the collaborative workstation. In the AR-based CPS system [106], the applied AR system provides cognitive feedback for the human operator during robot lead-through programming, and control parameters and commands generated in the collaborative workstation are used to control the industrial robot. The physical part in such a system includes the AR hardware, human operators, and robots. The video streaming of the remote workstation captured by the physical vision system is sent to the cyber part of the system. In the cyber part of the system, the embedded real-time objects recognition module is used to identify and recognise the diverse assembly objects. The recognised parts are further forwarded for AR models registration. The recognised assembly parts are tracked and registered in the right positions. In the physical part of the collaborative workstation, the relevant virtual models are displayed for the human operators.

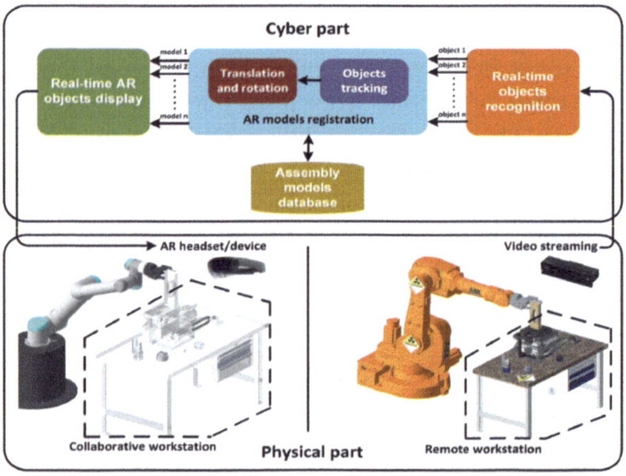

Fig. 13. AR-based CPS for remote HRC (adapted from [103]).

3.3.3 Remote HRC: A CPS Application in Hazard Manufacturing Environment

Human safety and health in any manufacturing setting is paramount. A hazard manufacturing environment for HRC, including radioactive component production and dangerous chemical product assembly, is not inevitable in the physical manufacturing activities. Therefore, to address such issues, a CPS application for the hazard manufacturing environment is proposed to implement a remote HRC, and the overview of the proposed remote HRC system is shown in Fig. 14, where the remote HRC system is comprised of a cyber part and a physical part. The physical part in the remote HRC includes two workstations: a collaborative workstation which includes a collaborative robot operated by a human operator and a remote workstation including an industrial robot and a collaborative working environment. An AR system designed for data exchange is shared by these two workstations in the physical part.

A collaborative ROS (Robot Operating System [107]) server and a remote ROS server included in the cyber part are used to support remote HRC control where the embedded industrial network takes the responsibility for data exchange and communication between the two servers. Both of the servers are installed with a ROS system that can support robot control and information processing functionalities. ROS provides a structural communication layer above the hosting computer operating system and brand-specific robot control system which is characterised with sensor connectivity, cross-brand robot control capability and software architecture flexibility. The robot action server in the collaborative ROS server is used to monitor and acquire the joint position of the collaborative robot in real time. The kinematic calculation node is then used to convert the joint position of the robot into the end-effector's position in the

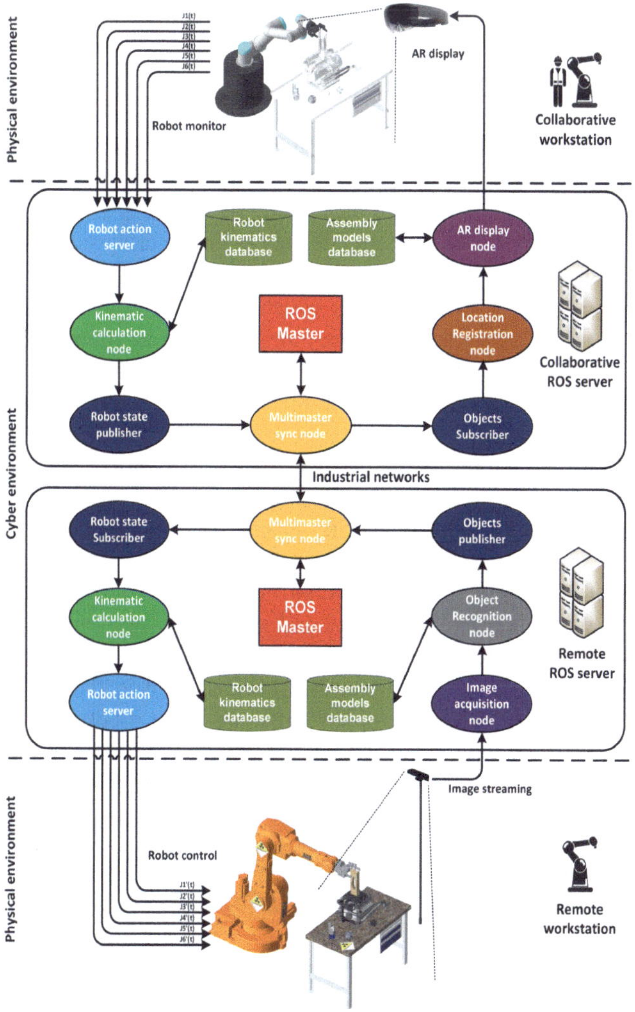

Fig. 14. Remote HRC: a CPS application in hazard environment (adapted from [103]).

Cartesian space, which is transmitted to a remote ROS server through Multimaster sync nodes [108]. In addition, the objects acquired from the remote workstation via Kinect sensors are recorded by the location registration node and displayed by the AR display node for the human operator in the collaborative workstation.

The Multimaster sync node in the remote ROS server is used for connectivity with the local server and data exchange (i.e. end-effector's position and robotic state). The kinematic calculation node converts the end-effector's position into the joint values of the industrial robot in the remote workstation. The image streaming of the assembly environment in the remote workstation is sent to the remote ROS server via the image acquisition node. Then, the object recognition node can recognise, localise, and track the assembly objects after the image processing. The recognised objects are further sent to a collaborative ROS server for AR display. Finally, the industrial robot in the remote workstation is controlled respectively by the relevant robot control nodes.

4 Multimodal Robot Control

In the context of manufacturing, today's robots are controlled by rigid native codes which in general are diverse due to the different robot manufacturers. Robot (re) programming with the requirements of the expert knowledge is also a big challenge for the wide use of industrial robots, especially for manufacturing companies. The conventional (re)programming of the industrial robots (i.e. manual online programming) is tedious, time-consuming and error-prone, which turns out to be a big bottleneck for robotic applications [109]. In addition, the training for programming of the industrial robots also requires high effort for users [110]. Although some of the control approaches attempted to implement the intuitive programming of the industrial robots such as the *tech-in* programming and programming by demonstration, current robot control approaches based on the rigid codes can no longer support the symbiotic HRC [111]. Therefore, a symbiotic HRC system needs to be enabled by a new means that is multimodal and symbiotic to facilitate any change during collaboration. To address such challenges, multimodal intuitive programming offers the potential programming-free approach for robot programming without the expert knowledge in which diverse communication channels between humans and robots (i.e. gesture, poster, voice, and haptic) can be used for HRC control in a shared working environment [1]. In recent years, multimodal intuitive programming has been applied in the design and implementation of natural HRC and control systems, and these robotic applications show high overall productivity with better product quality.

This section presents the robotic applications of the multimodal intuitive programming for HRC control. They are multimodal programming control, sensorless haptic control, and deep learning-based multimodal control.

4.1 Multimodal Programming Control

In HRC, robots that are allowed to work in the same work cell as human beings are required to adapt the change of pre-planned tasks for high-level flexibility and adaptability. However, most of the robots are controlled by pre-programmed control codes, which can no longer meet the demands. Conventional programming/reprogramming for robot control needs extra efforts of users when lot sizes of production are small. This motivates robotic engineers to develop robotic programming and control methods with a higher degree of adaptability. Consequently, many approaches have attempted to tackle such challenges towards an intuitive and programming-free HRC control.

Often, a robot is programmed for repetitive tasks. Once pre-planned tasks are changed, reprogramming of the robot is required. This operation is time-consuming and costly for industrial companies. Additionally, robot programming that requires expert knowledge is one of the big challenges. Many types of robots have been designed and developed for dedicated applications including medical robots, collaborative robots, humanoid robots, and industrial robots. While these robots are controlled by their own programming languages (i.e. KUKA Robot Language for KUKA Robot), they are not user-friendly. For the users of robots, the training for programming of the robots typically takes from several days up to weeks.

In HRC, it is important to have an intuitive way of communication with robots. Robot programming in Teach mode [112], as an intuitive method, allows the human operators without programming skills to teach the robot. In the *teach-in* mode, the robot is moved to each position by hands while pressing buttons on a teach pendant, and the reach positions are stored and integrated into a textual programme sequence. Finally, the robot can playback the points at full speed. Play-back programming offers a more direct approach of interaction through haptic guidance of the robot [113]. The approach, however, is limited by the poor transferability to robot systems in a large shared workspace. With *lead-through* programming, the robot is controlled to physically move through a task by an operator. Playback function can allow the robot to complete the task alone. Heavy-duty industrial robots widely used in the manufacturing industry make it difficult be adopted due to the efforts required from the operators and increased chances of hesitation or inaccuracies. Manual offline programming (i.e. programming with offline teaching software, scripting language, and offline-programming and simulation software), is a most often used programming approach in robotic research, while disadvantages of offline programming turn out to be a bottleneck for symbiotic HRC. Programming by demonstration [114] that end-users can teach robots new tasks without programming is a more intuitive programming paradigm which enables robots to autonomously perform new tasks. Meanwhile, this programming approach is interdisciplinary in a range of research fields including artificial intelligence (AI), sensor networks, image processing, automation control, and path planning. By the imitation learning of the human actions, a task or a programme used for imitating the actions can be automatically generated.

Recently, one promising approach of robot programming for HRC that has been actively explored is the multimodal intuitive programming where the specialised expert knowledge and high-level programming skills are not required [1, 115]. Multimodality refers to the use of diverse communication channels between a human and a robot which is important in the design of the natural HRI and control systems. Typically, speech, touch, gesture, and thoughts are communication methods natural to human, where speech and gesture recognition, and haptic interaction can be used to interpret theses communication methods [116]. A number of studies showed that multimodal design of industrial control and programming systems increases productivity.

As one of the most effective tools for human-human communication, speech commands have been extensively applied to robot intuitive programming and control [111]. Research results of the project MORPHA showed to be fundamental for numerous subsequent research activities in the field of intuitive control and programming for industrial and service robots with basic programming systems based on speech [117] and touch instructions [118, 119]. Perzanowski et al. [120] developed a multimodal human-robot interface design which incorporates both natural language understanding and gesture recognition as communication modes, and human operators interact with a mobile robot using natural language and gestures. In the project SMErobot [121], speech and gesture-based instructions were used for intuitive programming of industrial robots. Also, a demonstrator that combines both speech recognition and haptic control was developed to control a collaborative robot from Universal Robots which serves as an experimental platform and a pilot study [116]. Lastly, the EU project SYMBIO-TIC [17] developed a symbiotic HRC system where voice commands, gesture instructions, and haptic interactions are used for multimodal intuitive programming for HRC assembly tasks.

In the manufacturing context, the noisy background is a challenging factor for accurate recognition of the voice commands [111, 122]. Therefore, many researchers have explored the possibility to adopt human gestures as a form of nonverbal communication channel used for robot intuitive control and programming. According to Liu et al. [123], the gestures can be categorised into three types: body gestures, hand and arm gestures, and head and facial gestures. Meanwhile, In [124], Pavlovic stated that gestures describe intended movements which can be expressed by usually the arms and hands. Gestures can be divided into communicative and manipulative gestures in terms of the gesture's characters (as shown in Fig. 15) where communicative gestures can express the communicative intent while manipulative gestures can be used to manipulate the objects (i.e. translation, rotation, and deformation). The gesture-based instruction information can be transmitted and represented in three forms for control systems: pragmatic level (linguistic action), semantic level (meaning of the signs) and syntactic level (signs and rules). In multimodal HRC control, communicative gestures that are subdivided into symbols or actions are mostly combined with voice commands to implement the intuitive control and programming. The symbol gestures are labelled with linguistic functions by referential characters or modal function. For action gestures, gestures describe the movement itself. For example, mimetic gestures call for the imitation of the suggested action, whereby deictic gestures (pointing gestures) represent a local reference to the content of the action.

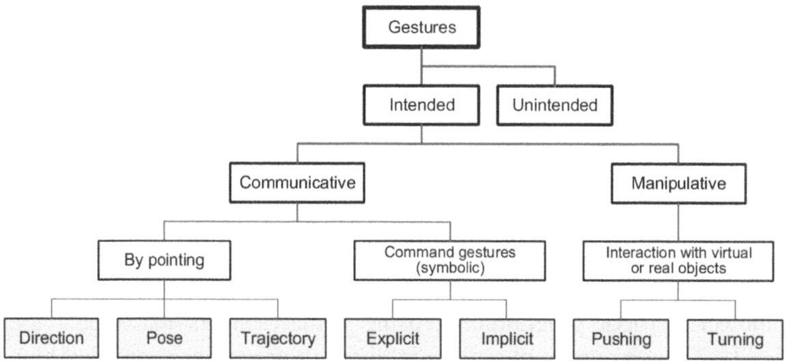

Fig. 15. Taxonomy of gestures in control and programming of industrial robots (adapted from [124]).

Figure 15 summarised the gesture-based applications in controlling and programming industrial robots. For communicative gestures, they are mainly defined as instructions for the robot motion control and motion specification. Meanwhile, the gesture-based manipulation of the objects can be translated into control commands in the intuitive programming. A comprehensive review of the gesture recognition used in robot programming for HRC manufacturing can be found in [123]. In HRC assembly, lightweight collaborative robots emerging in recent years make it possible for human operators in a shared workspace to haptically guide and control the robot in a more intuitive way in the *teach-in* mode [125–128]. In the *teach-in* mode, individual points or entire trajectories in the joint space are recorded by the control system and then the recorded trajectories can be played back after finishing the tasks. Collaborative robots often use playback programming methods for robot control and collaboration operations. To flexibly guide the robot, the installation of force/torque sensors and a compliance control strategy are needed. However, offline programming environment for some of the heavy-duty industrial robots is not available. Intuitive online programming such as playback programming and teach-in programming offers alternative approaches. Additionally, an object-oriented programming approach was developed to make dual-arm robot compliance programming simple by formulating bimanual actions which can be used for assembly operations such as BI-Approach, BI-Hold or BI-Insert-Extract in a natural way [129]. Inspired by this, a method for the intuitive programming of dual-arm robots was studied in which a task-oriented programming procedure is described, including a dual-arm robotics library and the human language [130]. The robotics library aspires after human-like capabilities and implements bi-manual operations. This intuitive programming framework is based on a service-oriented architecture and is developed in ROS. The user can easily interact with a dual-arm robot platform through depth sensors, noise-cancelling microphones and GUIs [131].

Several recent approaches for robot intuitive programming assisted by the advanced technologies have been reported. Ong et al. [132, 133] explored the potential of using an AR environment to facilitate immersive robot programming in unknown environments and also the use of an AR environment for facilitating intuitive robot programming. Lastly, a spatial programming approach for industrial robots based on gestures, AR, and task demonstration can be the support for different phases of the programming process as well as different levels of robot programming [134, 135].

In the last decades, to satisfy the users' demands for intuitive programming, robot programming software tools have been expected to be intuitive, efficient, and user-friendly. A new interface architecture towards minimum programming in HRC was developed in [136] where user interfaces, function blocks, functional modules and hardware are included. This designed interface allows engineers to customise the functions and tasks by dragging/dropping and linking the relevant FBs with minimum programming. In addition, a high-level robot programming method was presented to simplify industrial robot programming using visual sensors that detect the human motions [137]. The movement of a robot in different directions is enabled by a defined vocabulary of body and hand gestures, and a decoder application embedded in the robot controller is developed for translating the human messages into robot motions. The easy extensibility with new functionalities can be optional by integrating this method within an ROS-based open communication architecture. Multimodal intuitive programming of robots also requires embedding of complex algorithms on different levels of abstraction. The results achieved in different scientific literature have shown many efforts on integration and implementation of high-level functions for intuitive robot programming and control, as well as symbiotic collaboration between humans and robots, such as the use of function blocks.

4.2 Sensorless Haptic Control

Robots play important roles in manufacturing, especially in the assembly field, but they are far behind humans in terms of adaptability and flexibility. Complex assembly has demanding requirements in flexibility and adaptability, while fully automated assembly by robots or purely manual assembly cannot satisfy alone. A robot-assisted but human-guided assembly solution shows promise in handling such complex assembly processes as compared to full automation, where the flexibility and adaptability of a human can be combined with the speed, accuracy, and controllability of a robot. However, most solutions towards HRC assembly heavily rely on extra assistive systems, e.g., force sensors or haptic devices, to control external forces applied to a robot [138, 139].

Unfortunately, most heavy-duty industrial robots do not have built-in force/torque sensors. The installation of sensors for the robot systems proves to be not feasible because of the cost of sensors and difficulty on the mechanical integration. Sensorless haptic-driven robotics offers opportunities for addressing such challenges [140]. Figure 16 shows the overview of a system design for a sensorless haptic control for an industrial robot.

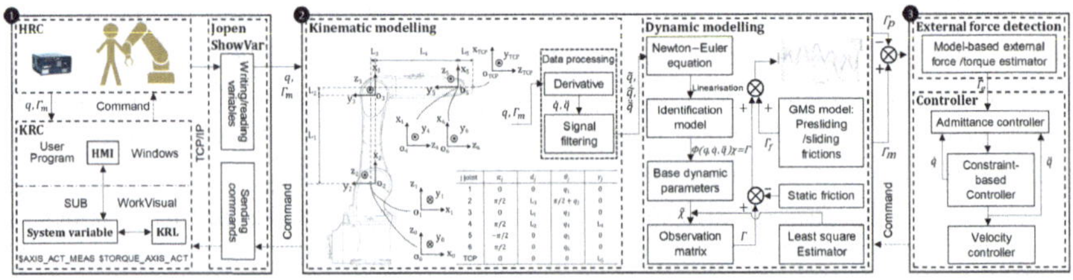

Fig. 16. System design of a sensorless haptic control for an industrial robot.

As shown in Fig. 16, the sensorless haptic control of an industrial robot incorporates an HRC and KRC (KUKA Robot Controller) module, a kinematics and dynamics module, a robotic control module, and a communication module.

In the HRC and KRC module, a sensorless haptic control strategy is developed to control the robot by a haptic command from a human. KRC is responsible for the execution of the motion and control commands. KRC is not open for robot users, and available feedback data from the robot controller are only joint position and joint torque. The axis-specific actual position of the robot and the current motor torque of an axis are stored in the system variable $AXIS_ACT_MEAS and $TORQUE_AXIS_ACT, respectively. To get access to the feedback data from and send movement commands to KRC, the communication between the human-machine interface (HMI) and KRC is built based on open-source cross-platform communication interface JOpenShowVar [141] which allows for reading and writing variables and data of the controlled manipulators.

The kinematics and dynamics module is responsible for kinematic and dynamic modelling of the industrial robot. The geometrical parameters of the industrial robot are acquired by the modified DH (Denavit–Hartenberg) based coordinate frames. The velocity and acceleration of the joints are obtained by the differential of the available joint positions. The acquired feedback data (joint positions and torques), velocity, and acceleration of the industrial robot are filtered to reduce the effect of the noise induced by the differential. The built-in kinematic model is used for the transfer between the joint positions in the joint space and the end-effector's positions in the Cartesian space, while the dynamic model is used to obtain the control output. The nonlinear Newton-Euler equation of the robotic dynamics is formulated as a linear identification model with certain inputs (joint positions, velocities, accelerations, and torques). Additionally, the friction in the robot's joints is also considered in the dynamic model for better control performance. The next step for the sensorless haptic control is the external force/torque detection which is the key and also a prerequisite of the robotic haptic control. The output of the dynamic model of the robot is defined as predicted torques, while the feedback data of the joint torques acquired from the robot controller are the measured torques. If a human operator does not apply any force on the end-effector of the robot, no external force is detected, and the predicted torques equal the measured ones. When the human operator applies the force on the robot, the external force induced by this can be detected by the external detection module where the external force is defined as the difference between the predicted force and measured force. The predicted force can be calculated by the Jacobian matrix based on the acquired joint torques.

The robotic control is to make the robot move when the external force is applied on the robot by a human operator. In the robotic control, an adaptive admittance controller is designed to transfer the external force into the reference position and velocity which are defined as the control output [142]. To make the interaction between the robot and human more natural and easier, adaptive admittance parameters are adopted in the controller. The control output commands are sent to the robot controller for the movement and task execution.

4.3 Deep Learning-Based Multimodal Control

In response to the needs for robot adaptability, a deep learning-based multimodal robot control approach for HRC, shown in Fig. 17 [111], is developed where three methods are integrated into the multimodality including voice recognition, human motion recognition, and body posture recognition. The proposed multimodal robot control approach allows a human operator to control a robot intuitively without programming rigid codes. Most of the existing multimodal robot control methods are not suitable or reliable for industrial applications because the feature representations are not shared across multiple modalities [37, 120, 143]. Therefore, the proposed multimodal control approach towards robust HRC is designed to deal with such challenges [111].

As shown in Fig. 17, the proposed multimodal control strategy consists of deep learning algorithms, sensors inputs, and an open-source robot control software. Body posture, hand motion, and voice commands that are acquired by cameras, Leap Motion sensors, and a microphone are defined as the control input of deep learning data processing block. The deep learning algorithms embedded for the model training are used to understand and classify the collected datasets, and the multi-channel communication signals are recognised and then these identified signals are analysed and fused to generate the robot control commands. Finally, these robot commands are sent to the robot controller in which the commands and task execution are performed.

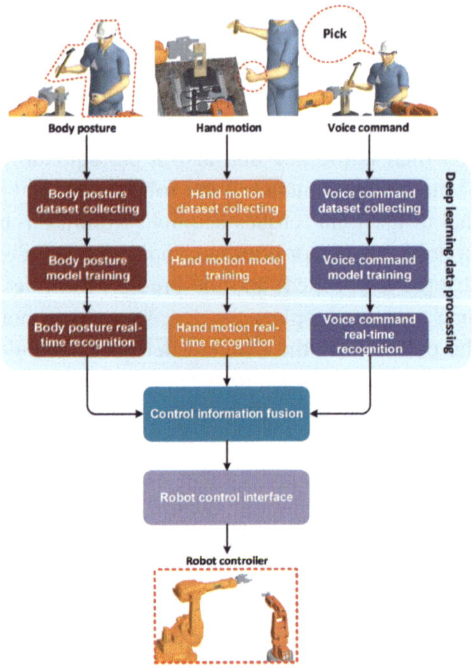

Fig. 17. Deep learning-based multimodal control for HRC (adapted from [111]).

Deep learning, as a branch of machine learning, has the capability of improving the performance of algorithms by learning through deep representations of the given dataset [144–146]. The network structure of deep learning algorithms consists of multiple network layers: one layer as input, several hidden layers to extract the deep features from the input layer, and one output layer for inference. In the network structure layers, the output of the previous layer is the input of the current layer. Different algorithms embedded in deep learning can be used for specified tasks. For example, Conventional Neural Network (CNN) is good at image processing [147–149], while long-short-term memory networks are for sequence-to-sequence modelling and auto-encoder feature learning [150–152]. Additionally, the combination of deep learning and multimodal fusion offers flexibility and capability in capturing hidden patterns from high-dimensional multimodal data, which can be used to analyse the multimodality to make decisions for robust multimodal HRC. Therefore, the architecture of the multimodal HRC is enabled by a microphone-based voice command recognition, Leap Motion-based hand motion recognition, and camera-based human body posture recognition as shown in Fig. 18 [153].

The multimodal recognition problem in Fig. 18 is defined as a type of machine learning problem, and the problem is formulated to facilitate a machine learning solution which is solved by three separate unimodal models. Lastly, the multimodal fusion is performed. Here, three deep learning models that are selected to recognise three multimodalities collected from sensors are CNN for voice command recognition, LSTM for hand motion recognition, and transfer learning-enabled human body motion recognition. The processes of modality recognition and multimodality fusion are introduced as Fig. 19.

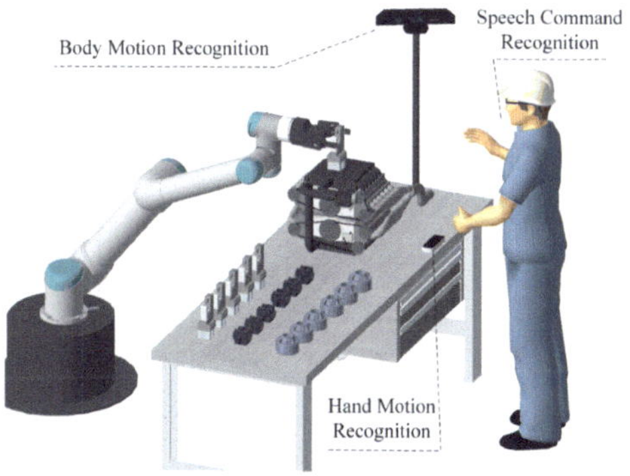

Fig. 18. Multimodal HRC enabled by voice command recognition, hand motion recognition, and human body motion recognition (adapted from [153]).

CNN for Voice Recognition: Before using CNN to process the voice commands, the voice command dataset is transformed into 2D spectrograms by Fast Fourier transform [154]. The convolution operation is formulated as a function of input maps and feature maps and feature map in which a *ReLU* activation function is adopted [147]. In the formulated convolution function, the weights of the input map are shared among each convolution neuron, and max-pooling outputs the maximum value of each of the local neighbour. Max-pooling makes each feature map invariant to local translations in the input map, which is also proven to be useful in CNN [155, 156]. In model training, the categorical cross-entropy is defined as the cost function.

The voice command dataset created by TensorFlow and AIY teams at Google is selected as the training dataset of voice command recognition, which consists of over 65,000 one-second audio recordings of 30 short words [157]. The voice commands used in the multimodal HRC are classified and labelled as *left*, *right*, *on*, *off*, *up*, and *down*. The recorded audio that is essentially a 1D vector of strength signals is transformed into a 2D matrix that can be treated as a single-channel image.

Hand Motion Recognition: Leap Motion as a non-contact sensor has the capacity of capturing hand motions by locating the fingers and hands within an interaction box. The Leap Motion can be connected with the mobile device (i.e. personal computer) for the data acquisition and the data output of Leap Motion offers a real-time representation of human hands with a series of timestamps, the finger positions, and hand position. To implement the human hand motion recognition in HRC, the Leap Motion dataset is built and multi-category classification is developed. A number of human-waving motions are generated and the recorded dataset from the Leap Motion is labelled as positive, while other random motion datasets collected from the sensors are labelled as negative. The labelled dataset is trained by the algorithm to identify the hand-waving motion. The hand motion dataset consists of a number of sequences of hand motions with six different categorical labels. The Leap Motion Controller [158] captures the direction and orientation of key hand joints and bones frequency of 100 Hz. The hand skeleton models can be built by tracking and capturing hand bones and joints. In each

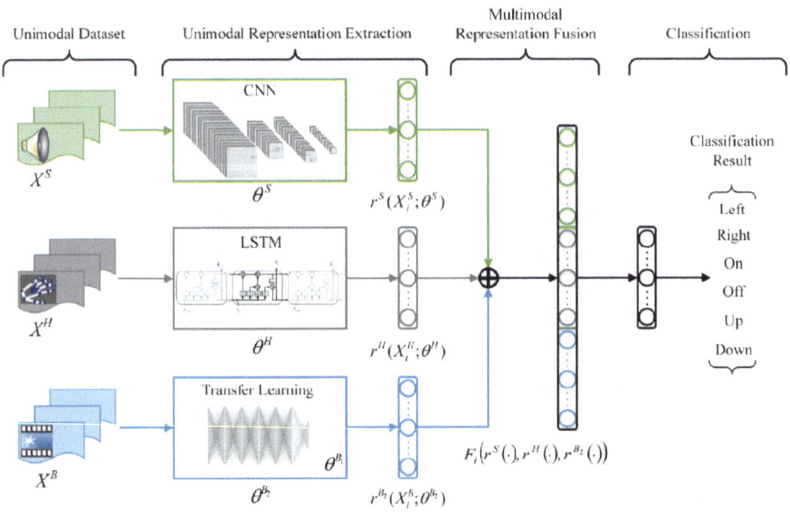

Fig. 19. Workflow of three unimodality recognition and multimodality fusion (adapted from [153]).

timestamp of the hand motion, 64 hand motion features which are defined by the parameters from the skeletons are captured. In addition, the same categories as the voice commands are defined for human motion recognition and body motion recognition.

Human Body Posture Recognition: A transfer learning-enabled body motion recognition method [159–161] is adopted to recognise body motion from video steams where transfer learning is proven to be an effective approach for image recognition. The training time of a new model can be reduced by reusing the knowledge from other networks. Here, transfer learning can be defined by a domain, a task, a learning source, and a target source where a domain consists of a feature space and a marginal probability distribution. When the domain is given, the task can be represented by a label space and a predictive function which can be learned from the training dataset. In the specific case of body motion recognition, a function is trained from the source domain and the source task with a large amount of labelled images. Since the source domain is in image format and the target domain is in video format, the data of the target domain can be sampled as sequences of images where the pre-trained network can act as a generic feature extractor to transfer the knowledge represented by parameters learned from the source domain. After the transfer learning-enabled feature extraction, the representation of the training dataset can be denoted a function of parameters transferred from the target predictive function and the training samples from body motion dataset. The network is further trained by minimising the cross-entropy loss, before being fed into a softmax layer for normalisation. To facilitate the training of the transfer learning, the collected video clips are sampled into sequences of images where the features are extracted from the image sequences by Inception-v3 [162] pre-trained model. The processed image sequences are stored and prepared as the input data for training multimodal fusion model [163].

Multimodal Fusion: These three modalities: voice command, hand motion, and body motion without the last layer are formulated as a function of the training samples from the corresponding datasets and the network parameters from these three models [153]. The three trained models are then fused by a concatenate function. The fused model can be further trained by minimising a loss function defined by cross-entropy. Finally, the optimised network is connected to a softmax function to normalise the output result. Here, data from the above three datasets are sampled randomly without semantic change. The representation trained from the unimodal models is concatenated and fed into a designed multilayer perceptron classifier which is used for identifying the commands.

The classification results in [153] show that the deep learning models outperform the traditional machine learning models (i.e. Support Vector Machine and Random Forest baseline models). The accuracy of the voice command recognition is improved significantly compared with the other two modalities. The deep learning models can capture some of the hidden patterns that cannot be recognised by the traditional machine learning models. Knowledge representations learned by the multimodal representations fusion process can contribute to the accuracy improvement of the identification. In the HRC enabled by the proposed multimodal fusion, the human operators are expected to control the robot intuitively with user-friendly modalities. Meanwhile, the robot control commands can be active once all the three modalities are in place and a right output is obtained from the fusion model.

5 Conclusions and Future Research Directions

This paper presents an overview of human-robot collaboration in manufacturing. Research on HRC has been active for many years. Despite the advancement and applications of HRC in recent years, confusions exist in different forms of human-robot relationships. These relationships and the unique characteristics with clear definitions need to be classified and analysed. This paper is aimed to provide detailed treatments with a focus on HRC in manufacturing together with existing challenges and recent technological advancements. Within the context of HRC, topics covered consist of scenarios of HRC in manufacturing and multimodal robot control. In the industrial scenarios, symbiotic HRC assembly, safety of HRC in manufacturing, and HRC in cyber-physical systems are introduced, and also multimodal robot control enabled by multimodal programming, sensorless haptic interaction, and deep learning-based multimodal control with voice commands, hand motion, and body motion is presented.

In [1], a comprehensive summary of the future research directions and challenges of HRC in manufacturing was presented with 12 research topics. With the support of the latest technologies of sensing, communication, AI, AR and robot control, HRC will find its way to practical applications on shop floors in factories of the future.

References

1. Wang, L., Gao, R., Váncza, J., Krüger, J., Wang, X.V., Makris, S., Chryssolouris, G.: Symbiotic human-robot collaborative assembly. CIRP Ann. **68**(2), 701–726 (2019)
2. Wang, X.V., Kemény, Z., Váncza, J., Wang, L.: Human–robot collaborative assembly in cyber-physical production: classification framework and implementation. CIRP Ann. Manuf. Technol. **66**(1), 5–8 (2017)
3. Schmidtler, J., Knott, V., Hölzel, C., Bengler, K.: Human centered assistance applications for the working environment of the future. Occup. Ergon. **12**(3), 83–95 (2015)
4. Wang, X.V., Seira, A., Wang, L.: Classification, personalised safety framework and strategy for human-robot collaboration. In: Proceedings of International Conference on Computers & Industrial Engineering, CIE 2018 December (2018)
5. ISO 10218-1:2011 Robots and robotic devices — Safety requirements for industrial robots — Part 1: Robots
6. Lien, T.K., Verl, A.: Cooperation of human and machines in assembly lines. CIRP Ann. Manuf. Technol. **58**, 628–646 (2009)
7. Yanco, H.A., Drury, J.: Classifying human-robot interaction: an updated taxonomy. In: Conference Proceedings - IEEE International Conference on Systems, Man and Cybernetics, vol. 3, pp. 2841–2846 (2004)
8. De Santis, A., Siciliano, B., De Luca, A., Bicchi, A.: An atlas of physical human–robot interaction. Mech. Mach. Theory **43**(3), 253–270 (2008)
9. Krüger, J., Lien, T.K., Verl, A.: Cooperation of human and machines in assembly lines. CIRP Ann. **58**(2), 628–646 (2009)

10. Pellegrinelli, S., Moro, F.L., Pedrocchi, N., Molinari Tosatti, L., Tolio, T.: A probabilistic approach to workspace sharing for human–robot cooperation in assembly tasks. CIRP Ann. **65**(1), 57–60 (2016)
11. Wang, L., Haghighi, A.: Combined strength of holons, agents and function blocks in cyber-physical systems. J. Manuf. Syst. **40**, 25–34 (2016)
12. Leitão, P.: Agent-based distributed manufacturing control: a state-of-the-art survey. Eng. Appl. Artif. Intell. **22**(7), 979–991 (2009)
13. Monostori, L., Váncza, J., Kumara, S.R.T.: Agent-based systems for manufacturing. CIRP Ann. **55**(2), 697–720 (2006)
14. Bi, Z.M., Wang, L., Lang, S.Y.T.: Current status of reconfigurable assembly systems. Int. J. Manuf. Res. **2**(3), 303–328 (2007)
15. Musić, S., Hirche, S.: Control sharing in human-robot team interaction. Annu. Rev. Control **44**, 342–354 (2017)
16. Janni, P.: Human-robot collaboration: a survey. Lingua Nostra **67**(3–4), 122–124 (2006)
17. EU project: SYMBIO-TIC. http://www.symbio-tic.eu/
18. Sadrfaridpour, B., Wang, Y.: Collaborative assembly in hybrid manufacturing cells: an integrated framework for human-robot interaction. IEEE Trans. Autom. Sci. Eng. **15**(3), 1178–1192 (2018)
19. Elena, R., Brian, A.: Levels of human and robot collaboration for automotive manufacturing. In: Proceedings of the Workshop on Performance Metrics for Intelligent Systems, March 2012 (2012)
20. Tan, J.T.C., Duan, F., Zhang, Y., Watanabe, K., Kato, R., Arai, T.: Human-robot collaboration in cellular manufacturing: design and development. In: 2009 IEEE/RSJ International Conference on Intelligent Robots and Systems, IROS 2009, pp. 29–34 (2009)
21. Sadrfaridpour, B., Saeidi, H., Wang, Y.: An integrated framework for human-robot collaborative assembly in hybrid manufacturing cells. In: International Conference on Automation Science and Engineering, November 2016, pp. 462–467 (2016)
22. Wang, L.: From intelligence science to intelligent manufacturing. Engineering **5**(4), 615–618 (2019)
23. Thomas, C., Matthias, B., Kuhlenkötter, B.: Human - robot collaboration – new applications in industrial robotics. In: International Conference in Competitive Manufacturing, January, pp. 293–299 (2016)
24. Kardos, C., Kovács, A., Váncza, J.: Decomposition approach to optimal feature-based assembly planning. CIRP Ann. **66**(1), 417–420 (2017)
25. Kardos, C., Váncza, J.: Mixed-initiative assembly planning combining geometric reasoning and constrained optimization. CIRP Ann. **67**(1), 463–466 (2018)
26. Michalos, G., Makris, S., Tsarouchi, P., Guasch, T., Kontovrakis, D., Chryssolouris, G.: Design considerations for safe human-robot collaborative workplaces. Procedia CIRP **37**, 248–253 (2015)
27. Rosenthal, S., Biswas, J., Veloso, M.: An effective personal mobile robot agent through symbiotic human-robot interaction. In: Proceedings of International Joint Conference on Autonomous Agents and Multiagent Systems, AAMAS, vol. 2, pp. 915–922 (2010)
28. Monostori, L., Kádár, B., Bauernhansl, T., Kondoh, S., Kumara, S., Reinhart, G., Sauer, O., Schuh, G., Sihn, W., Ueda, K.: Cyber-physical systems in manufacturing. CIRP Ann. **65** (2), 621–641 (2016)
29. Wang, L., Balasubramanian, S., Norrie, D.H,, Brennan, R.W.: Agent-based control system for next generation manufacturing (1998)
30. Shen, W., Wang, L., Hao, Q.: Agent-based distributed manufacturing process planning and scheduling: a state-of-the-art survey. IEEE Trans. Syst. Man Cybern. Part C Appl. Rev. **36** (4), 563–577 (2006)

31. Cherubini, A., Passama, R., Crosnier, A., Lasnier, A., Fraisse, P.: Collaborative manufacturing with physical human-robot interaction. Robot. Comput. Integr. Manuf. **40**, 1–13 (2016)
32. Ji, W., Yin, S., Wang, L.: A virtual training based programming-free automatic assembly approach for future industry. IEEE Access **6**, 43865–43873 (2018)
33. International Electrotechnical Commission: International Standard of Function Blocks – Part 1: Architecture, IEC 61499, pp. 1–111 (2005)
34. Tsarouchi, P., Michalos, G., Makris, S., Athanasatos, T., Dimoulas, K., Chryssolouris, G.: On a human–robot workplace design and task allocation system. Int. J. Comput. Integr. Manuf. **30**(12), 1272–1279 (2017)
35. Liu, S., Wang, Y., Wang, X.V., Wang, L.: Energy-efficient trajectory planning for an industrial robot using a multi-objective optimisation approach. Procedia Manuf. **25** (August), 517–525 (2018)
36. Ranz, F., Hummel, V., Sihn, W.: Capability-based task allocation in human-robot collaboration. Procedia Manuf. **9**, 182–189 (2017)
37. Tsarouchi, P., Makris, S., Chryssolouris, G.: Human–robot interaction review and challenges on task planning and programming. Int. J. Comput. Integr. Manuf. **29**(8), 916–931 (2016)
38. Tsarouchi, P., Makris, S., Chryssolouris, G.: On a human and dual-arm robot task planning method. Procedia CIRP **57**, 551–555 (2016)
39. Wang, L., Givehchi, M., Schmidt, B., Adamson, G.: Robotic assembly planning and control with enhanced adaptability. Procedia CIRP **3**(1), 173–178 (2012)
40. Wang, L., Schmidt, B., Givehchi, M., Adamson, G.: Robotic assembly planning and control with enhanced adaptability through function blocks. Int. J. Adv. Manuf. Technol. **77**(1–4), 705–715 (2015)
41. ISO 10218-2:2011 Robots and robotic devices — Safety requirements for industrial robots — Part 2: Robot systems and integration
42. Lasota, P.A., Fong, T., Shah, J.A.: A survey of methods for safe human-robot interaction. Found. Trends Robot. **5**(3), 261–349 (2017)
43. Haddadin, S., De Luca, A., Albu-Schäffer, A.: Robot collisions: a survey on detection, isolation, and identification. IEEE Trans. Robot. **33**(6), 1292–1312 (2017)
44. Schmidt, B., Wang, L.: Active collision avoidance for human-robot collaborative manufacturing. In: The 5th International Swedish Production Symposium, pp. 81–86 (2012)
45. Vasic, M., Billard, A.: Safety issues in human-robot interactions. In: Proceedings - IEEE International Conference on Robotics and Automation, pp. 197–204 (2013)
46. Carlson, J., Murphy, R.R.: How UGVs physically fail in the field. IEEE Trans. Robot. **21** (3), 423–437 (2005)
47. Schmidt, B., Wang, L.: Depth camera based collision avoidance via active robot control. J. Manuf. Syst. **33**(4), 711–718 (2014)
48. KUKA System Software 8.3 Operating and Programming Instructions for End Users (2019)
49. ISO 13855:2010 Safety of machinery — Positioning of safeguards with respect to the approach speeds of parts of the human body
50. Kosuge, K., Yoshida, H., Taguchi, D., Fukuda, T., Hariki, K., Kanitani, K., Sakai, M.: Robot-human collaboration for new robotic applications. In: Proceedings of IECON - Industrial Electronics Conference, vol. 2, pp. 713–718 (1994)
51. ISO/TS 15066:2016 Robots and robotic devices — Collaborative robots
52. Pfitzner, C., Antal, W., Hess, P., May, S., Merkl, C., Koch, P., Koch, R., Wagner, M.: 3D multi-sensor data fusion for object localization in industrial applications. In: ISR/Robotik 2014; 41st International Symposium on Robotics Proceedings, pp. 1–6. VDE (2014)

53. Michalos, G., Kousi, N., Karagiannis, P., Gkournelos, C., Dimoulas, K., Koukas, S., Mparis, K., Papavasileiou, A., Makris, S.: Seamless human robot collaborative assembly – an automotive case study. Mechatronics **55**, 194–211 (2018)
54. Rocco, P., Zanchettin, A.M., Matthias, B., Ding, H., Ceriani, N.M.: Safety in human-robot collaborative manufacturing environments: metrics and control. IEEE Trans. Autom. Sci. Eng. **13**(2), 882–893 (2015)
55. Heinzmann, J., Zelinsky, A.: Quantitative safety guarantees for physical human-robot interaction. Int. J. Robot. Res. **22**(7-8 Special Issue), 479–504 (2003)
56. Calinon, S., Sardellitti, I., Caldwell, D.G.: Learning-based control strategy for safe human-robot interaction exploiting task and robot redundancies. In: IEEE/RSJ 2010 International Conference on Intelligent Robots and Systems, IROS 2010 - Conference Proceedings, pp. 249–254 (2010)
57. Laffranchi, M., Tsagarakis, N.G., Caldwell, D.G.: Safe human robot interaction via energy regulation control. In: 2009 IEEE/RSJ International Conference on Intelligent Robots and Systems, IROS 2009, pp. 35–41 (2009)
58. Geravand, M., Shahriari, E., De Luca, A., Peer, A.: Port-based modeling of human-robot collaboration towards safety-enhancing energy shaping control. In: Proceedings - IEEE International Conference on Robotics and Automation, June 2016, pp. 3075–3082 (2016)
59. Meguenani, A., Padois, V., Da Silva, J., Hoarau, A., Bidaud, P.: Energy based control for safe human-robot physical interaction (2017). https://doi.org/10.1007/978-3-319-50115-4
60. Yamada, Y., Hirasawa, Y., Huang, S., Umetani, Y., Suita, K.: Human-robot contact in the safeguarding space. IEEE/ASME Trans. Mechatron. **2**(4), 230–236 (1997)
61. Suita, K., Yamada ,Y., Tsuchida, N., Imai, K., Sugimoto, N.: A failure-to-safety "Kyozon" system with simple contact detection and stop capabilities for safe human-autonomous robot coexistence, pp. 3089–3096 (1965)
62. Wang, L., Wang, X.V., Wang, L., Wang, X.V.: Safety in human-robot collaborative assembly. Cloud-Based Cyber-Phys. Syst. Manuf. (2018). https://doi.org/10.1007/978-3-319-67693-7_9
63. Dombrowski, U., Stefanak, T., Reimer, A.: Simulation of human-robot collaboration by means of power and force limiting. Procedia Manuf. **17**, 134–141 (2018)
64. Kokkalis, K., Michalos, G., Aivaliotis, P., Makris, S.: An approach for implementing power and force limiting in sensorless industrial robots. Procedia CIRP **76**, 138–143 (2018)
65. Takakura, S., Murakami, T., Ohnishi, K.: Approach to collision detection and recovery motion in industrial robot. In: Proceedings of IECON - Industrial Electronics Conference, vol. 2, pp. 421–426 (1989)
66. De Luca, A., Flacco, F.: Integrated control for pHRI: collision avoidance, detection, reaction and collaboration. In: Proceedings IEEE RAS EMBS International Conference on Biomedical Robotics and Biomechatronics, pp. 288–295 (2012)
67. De Luca, A., Mattone, R.: Sensorless robot collision detection and hybrid force/motion control. In: Proceedings - IEEE International Conference on Robotics and Automation, pp. 999–1004 (2005)
68. Kosuge, K., Matsumoto, T., Morinaga, S.: Collision detection system for manipulator based on adaptive control scheme. Trans. Soc. Instrum. Control Eng. **39**(6), 552–558 (2003)
69. Wu, D., Liu, Q., Xu, W., Liu, A., Zhou, Z., Pham, D.T.: External force detection for physical human-robot interaction using dynamic model identification. In: International Conference on Intelligent Robotics and Applications, October, vol. 10462, pp. 581–592 (2017)
70. Krüger, J., Nickolay, B., Heyer, P., Seliger, G.: Image based 3D surveillance for flexible man-robot-cooperation. CIRP Ann. **54**(1), 19–22 (2005)

71. Corrales, J.A., Candelas, F.A., Torres, F.: Safe human–robot interaction based on dynamic sphere-swept line bounding volumes. Robot. Comput. Integr. Manuf. **27**(1), 177–185 (2011)
72. Ebert, D.M., Henrich, D.D.: Safe human-robot-cooperation: image-based collision detection for industrial robots. In: IEEE International Conference on Intelligent Robots and Systems, October, vol. 2, pp. 1826–1831 (2002)
73. Henrich, D., Gecks, T.: Multi-camera collision detection between known and unknown objects. In: 2008 2nd ACM/IEEE International Conference on Distributed Smart Cameras, ICDSC 2008 (2008). https://doi.org/10.1109/icdsc.2008.4635717
74. Vogel, C., Walter, C., Elkmann, N.: A projection-based sensor system for ensuring safety while grasping and transporting objects by an industrial robot. In: 2015 IEEE International Symposium on Robotics and Intelligent Sensors, pp. 271–277 (2016)
75. Vogel, C., Walter, C., Elkmann, N.: A projection-based sensor system for safe physical human-robot collaboration. In: IEEE International Conference on Intelligent Robots and Systems, pp. 5359–5364 (2013)
76. Tan, J.T.C., Arai, T.: Triple stereo vision system for safety monitoring of human-robot collaboration in cellular manufacturing. In: Proceedings - 2011 IEEE International Symposium on Assembly and Manufacturing, ISAM 2011, pp. 1–6 (2011)
77. Schiavi, R., Bicchi, A., Flacco, F.: Integration of active and passive compliance control for safe human-robot coexistence. In: Proceedings - IEEE International Conference on Robotics and Automation, pp. 259–264 (2009)
78. Fischer, M., Henrich, D.: 3D collision detection for industrial robots and unknown obstacles using multiple depth images. In: Advances in Robotics Research: Theory, Implementation, Application, pp. 111–122 (2009)
79. Rybski, P., Anderson-Sprecher, P., Huber, D., Niessl, C., Simmons, R.: Sensor fusion for human safety in industrial workcells. In: IEEE International Conference on Intelligent Robots and Systems, pp. 3612–3619 (2012)
80. Dániel, B., Korondi, P., Thomessen, T.: Joint level collision avoidance for industrial robots. IFAC Proc. **45**(22), 655–658 (2012)
81. Morato, C., Kaipa, K.N., Zhao, B., Gupta, S.K.: Toward safe human robot collaboration by using multiple kinects based real-time human tracking. J. Comput. Inf. Sci. Eng. (2014). https://doi.org/10.1115/1.4025810
82. Mohammed, A., Schmidt, B., Wang, L.: Active collision avoidance for human–robot collaboration driven by vision sensors. Int. J. Comput. Integr. Manuf. **30**(9), 970–980 (2017)
83. Wang, L., Schmidt, B., Nee, A.Y.C.: Vision-guided active collision avoidance for human-robot collaborations. Manuf. Lett. **1**(1), 5–8 (2013)
84. Graham, J.H.: A fuzzy logic approach for safety and collision avoidance in robotic systems. Int. J. Hum. Factors Manuf. **5**(4), 447–457 (1995)
85. Haddadin, S., Albu-sch, A., De Luca, A., Hirzinger, G.: Collision detection and reaction: a contribution to safe physical human-robot interaction, pp. 22–26 (2008)
86. De Luca, A., Ferrajoli, L.: Exploiting robot redundancy in collision detection and reaction. In: 2008 IEEE/RSJ International Conference on Intelligent Robots and Systems, IROS, pp. 3299–3305 (2008)
87. Parusel, S., Haddadin, S., Albu-Schäffer, A.: Modular state-based behavior control for safe human-robot interaction: a lightweight control architecture for a lightweight robot. In: Proceedings - International Conference on Robotics and Automation, pp. 4298–4305 (2011)

88. Flacco, F., Kroeger, T., De Luca, A., Khatib, O.: A depth space approach for evaluating distance to objects: with application to human-robot collision avoidance. J. Intell. Robot. Syst. Theory. Appl. **80**, 7–22 (2015)
89. Balan, L., Bone, G.M.: Real-time 3D collision avoidance method for safe human and robot coexistence, pp. 276–282 (2006)
90. Seto, F., Kosuge, K., Hirata, Y.: Self-collision avoidance motion control for human robot cooperation system using RoBE. In: 2005 IEEE/RSJ International Conference on Intelligent Robots and Systems, IROS, pp. 50–55 (2005)
91. Ratsamee, P., Mae, Y., Ohara, K., Takubo, T., Arai, T.: Human-robot collision avoidance using a modified social force model with body pose and face orientation. Int. J. Humanoid Robot. **10**(1), 1–24 (2013)
92. Polverini, M.P., Zanchettin, A.M., Rocco, P.: Real-time collision avoidance in human-robot interaction based on kinetostatic safety field. In: IEEE International Conference on Intelligent Robots and Systems (IROS), pp. 4136–4141 (2014)
93. Tamura, Y., Fukuzawa, T., Asama, H.: Smooth collision avoidance in human-robot coexisting environment. In: IEEE/RSJ 2010 International Conference on Intelligent Robots and Systems, IROS 2010 - Conference Proceedings, pp. 3887–3892 (2010)
94. Wang, L., Törngren, M., Onori, M.: Current status and advancement of cyber-physical systems in manufacturing. J. Manuf. Syst. **37**, 517–527 (2015)
95. Wang, L., Wang, X.V., Wang, L., Wang, X.V.: Latest advancement in CPS and IoT applications. In: Cloud-Based Cyber-Physical Systems in Manufacturing, pp. 33–61. Springer, Cham (2018)
96. Zhang, Y., Liu, S., Liu, Y., Yang, H., Li, M., Huisingh, D., Wang, L.: The 'Internet of Things' enabled real-time scheduling for remanufacturing of automobile engines. J. Clean. Prod. **185**, 562–575 (2018). https://doi.org/10.1016/j.jclepro.2018.02.061
97. Liu, S., Zhang, Y., Liu, Y., Wang, L., Wang, X.V.: An 'Internet of Things' enabled dynamic optimization method for smart vehicles and logistics tasks. J. Clean. Prod. **215**, 806–820 (2019)
98. Liu, S., Zhang, G., Wang, L.: IoT-enabled dynamic optimisation for sustainable reverse logistics. Procedia CIRP **69**, 662–667 (2018)
99. Wang, L., Wang, X.V., Wang, L., Wang, X.V.: Cloud robotics towards a CPS assembly system. In: Cloud-Based Cyber-Physical Systems in Manufacturing, pp. 243–259. Springer, Cham (2018)
100. Weidner, R., Kong, N., Wulfsberg, J.P.: Human hybrid robot: a new concept for supporting manual assembly tasks. Prod. Eng. **7**(6), 675–684 (2013)
101. Adamson, G., Wang, L., Moore, P.: Feature-based control and information framework for adaptive and distributed manufacturing in cyber physical systems. J. Manuf. Syst. **43**, 305–315 (2017)
102. Wang, X.V., Wang, L., Mohammed, A., Givehchi, M.: Ubiquitous manufacturing system based on Cloud: a robotics application. Robot. Comput. Integr. Manuf. **45**, 116–125 (2017)
103. Liu, H., Wang, L.: Remote human-robot collaboration: a cyber-physical system application for hazard manufacturing environment. J. Manuf. Syst. **54**, 24–34 (2019)
104. Kebria, P.M., Al-Wais, S., Abdi, H., Nahavandi, S.: Kinematic and dynamic modelling of UR5 manipulator. In: 2016 IEEE International Conference on Systems, Man, and Cybernetics, SMC 2016 - Conference Proceedings, pp. 4229–4234 (2017)
105. Product specification IRB 120
106. Hashimoto, S., Ishida, A., Inami, M., Igarash, T.: TouchMe: an augmented reality based remote robot manipulation. In: 21st International Conference on Artificial Reality and Telexistence, pp. 1–6 (2011)

107. Quigley, M., Gerkey, B., Conley, K., Faust, J., Foote, T., Leibs, J., Berger, E., Wheeler, R., Ng, A.Y.: ROS: an open-source robot operating system. In: IECON 2015 - 41st Annual Conference of the IEEE Industrial Electronics Society (Figure 1), pp. 4754–4759 (2015)
108. Hernandez, S., Fernando, J., Cotarelo, H.: Multi-master ROS systems (2015)
109. Pan, Z., Polden, J., Larkin, N., Van Duin, S., Norrish, J.: Recent progress on programming methods for industrial robots. In: Joint International Symposium on Robotics and 6th German Conference on Robotics 2010, ISR/ROBOTIK 2010, vol. 1, pp. 619–626 (2010)
110. Denkena, B., Denkena, B., Wörn, H., Apitz, R., Bischoff, R., Hein, B., Kowalski, P., Mages, D., Schuler, H.: Roboterprogrammierung in der Fertigung, vol. 9, pp. 656–60. Springer (2005)
111. Liu, H., Fang, T., Zhou, T., Wang, Y., Wang, L.: Deep learning-based multimodal control interface for human-robot collaboration. Procedia CIRP **72**, 3–8 (2018)
112. Perzylo, A., Somani, N., Profanter, S., Rickert, M., Knoll, A.: Toward efficient robot teach-in and semantic process descriptions for small lot sizes. In: Proceedings of Robotics: Science and Systems (RSS), Workshop on Combining AI Reasoning and Cognitive Science with Robotics, pp. 1–7 (2015)
113. Cevzar, M., Petrič, T., Babič, J.: Sensor-based loops and branches for playback-programmed robot systems. Mech. Mach. Sci. **49**(Raad), 797–804 (2018)
114. Nippun Kumaar, A.A., Sudarshan, T.S.B.: Mobile robot programming by demonstration. In: International Conference on Emerging Trends on Engineering Science, Technology, ICETET, pp. 206–209 (2011)
115. Akan, B., Ameri, A., Cürüklü, B., Asplund, L.: Intuitive industrial robot programming through incremental multimodal language and augmented reality. In: Proceedings – IEEE International Conference on Robotics and Automation, pp. 3934–3939 (2011)
116. Gustavsson, P., Syberfeldt, A., Brewster, R., Wang, L.: Human-robot collaboration demonstrator combining speech recognition and haptic control. Procedia CIRP **63**, 396–401 (2017)
117. Box, P.O., Izoellner, I.: Using gesture and speech control for commanding a robot assistant, pp. 454–459 (2002)
118. Albu-Schäffer, A., Hirzinger, G.: Cartesian impedance control techniques for torque controlled light-weight robots. In: Proceedings – IEEE International Conference on Robotics and Automation, May, vol. 1, pp. 657–663 (2002)
119. Grunwald, G., Schreiber, G., Hirzinger, G.: Touch: the direct type of human interaction with a redundant service robot, pp. 347–352 (2001)
120. Perzanowski, D., Schultz, A.C., Adams, W., Marsh, E., Bugajska, M.: Building a multimodal human-robot interface. IEEE Intell. Syst. Their Appl. **16**(1), 16–21 (2001)
121. Neto, P., Norberto Pires, J., Paulo Moreira, A.: High-level programming and control for industrial robotics: using a hand-held accelerometer-based input device for gesture and posture recognition. Ind. Rob. **37**(2), 137–147 (2010)
122. Kardos, C., Kemény, Z., Kovács, A., Pataki, B.E., Váncza, J.: Context-dependent multimodal communication in human-robot collaboration. Procedia CIRP **72**, 15–20 (2018)
123. Liu, H., Wang, L.: Gesture recognition for human-robot collaboration: a review. Int. J. Ind. Ergon. **68**, 355–367 (2018)
124. Pavlovic, V.I., Sharma, R., Huang, T.S.: Visual interpretation of hand gestures for human-computer interaction a review. IEEE Trans. Pattern Anal. Mach. Intell. **19**(7), 677–695 (1997)
125. Krüger, J., Wang, L., Verl, A., Bauernhansl, T., Carpanzano, E., Makris, S., Fleischer, J., Reinhart, G., Franke, J., Pellegrinelli, S.: Innovative control of assembly systems and lines. CIRP Ann. **66**(2), 707–730 (2017)

126. Reed, K.B., Peshkin, M.A.: Physical collaboration of human-human and human-robot teams. IEEE Trans. Haptics **1**(2), 108–120 (2008)
127. Huang, S., Ishikawa, M., Yamakawa, Y.: An active assistant robotic system based on high-speed vision and haptic feedback for human-robot collaboration. In: Proceedings IECON 2018 - 44th Annual Conference of the IEEE Industrial Electronics Society, vol. 1, pp. 3649–3654 (2018)
128. Feth, D., Groten, R., Peer, A., Buss, M.: Haptic human-robot collaboration: Comparison of robot partner implementations in terms of human-likeness and task performance. Presence Teleoperators Virtual Environ. **20**(2), 173–189 (2011)
129. Surdilovic, D., Yakut, Y., Nguyen, T.M., Pham, X.B., Vick, A., Martin Martin, R.: Compliance control with dual-arm humanoid robots: design, planning and programming. In: 2010 10th IEEE-RAS International Conference on Humanoid Robots, pp. 275–281 (2010)
130. Makris, S., Tsarouchi, P., Surdilovic, D., Krüger, J.: Intuitive dual arm robot programming for assembly operations. CIRP Ann. **63**(1), 13–16 (2014)
131. Kousi, N., Michalos, G., Aivaliotis, S., Makris, S.: An outlook on future assembly systems introducing robotic mobile dual arm workers. Procedia CIRP **72**, 33–38 (2018)
132. Ong, S.K., Chong, J.W.S., Nee, A.Y.C.: Methodologies for immersive robot programming in an augmented reality environment. In: Proceedings - Graph 2006 International Conference on Computer Graphics and Interactive Techniques in Australasia and South East Asia, January, pp. 237–244 (2006)
133. Chong, J.W.S., Ong, S.K., Nee, A.Y.C., Youcef-Youmi, K.: Robot programming using augmented reality: an interactive method for planning collision-free paths. Robot. Comput. Integr. Manuf. **25**(3), 689–701 (2009)
134. Lambrecht, J., Kleinsorge, M., Rosenstrauch, M., Krüger, J.: Spatial programming for industrial robots through task demonstration. Int. J. Adv. Robot. Syst. (2013). https://doi.org/10.5772/55640
135. Lambrecht, J., Kruger, J.: Spatial programming for industrial robots based on gestures and Augmented Reality. In: IEEE International Conference on Intelligent Robots and Systems, pp. 466–472 (2012)
136. Ji, W., Wang, Y., Liu, H., Wang, L.: Interface architecture design for minimum programming in human-robot collaboration. Procedia CIRP **72**, 129–134 (2018)
137. Tsarouchi, P., Athanasatos, A., Makris, S., Chatzigeorgiou, X., Chryssolouris, G.: High level robot programming using body and hand gestures. Procedia CIRP **55**, 1–5 (2016)
138. Rozo, L., Calinon, S., Caldwell, D.G., Jiménez, P., Torras, C.: Learning physical collaborative robot behaviors from human demonstrations. IEEE Trans. Robot. **32**(3), 513–527 (2016)
139. Lamy, X., Collédani, F., Geffard, F., Measson, Y., Morel, G.: Human force amplification with industrial robot: study of dynamic limitations. In: IEEE/RSJ 2010 International Conference on Intelligent Robots and Systems, IROS 2010 - Conference Proceedings, pp. 2487–2494 (2010)
140. Yao, B., Zhou, Z., Wang, L., Xu, W., Liu, Q., Liu, A.: Sensorless and adaptive admittance control of industrial robot in physical human−robot interaction. In: Robotics and Computer-Integrated Manufacturing, November 2017, vol. 51, pp. 158–168 (2018)
141. Sanfilippo, F., Hatledal, L.I., Zhang, H., Fago, M., Pettersen, K.Y.: Controlling Kuka industrial robots: flexible communication interface JOpenShowVar. IEEE Robot. Autom. Mag. **22**(4), 96–109 (2015)
142. Landi, C.T., Ferraguti, F., Sabattini, L., Secchi, C., Fantuzzi, C.: Admittance control parameter adaptation for physical human-robot interaction, pp. 2911–2916 (2017)

143. Cherubini, A., Passama, R., Meline, A., Crosnier, A., Fraisse, P.: Multimodal control for human-robot cooperation. In: IEEE International Conference on Intelligent Robots and Systems, pp. 2202–2207 (2013)
144. Lecun, Y., Bengio, Y., Hinton, G.: Deep learning. Nature **521**(7553), 436–444 (2015)
145. Silver, D., Schrittwieser, J., Simonyan, K., et al.: Mastering the game of Go without human knowledge. Nature **550**(7676), 354–359 (2017)
146. Deng, L., Yu, D., Dahl, G.E., Mohamed, A., Jaitly, N., Senior, A., Vanhoucke, V., Nguyen, P., Sainath, T.N., Kingsbury, B.: IEEE Signal Process. Mag. **24**(99), c1–c1 (2008)
147. Gonzalez, T.F.: ImageNet classification with deep convolutional neural networks. In: Handbook of Approximation Algorithms and Metaheuristics, pp. 1–1432 (2007)
148. Hershey, S., Chaudhuri, S., Ellis, D.P.W., et al.: CNN architectures for large-scale audio classification. In: ICASSP, IEEE International Conference on Acoustics, Speech, and Signal Processing - Proceedings, pp. 131–135 (2017)
149. Abdel-Hamid, O., Mohamed, A.R., Jiang, H., Penn, G.: Applying convolutional neural networks concepts to hybrid NN-HMM model for speech recognition. In: ICASSP, IEEE International Conference on Acoustics, Speech, and Signal Processing – Proceedings, July 2015, pp. 4277–4280 (2012)
150. Hochreiter, S.: Long short-term. Memory **1780**, 1735–1780 (1997)
151. Sundermeyer, M., Schlüter, R., Ney, H.: LSTM neural networks for language modeling. In: 13th Annual Conference of the International Speech Communication Association 2012, INTERSPEECH 2012, vol. 1, pp. 194–197 (2012)
152. Zhu, W., Lan, C., Xing, J., Zeng, W., Li, Y., Shen, L., Xie, X.: Co-occurrence feature learning for skeleton based action recognition using regularized deep LSTM networks. In: 30th AAAI Conference on Artificial Intelligence, AAAI 2016, vol. ii, pp. 3697–3703 (2016)
153. Liu, H., Fang, T., Zhou, T., Wang, L.: Towards robust human-robot collaborative manufacturing: multimodal fusion. IEEE Access **6**, 74762–74771 (2018)
154. Zhang, H., McLoughlin, I., Song, Y.: Robust sound event recognition using convolutional neural networks. In: ICASSP, IEEE International Conference on Acoustics, Speech, and Signal Processing – Proceedings, 2015 August, pp. 559–563 (2015)
155. Jarrett, K., Kavukcuoglu, K., Ranzato, M., LeCun, Y.: What is the best multi-stage architecture for object recognition? In: Proceedings of the IEEE International Conference on Computer Vision, pp. 2146–2153 (2009)
156. Boureau, Y., Bach, F.: Learning mid-level features for recognition (2010)
157. van der Maaten, L., Hinton, G.: Visualizing data using t-SNE. J. Mach. Learn. Res. **1**, 1–48 (2008)
158. Weichert, F., Bachmann, D., Rudak, B., Fisseler, D.: Analysis of the accuracy and robustness of the Leap Motion Controller. Sensors (Switzerland) **13**(5), 6380–6393 (2013)
159. Oquab, M., Bottou, L., Laptev, I., Sivic, J.: Learning and transferring mid-level image representations using convolutional neural networks. In: Proceedings IEEE Conference on Computer Vision and Pattern Recognition, pp. 1717–1724 (2004)
160. Pan, S.J., Yang, Q.: A survey on transfer learning. IEEE Trans. Knowl. Data Eng. **22**(10), 1345–1359 (2010)
161. Wang, P., Liu, H., Wang, L., Gao, R.X.: Deep learning-based human motion recognition for predictive context-aware human-robot collaboration. CIRP Ann. **67**(1), 17–20 (2018)
162. Donahue, J., Jia, Y., Vinyals, O., Hoffman, J., Zhang, N., Tzeng, E., Darrell, T.: DeCAF: a deep convolutional activation feature for generic visual recognition. 31st International Conference on Machine Learning, ICML 2014, pp. 988–996 (2014)
163. Ngiam, J., Khosla, A., Kim, M., Nam, J., Lee, H., Ng, A.Y.: Multimodal deep learning. In: Proceedings 28th International Conference on Machine Learning, ICML 2011, pp. 689–696 (2011)

A Bending Test of the Additively Produced Porous Sample

Katarina Monkova[1,2(✉)], Peter Pavol Monka[1,2], Jozef Tkac[1], and Jan Vanca[1]

[1] Faculty of Manufacturing Technologies, Technical University in Kosice, Presov, Slovakia
katarina.monkova@tuke.sk
[2] Faculty of Technology, UTB Zlin, Vavreckova 275, Zlin, Czech Republic

Abstract. With Industry 4.0, additive-manufacturing methods will be widely used to produce small batches of customized products that offer construction advantages, such as complex, lightweight designs. One type of products that belong to the sophisticated components, producible by 3D printing technology, is a lattice structure. The article deals with the three points bending test of the cylindrical samples. FDM (Fused Deposition modelling) technique and ABS (Acrylonitrile Butadiene Styrene) material were selected for the samples production. Two types of samples (with a simple lattice structure and fully filled by material) were analyzed experimentally and numerically. The results showed that in spite of the material saving at the porous structure production, the lattice structure with body-centred cube cell is not very suitable for the components loaded by bending, because a unit of material at this type of sample is able to carry down less stress compared to the sample fully filled by the material.

Keywords: Additive technology · Sophisticated component · Cellular sample · Bending test

1 Introduction

The current situation in manufacturing requires to be the components manufactured quickly and flexibly not only from the view of production efficiency but also from the view of enterprise competitiveness. All the companies are looking for better responsiveness, because - what is important - is a customer, and what is important for the customer is to get what they ask for (including function, aesthetics or competitive prices but also, for example, recyclability or energetically efficient products). Companies seek to find better methods and improvements. In the competitive market, customer-based production is what companies have in mind.

A new wave within a current trend in automation and data exchange in manufacturing technologies that include a combination of Cyber-physical systems, the Internet of Things, Cloud computing and Cognitive computing offers Industry 4.0. A vital part of Industry 4.0 is also 3D printing technology. While 3D printers hit the market way back in the 80 s, commercially viable 3D printing has been possible only in the last

L. Wang et al. (Eds.): AMP 2020, LNME, pp. 59–68, 2020.
https://doi.org/10.1007/978-3-030-46212-3_3

decade [1]. 3D printing technology today is at a stage where companies are starting to realize significant, tangible new value for themselves and their customers using them.

Modern method of 3D printing allows creating complex features of parts geometry by their direct integration into the production process. When using this technology, designers and engineers have more freedom at the design creating, since practically all components that can be modelled in a virtual environment of 3D software it is possible to produce by this technology. There are no limitations relate e.g. to the production of cavities as it is in the case of conventional technologies, because metallic material is applied only into the places for which it is intended. The additive manufacturing can be with the great advantage used for example for the manufacturing of moulds for plastic injection, in which cooling channels are passed in different ways, along different trajectories, with variable cross-section, etc. Despite the fact that the process of additive manufacturing appears to be simple, experience of foreign producers using this technology say it is not true.

2 Porous Structures and Additive Technology

The world develops permanently. According to [2], the progress of engineering disciplines is indisputably bounded with the continuous learning and production of knowledge and vice versa. According to Aristotle, knowledge is developed through experience and empirical observations which are the foundations of intellectual knowledge. Acquisition of information coming from experience interacted by intellectual cognition can be transformed to the formulation of conclusions and statements that can be considered as "elements" of knowledge.

In recent years, two new fields which change approach to the manufacture might be observed. In the area of manufacturing, there are additive manufacturing techniques which develop fast. A few years ago, these techniques allowed only to execute elements only for spatial visualizations of designed parts. Currently, we can print 3D elements made of different types of plastics or metals. Additive printing technologies allow the production of elements which have hitherto been considered as the non-technological and not possible to made from the point of view of production technology, capacity and cost-effectiveness.

Additive manufacturing, also commonly referred to as 3D printing, is the process of building a three-dimensional (3D) structure or component from the ground up, generally, layer by layer. The CAD model, which represents the part to be manufactured, is converted into STL format. Nearly every AM technology uses the STL file format. The term STL was derived from STereoLithograhy, which was the first commercial AM technology from 3D Systems in the 1990 s. Considered a de facto standard, STL is a simple way of describing a CAD model in terms of its geometry alone. It works by removing any construction data, modelling history, etc., and approximating the surfaces of the model with a series of triangular facets. The minimum size of these triangles can be set within the most CAD software and the objective is to ensure the models created do not show any obvious triangles on the surface. This STL file is then transferred into the AM machine software which slices the model into 2D slices with specified thicknesses. Different AM machines use different slicing formats such as SLI

(System Layer Interface) by 3D systems. This is followed by setting-up the processing parameters and preparing the materials [3, 4].

A build preparation of the part begins with its digital positioning onto the build platform and the necessary support structures are strategically placed onto the part. Technicians then go through a layer-by-layer review of the build to ensure that all of the features are adequately supported for a successful build. Once the review is complete, the technicians will digitally create the required amount of part copies needed for a particular order before it is ultimately sent to the printing machine.

The building process depends on the material and the selected technology. While the additive technology for metal parts is commonly based on powder bed fusion using a laser or electron beam, the Plastic Additive Manufacturing (AM) uses primarily the FDM (Fused Deposition Modelling) or FFF (Fused Filament Fabrication) methods. There is no difference between FFF and FDM, only a trademark issue [5].

The special area of produced components incorporates the parts with internal structures. Their design is considered in all industries as the most sophisticated factor of the manufacturing process. The conventional production of these types of components is often costly, time-consuming and technically very difficult. Mechanical properties of porous structures depend primary on material, from which they are produced, and on topology including geometry and relative density. The relative density or "Volume ratio *Vr*" is a parameter that expresses how many percentages of the cell space is filled by the material. It is given by equation [6]

$$V_r = \frac{Volume\ of\ solid\ phase}{Total\ volume} 100[\%] \tag{1}$$

Two common types of cellular structures include stochastic cellular solids and non-stochastic periodic cellular solids – called as regular porous structures. Mechanical properties of the regular porous structure are better controlled [7].

Basic building units of geometrically defined porous structures are cells that are created by simple bodies, e.g. cube, cuboid, cylinder or sphere. These geometries are usually well modelled in CAD applications. Regular structures are those which have a characteristic 2D or 3D periodicity (i.e. repeating and ordered unit cells). The honeycomb structure is a good example of a prismatic cellular solid [8]. Figure 1 shows a schematic profile representation of some common and less common prismatic cellular solids.

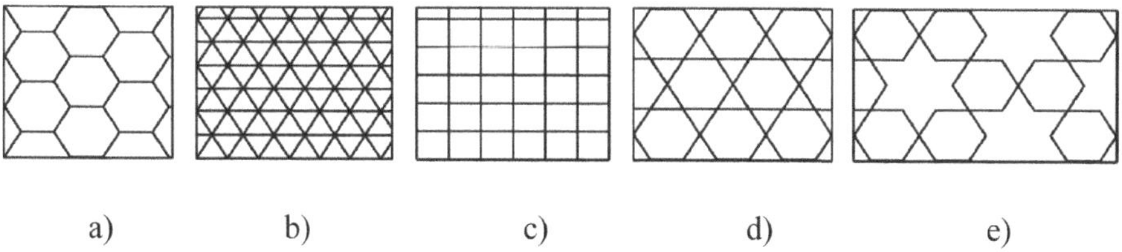

Fig. 1. Five samples of prismatic cellular topologies (a) hexagonal honeycomb, (b) triangulated, (c) square, (d) Kagomé, e) star-hex

There is also another class of sophisticated material that is based upon 3-dimensional lattices of trusses. 3-dimensional periodic lattice solids have uniform structures that are based on repeating unit cells in three co-ordinates. These 3D periodic lattice structures consist of strait prismatic strut making up the unit cell with specific cell size and thickness They have been shown to have superior mechanical properties including energy absorption, strength and stiffness, as well as easier control of structure properties, better load sustaining capabilities and higher surface area densities than stochastic and prismatic cellular solids [9]. Some of these structures including Hexagon, Sphere, Octagon, Cube, Tower, Rhombic, Diamond, Diagonal cross, X-cross, Circular-pipe, and Triangular etc. are shown in Fig. 2.

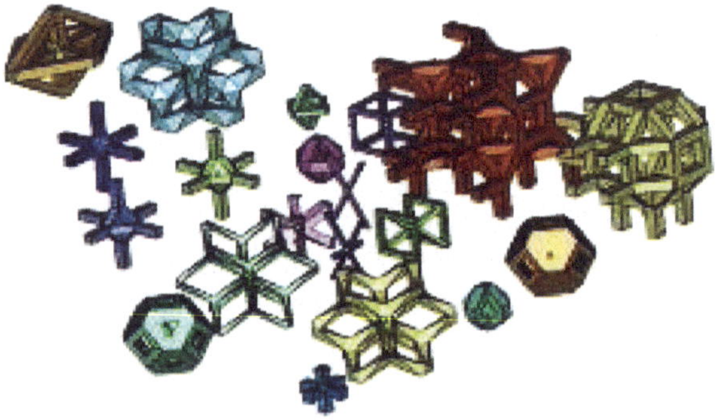

Fig. 2. Unit cell types in Magics software

Lattice truss structures used for the cores of sophisticated components has been designed as a means to achieve efficient load support and other functionality such as cross-flow heat exchange or high-intensity dynamic load mitigation. The emerging applications for metal periodic lattice structures range from ultra-lightweight multi-functional structures to automobile, medical and aerospace components. A new challenge lies in looking for the applications of the structures in other industrial and commercial components. That is why the authors decided to investigate the possibility to use a simple cellular structure in such a conventional industrial component as it is a shaft. Within this goal, preliminary research has been already done and it is described in the article.

3 Bending Stress Investigation

As it has been already said, the internal cellular structure can give to a final product extraordinary combination of properties such are high strength, stiffness along with low weight and good absorption of energy. The goal of the presented research was to compare the bending stress of a shaft, which core is filled be a lattice structure, with a shaft that is fully filled by a material, and consequently to evaluate achieved results in relation to the volume of a material that was used for the production both of these shafts. FDM (Fused Deposition modelling) technique and ABS (Acrylonitrile Butadiene Styrene) material were selected for the samples production because of their good availability at the authors' workplaces and due to less financial expenses compared with a metal printing method.

ABS is a part of the thermoplastic polymers' family. As its name implies, ABS is created from Acrylonitrile, Butadiene and Styrene polymers. It is popular in large part because it has great plastic properties. It is lightweight, has good impact strength, it is abrasion resistant and affordable. Moreover, ABS polymers withstand a lot of chemical formulas. The melting temperature of ABS plastic is 200 °C (392 °F), making it ideal for use in relatively safe machines that are easy to operate (the safety of household machines is important) [10]. This is an ideal material to manufacture low-cost prototypes and architectural models for engineers or research departments, as well as to create low-cost medical prostheses or material handling equipment. The physical properties of this type of plastic, like its tensile strength and stiffness, and its heat deflection temperature, are real advantages. It can also be used for mechanical purpose, or for its electrical properties. In addition to its chemical resistance and mechanical properties, ABS material has a good surface quality and is flame retardant. There have been recent concerns about the toxicity of the ABS material used in printing when it is brought to its melting point. In fact, some studies indicate that ABS emit fumes when it exits the extrusion head at its melting temperature [11, 12].

ABS is a material, which is done using primarily FDM (Fused Deposition modelling) or FFF (Fused Filament Fabrication) 3D printers. Plastics come in the form of a long filament wound around a spool. By means of a print head, a molten layer of plastic is deposited on the print bed, which then adheres. Once the first layer has been drawn, the print bed drops and a new layer is built on the previous layer. This is repeated several times, ultimately resulting in a 3D printed model. The principle of FDM technique is shown in Fig. 3 [13].

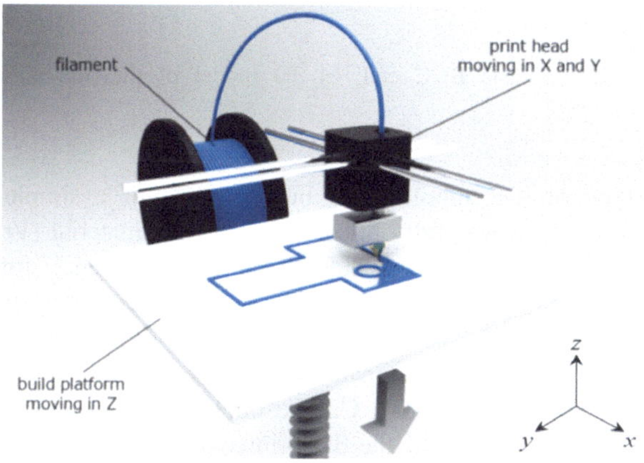

Fig. 3. The principle of FDM technique.

Within the preliminary research, two types of samples were modelled, produced and tested. The first type was designed in the tube shape with a structure at the volume ratio of $Vr = 44\%$.

The samples sizes were: diameter of ϕ29 mm and length of 200 mm (based on testing device requirements), the thickness of outside cylindrical shell was 2 mm. A simple Body-Centred Cubic (BCC) unit cell sizes 5 × 5 × 5 mm with strengthened linear edges and diameter of a strut 1 mm was selected for this research. The basic cell has been regularly patterned in all three directions. The basic BCC cell, the virtual model of a sample with the in structure, as well as the basic dimensions of the sample are presented in Fig. 4.

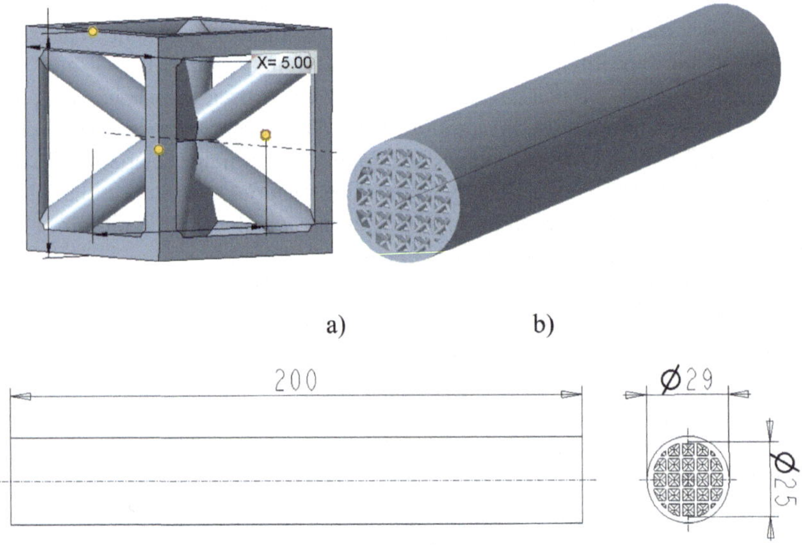

c)

Fig. 4. Tested sample, a) basic BCC cell, b) 3D model of the porous sample, c) basic dimensions of the sample.

The second type of the samples was in the shape of a simple cylinder ϕ29 200 mm, volume of which was fully filled with the ABS material ($Vr = 100\%$).

Six samples of both types (porous and fully filled by material) for pressure tests were produced from the material ABSplus-P430 Ivory plastics using FDM technique. The 3D printer Prusa i3 Mk2 was selected for the samples production. The melting temperature of the filament is 300 °C, while the nozzle temperature was 255 °C. The temperature of the basement was 100 °C. The thickness of the deposited layer was 0,254 mm given by the producer of the 3D printer device and the diameter of ABS filament was ϕ 1.75 mm. The speed of solid infill was 40 mm/s and at external perimeters, it was 30 mm/s.

Bending tests were carried out according to the standard EN ISO 604 using the ZWICK 1456 testing machine at an ambient temperature of 22 °C and at a relative humidity of 60%. The distance between supports was 170 mm. Crossbeam feed at experiments was 20 mm/min, a round of push thorn was 5 mm. The orientation of cross-section was the same at all samples with an internal lattice structure. The sample position at a testing machine at three points bending test, orientation at a force load and detail of a sample after damage are presented in Fig. 5.

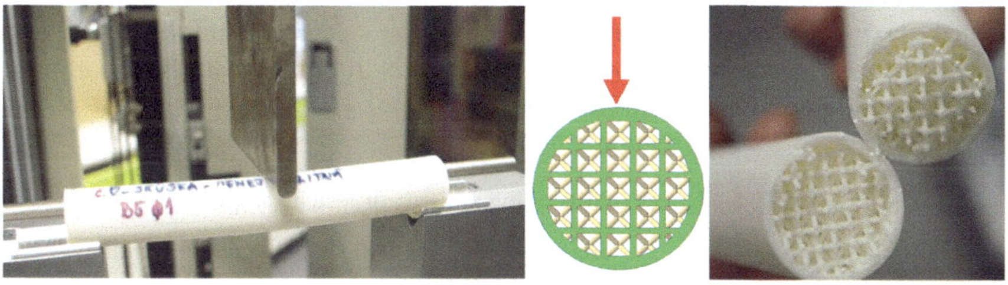

a) b) c)

Fig. 5. Bending test, a) position of a sample, b) orientation of sample cross-section, c) sample after damage

Results were evaluated using TestXpert software. Obtained dependences of the force on deformation for the samples with the material in full-volume are presented in Fig. 6.

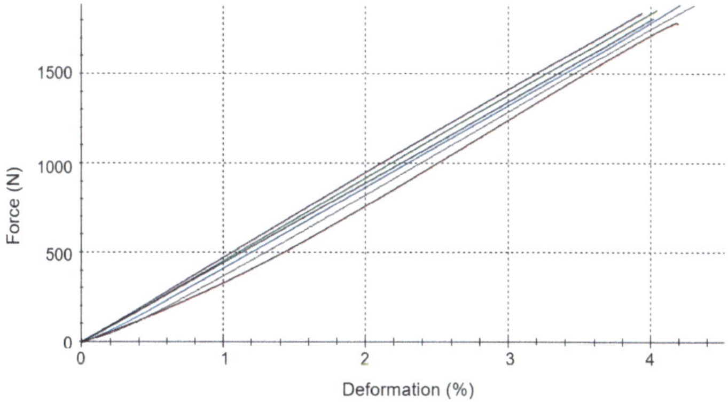

Fig. 6. The dependencies of loading force on deformation for full-volume samples.

To compare the results of both approach - experimental and numerical, the bending stresses were calculated based on measured forced $F_{\max}$ at the damage, while a sectional bending modulus W_o has been specified by means of the 3D model within the PTC Creo software. They are

W_o = 1212.29 mm^3 for the samples with the lattice structure, and
W_o = 2394.38 mm^3 for the samples fully filled with material.
The following equation for the bending stress computation was used [14]:

$$\sigma = \frac{F_{\mathrm{Max}} l}{Wo} = \frac{Mo_{\mathrm{Max}}}{Wo} \tag{2}$$

where $F_{\max}$ is maximal force; l is the arm on which the force acts; $Mo_{\max}$ is maximal bending moment; σ is bending stress and W_o is sectional bending modulus. The recorded maximal forces at the samples damage, the calculated stresses at individual measures, as well as the average bending stress for both types of samples (porous and fully filled with the material), are in Table 1.

Table 1. Forces and bending stresses of tested samples

Sample type	Sample number	F_{max} (N)	Bending moment Mo (Nmm)	Bending stress σ_o (MPa)	Average bending stress σ_o (MPa)
Vr_44	1	259	11 007.5	9.08	8.89
	2	273	11 602.5	9.57	
	3	258	10 965.0	9,04	
	4	232	9 860.0	8.13	
	5	256	10 880.0	8.97	
	6	244	10 370.0	8.55	
Vr_100	1	1890	80 325.0	33.55	32.86
	2	1845	78 412.5	32.75	
	3	1856	78 880.0	32.94	
	4	1848	78 540.0	32.80	
	5	1830	77 775.0	32.48	
	6	1840	78 200.0	32.66	

Numerical simulation was done in the software PTC Creo. The constraints and the force load based on measured values were defined by means of surface regions. The number of elements of the tetrahedrons type was 758 248. An example of created mesh and the simulation of a sample behaviour are presented in Fig. 7.

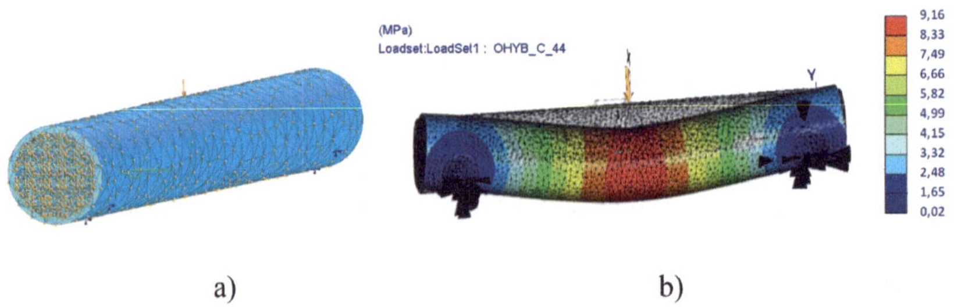

Fig. 7. A numerical analysis of the porous sample, a) mesh, b) bending stress behavior

Comparison of the results obtained by experimental testing and by numerical analysis are shown in Table 2. It can be said that the results are comparable and that the experimental tests confirmed the boundary conditions of the numerical analysis.

Due to the possibility to compare achieved results, the obtained stresses were recalculated on the stress per volume unit of material. It was found out that one cubic centimetre of the material at the sample with a porous structure is able to carry down 0.158 MPa, while the same unit of the material of a sample with 100% volume ratio of the material is able to carry down the bending stress 0.249 MPa.

Table 2. The comparison of bending stresses

Sample type	Bending stress	
	Experimental testing (MPa)	Numerical analysis (MPa)
Vr_44	8.89	9.16
Vr_100	32.86	35.14

4 Conclusions

The companies are getting to be part of another industrial revolution called Industry 4.0. As the speed, reliability, safety and quality of 3D printers improve, and the cost reduces, 3D printers are set to play an important role in this digital transformation of the industry. As the performance of 3D printers improves rapidly and the cost decreases, new opportunities will arise that will take 3D printing ever closer to mass production. As 3D printing develops, the range of products that can be manufactured is also set to grow. Rate of development of specialized printing materials, integration of digital security to protect IP and certification of 3D products by regulatory agencies will boost adoption of 3D printers in Industry 4.0 But of course, it is the willingness of innovative manufacturers who choose to embrace the tenets of Industry 4.0 and digitalize their businesses fast who will benefit the most [15].

Within the presented research, the bending stresses of two types of cylindrical samples were evaluated. The first type of samples was designed with a simple lattice structure created by body-centred cube cell. Samples of the second type were fully filled by the material. Numerically obtained results showed good compliance with the values of stresses achieved experimentally through three points bending test. Based on the results it can be said that in spite of the material saving at the porous structure production, this type is not very suitable for the components loaded by bending, because a unit of material at this type of sample is able to carry down less stress compared to the sample fully filled by the material.

Acknowledgement. The article was prepared thanks to supporting by the Ministry of Education of the Slovak Republic through the grants KEGA 007TUKE-4/2018 and APVV-19-0550.

References

1. Majstorovic, V., et al.: Industry 4.0 – a national program of Serbia. Mechanical Engineering Faculty, Belgrade (2019)
2. Pantazopoulos, G.A.: The value of engineering literature and the complex role of its contributors. J Fail. Anal. Preven. **16**, 695–696 (2016)
3. Huang, Y., Leu, M.C., Mazumder, J., Donmez, A.: Additive manufacturing: current state, future potential, gaps and needs, and recommendations. J. Manuf. Sci. Eng. **137**, 014001–0140010 (2015)
4. Baron, P., Dobransky, J., Pollak, M., Kocisko, M., Cmorej, T.: The parameter correlation of acoustic emission and high-frequency vibrations in the assessment process of the operating state of the technical system. Acta Mechanica et Automatica **10**(2), 112–116 (2016)
5. Pereira, A.C., Romero, F.: A review of the meanings and the implications of the Industry 4.0 concept. Proc. Manuf. **13**, 1206–1214 (2017)
6. Hao, L., Raymont, D., Yan, C., Hussiein, A., Young, P.: Design and Additive Manufacturing of Cellular Lattice Structures (2011). https://doi.org/10.1201/b11341-40
7. Brenne, F., Niendorf, T., Maier, H.J.: Additively manufactured cellular structures: Impact of microstructure and local strains on the monotonic and cyclic behavior under uniaxial and bending load. J. Mater. Process. Technol. **213**, 1558–1564 (2013)
8. Ungureanu, M., et al.: Innovation and technology transfer for business development. Proc. Eng. **149**, 495–500 (2016)
9. Beno, P., Kozak, D., Konjatic, P.: Optimization of thin-walled constructions in CAE system ANSYS. Technicki Vjesnik **21**(5), 1051–1055 (2014)
10. Stansbury, J.W., Idacavage, M.J.: 3D printing with polymers: challenges among expanding options and opportunities. Dent. Mater. **32**, 54–64 (2016)
11. Bourell, D., Pierre, J., Leu, M., Levy, G., Rosen, D., Beese, A.M., Clare, A.: Materials for additive manufacturing. CIRP Ann. Manuf. Technol. **66**, 659–681 (2017)
12. Miller, A.T., Safranski, D.L., Wood, C., Guldberg, R.E., Gall, K.: Deformation and fatigue of tough 3D printed elastomer scaffolds processed by fused deposition modeling and continuous liquid interface production. J. Mech. Behav. Biomed. Mater. **75**, 1–13 (2017)
13. Guo, N., Leu, M.C.: Additive manufacturing: technology, applications and research needs. Front. Mech. Eng. **8**, 215–243 (2013)
14. Vychytil, J., Holecek, M.: The simple model of cell prestress maintained by cell incompressibility. Math. Comput. Simul. **80**(6), 1337–1344 (2010)
15. Monkova, K., et al.: Study of 3D printing direction and effects of heat treatment on mechanical properties of MS1 maraging steel. Arch. Appl. Mech. **89**, 791–804 (2018)

Assessing Industry 4.0 Readiness in Manufacturing Companies from Serbia

Vidosav D. Majstorović(✉), Radivoje M. Mitrović, and Žarko Z. Mišković

Faculty of Mechanical Engineering, University of Belgrade, Belgrade, Serbia
vidosav.majstorovic@sbb.rs

Abstract. Industry 4.0 has become a global programme of scientific and technological development, which covers all economic activities of today. The most developed countries have adopted their own programmes and they implement them for Industry 4.0. Our country is also intensively working on this Programme. This paper presents a model for assessing maturity and readiness of manufacturing organizations (industry branches) to operate and implement the Project I4.0 in their environment.

Keywords: Industry 4.0 · Manufacturing companies · Serbia

1 Introduction

Ever since it was officially defined as a new, Internet-based model of automatization of technological systems in 2011, the concept of Industry 4.0 has evolved to a great extent. At the time, an advanced manufacturing industry model was discussed, based on the new concept of connecting machines and computers (cyber-physical systems (CFS)), their networking (cloud computing and Internet of Things (IoT)) and the widespread use of artificial intelligence in these systems (AI). This is how the concept of smart manufacturing was created, and nowadays we speak about smart vehicles, roads, networks, cities, services, basically, everything.

One decade of development of the Industry 4.0 concept has raised a whole array of issues from the standpoint of its implementation in practice, especially in the industry: digitalization and its model in Industry 4.0 (formats and secure use of data, large volume of data), assessing the level of readiness of an organization for digital transformation (the level of maturity, business model), the achieved level of digital readiness, best practices of I4.0 in industry, knowledge and skills in I4.0 jobs, etc.

The following facts are relevant for Serbia when it comes to the implementation of I4.0 in industry: industrial policy for I4.0, education for I4.0 (higher education and secondary education), research for I4.0, especially applied research, and the readiness of SMEs, since these form the dominant structure of the Serbian industry today.

This paper focuses on the development of the model for assessing the readiness of the Serbian manufacturing industry for the implementation of the I4.0 project in Serbia [1–6].

L. Wang et al. (Eds.): AMP 2020, LNME, pp. 69–79, 2020.
https://doi.org/10.1007/978-3-030-46212-3_4

2 Literature Review

A detailed analysis of the literature starting from the year 2011 [7], from the point of view of the readiness of an organization to start with digital transformation and in order to implement the concept of I4.0 in a certain environment, shows that three approaches are currently dominant in the world: (a) analysis and assessment of the organization for the implementation of the I4.0 project in a certain environment (the level of maturity), (b) the organization's readiness to implement good practice of I4.0 in its environment – readiness, and (iii) the context of the organization's business model with respect to the I4.0 context. The analysis of these elements is provided in further text.

2.1 Levels of Maturity for I4.0

The basic definition of maturity says that it is a change-management process. In this case, the changes are related to the context of implementation of the I4.0 project in an organization, and they are defined via several levels, most frequently five. In the last several decades, different models were developed to evaluate the maturity of technological systems from different aspects (e.g. technology, automation, doing business). In this paper, we are interested in those aspects referring to the implementation and integration of ICT technologies in manufacturing. A detailed analysis of these aspects reveals the following facts:

(a) the models related to assessing maturity for the implementation of national programmes for I4.0 [27, 32]. Both models were developed in Germany and serve as a framework for defining an integrated maturity index, which has six levels [32]: computerization, connectivity, visibility, transparency, predictive capacity and adaptability. The first two levels refer to digitalization, while the other four are related to the I4.0 structure. Due to the fact that the original I4.0 concept originated in Germany, this model is also one of the best solutions in this field.

(b) the generally developed models of maturity and their assessment of the existing business practices with respect to the requirements of I4.0 [23, 30], which are divided into scientific [23] and practical. Nowadays, in the context of I4.0, they define the path of excellence in digitalization as the basis for I4.0, in order to be applied in practice: CPS, IoT, Big Data analytics and machine learning; all of these are related to cloud computing. Scientific models were created as a result of research into good manufacturing practice from the aspect of automation, usually defined via five levels, from initial to digital-oriented. Currently, there are ten of them [30]. The practical ones emerged as the consulting models of the well-known companies around the world who are devoted to developing and implementing good management practice in the field of I4.0. As a rule, they are oriented towards the development and implementation of individual elements of I4.0 in practice (CPS, AI, etc.).

(c) maturity models developed for specific industries [13, 28]. The latter references provide an overview of two models for the defence and space sectors. They comply with the characteristic technological and ICT requirements stemming from the particularities of products of these industries. The I4.0 elements particularly important in these industries include additive manufacturing and e-value chains.

(d) the maturity model for SMEs [11, 26]. These are the models that draw the most attention on the part of users from the point of view of their implementation in practice. SMEs are the dominant form of production in the industry, and, hence, the maturity model in this structure features three elements: vision (resources and understanding of the I4.0 concept), a roadmap with the I4.0 elements for the organization and the I4.0 project for the organization (training, risk management and implementation of I4.0 elements in practice). For a specific organization, this model is defined via the maturity level, from the beginner to business model transformation for I4.0.

(e) the models that integrate maturity and readiness [16, 17]. This is the latest approach that allows us to connect these two elements at the same time: to define the level of maturity and to assess the level of readiness of an organization to implement the I4.0 project in its environment. Five models were developed for this approach [17], featuring four to six levels and five to nine dimensions of the model that is being assessed. The assessment itself is performed by calculating the weight coefficients for a certain dimension and the total score is displayed via a radar chart, ranging from 1 (the lowest level) to 5 (the highest level).

Therefore, we can conclude that there are different approaches in this area. The latter two approaches seem the most suitable for our country.

2.2 Organization Readiness for the I4.0 Concept

The readiness of an organization in the context of the implementation of the I4.0 project in its environment implies the ability to apply the planned steps in the process of implementation of the I4.0 concept, which is evaluated via different criteria and weight factors. Formerly, maturity and readiness of an organization were viewed as separate approaches, but today these are most often viewed as an integral concept. By analysing this approach in detail, we can conclude the following:

(a) complex models for assessing readiness [10, 14, 19, 33]. The most famous model is [33], defined via four dimensions of the I4.0 model: smart factory, smart products, smart operations and data-driven services. When the self-assessment model is applied, an organization can be at one of six levels of readiness: outsider, beginner, intermediate, experienced, expert and top performer. These models connect maturity and readiness.

(b) the models for assessing the readiness of manufacturing organizations [8, 9, 17, 18, 25]. The whole concept of I4.0 was developed and initially implemented in manufacturing companies. Several models for the assessment of readiness have been developed for this type of industry, the most well-known being the VDMA with six dimensions and 18 elements.

(c) general models for assessing readiness [20, 21, 24], most commonly of a country or a region. The most famous model in this group is the one developed within the World Economic Forum [24], used to assess countries' readiness via over 59 indicators.

(d) the models of readiness for SMEs [16, 22]. Finally, similar as for maturity, we also have models for SMEs. These bear similarities to the approach (a), but they are simpler.

Approach (b) seems to be the most suitable for our research, referring to our country, for this model.

2.3 Context of Organization for I4.0

This approach, basically used for analysing and assessing maturity and readiness, departs from the main elements of I4.0 (autonomous robots, simulation, horizontal and vertical system integration, the Industrial Internet of Things, cyber security, the Cloud, additive manufacturing, augmented reality and Big Data) and proceeds with their connecting, for example, with the metrics for technological performance: product traceability, process monitoring, reschedulability, cyber security and networking [31]. Reference [29] presents the I4.0 model for the development and assessment of maturity and readiness by connecting the architecture of the I4.0 concept, its basic attributes and main functions in the following way: (a) the connection level – communication – hardware connection, (b) the conversion level – information – discovery information, (c) the cyber level – control – automated system, (d) the cognition level - early awareness – predictive maintenance, (e) the configuration level – self-configure – intelligent manufacturing.

The remaining two examples refer to specific industries and their requirements for I4.0 [12, 15]. For instance, in the construction industry [15], the following key features are required for the implementation of Industry 4.0: (a) horizontal integration through value networks, (b) end-to-end digital integration of engineering across the entire value chain, and (c) vertical integration and networked manufacturing systems in the construction industry. The main features of this industry include: complexity, uncertainty, fragmented supply chain, short-term thinking and culture. Relying on these elements, a network is created for an organization, based on which the assessment of readiness for the I4.0 project is made.

As far as our research is concerned, this approach is not the priority at the moment.

3 Framework of Model

It has been said previously that the dominant structure of production organizations in Serbia includes SMEs. Starting from this fact and the analysis carried out above, a model for the assessment of maturity with four levels is proposed, Table 1 [3, 4].

Table 1. Levels of maturity.

	Level I4.0	Characteristics	Project
Level one	Beginners	An organization gets acquainted with I4.0	The idea to establish it is shaped
Level two	An established project	The defined vision and strategy for I4.0	The defined Project
Level three	Roadmap	Good practice for I4.0 is established	I4.0 elements are realized
Level four	All the Industry 4.0 elements for an organization have been implemented	The organization is working on the Platform for I4.0	The Project has been successfully completed

An organization at level one has a decision and a team to implement the Project I4.0 in its environment. It performs a system analysis of its business model (customers, suppliers, automation) and defines a business objective for I4.0. It also estimates costs and the time of Project implementation.

At level two, the Team defines a detailed Project with the required resources and provides a framework of the I4.0 module for specific functions.

At level three, a detailed Roadmap with all the elements of I4.0 for the organization is defined, according to which the Project is implemented and the achieved results are reviewed. Step by step, pilot projects are implemented for individual wholes.

Finally, at the last level, the organization has an established business model on the Platform I4.0.

The dimensions of the concept I4.0 readiness for an organization, which represent the framework of the business model, are comprehensive, with the structure shown in Table 2.

Table 2. Structure of the questionnaire for the assessment of readiness of an organization.

Dimension	Characteristics	Comments
Strategy	Implementation of the I4.0 roadmap, needed resources, design business model etc.	1
Culture of organization	Knowledge sharing, Innovations, cross-SME collaborations, value of ICT in the organization etc.	2
Leadership	Commitments and competences of leaders, Project leadership etc.	3
Customers	Customer database, digitalization of sale/services, Social networks etc.	4
Suppliers	Database of suppliers, digitalization of order and acceptance etc.	5
Products	Digitalization of the product, product integration in other systems	6

(continued)

Table 2. (*continued*)

Dimension	Characteristics	Comments
Technology	Existence of modern ICT, utilization of advanced machines etc.	7
People	Competences of employees in the SME, readiness of employees for new technology application etc.	8
I4.0 framework	Basic elements of I4.0, cloud technology etc.	9
Governance	Labour regulations for I4.0, utilization of ICT standards, protection of intellectual property etc.	10

The overall model of assessment has nine dimensions: from the organization's strategy for I4.0 to the legal framework of a digital factory. A separate questionnaire was constructed for each dimension, with the questions to be answered according to the Likert-scale model (1–5; 1 – not implemented, 2 – partially implemented, 3 – implemented in more than 50% of cases, 4 – implemented completely, 5 – improvements in the implementation are being made). That is why it is extremely important that the person who collects answers to the questions is extremely familiar with the concept of I4.0, especially from the point of view of implementation in the specific organization concerned.

The assessment is performed by determining the Mr coefficient for each dimension, using a formula that includes: the dimension, characteristic, weight factor within the dimension and the number of characteristics within the dimension. This provides a radar chart for each organization individually.

4 Results and Discussion

For the purposes of this research, Questionnaire 1 with 84 questions was developed for assessing readiness, and, in parallel, Questionnaire 2 was developed for analysing the general facts about I4.0 for SMEs in Serbia. Both questionnaires were distributed to 106 SMEs from the manufacturing sector, and 49 responses were obtained [3, 4].

The first group of questions from Questionnaire 2 addressed the organization's challenges regarding its overall readiness to become a part of the Project I4.0. The obtained responses are provided in Fig. 1. The most significant problem refers to the uncertainty surrounding the organization's required investment to implement and apply this concept within the organization (46%), and so on. These facts suggest that organizations must systematically approach this Project.

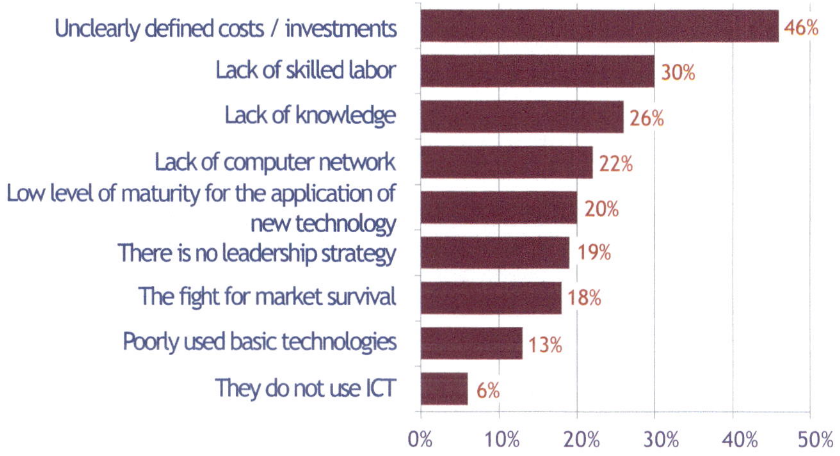

Questionnaire / 2019 - Industry 4.0 in Serbia for SMEs, n = 49.

Fig. 1. Challenges for successful implementation of Industry 4.0 in Serbia.

The second group of questions referred to the general facts about the possibility of implementing the I4.0 concept in the organization, Fig. 2. Almost half of the answers speak in favour of the fact that organizations in Serbia are seriously considering and working on the preparation of the I4.0 Project in their environment.

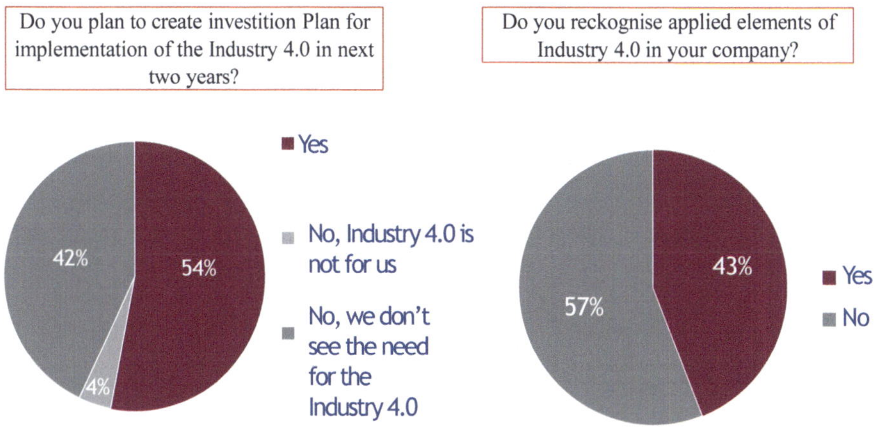

Questionnaire / 2019 - Industry 4.0 in Serbia for SMEs, n = 49.

Fig. 2. Facts on Industry 4.0 in SMEs.

Finally, the third group of questions referred to the functions of SMEs where I4.0 should be implemented first, Fig. 3. Our organizations understand the essence of the I4.0 concept and, hence, the logical answer is that production is the key element for the implementation of I4.0 in these enterprises. Supply chains come next, as an important element of I4.0 in real production.

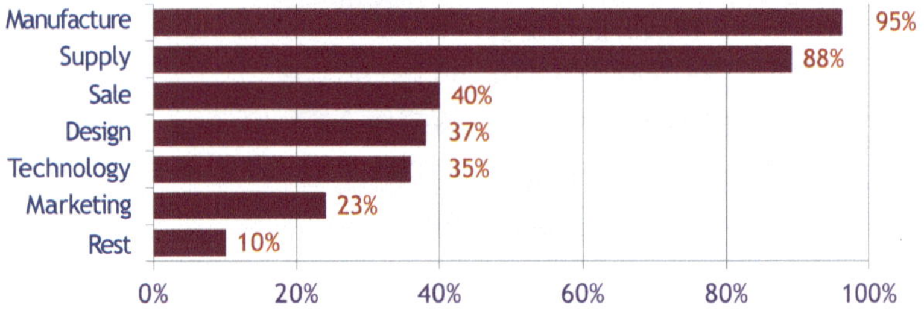

Questionnaire / 2019 - Industry 4.0 in Serbia for SMEs, n = 49.

Fig. 3. Application fields of the Industry 4.0.

Questionnaire 1 was used to perform the second experiment to assess the organization's readiness to implement the I4.0 elements in its environment, Fig. 4. The responses from 49 organizations were obtained and the next step is to cross-tabulate these responses to obtain readiness of all organizations in this branch of industry.

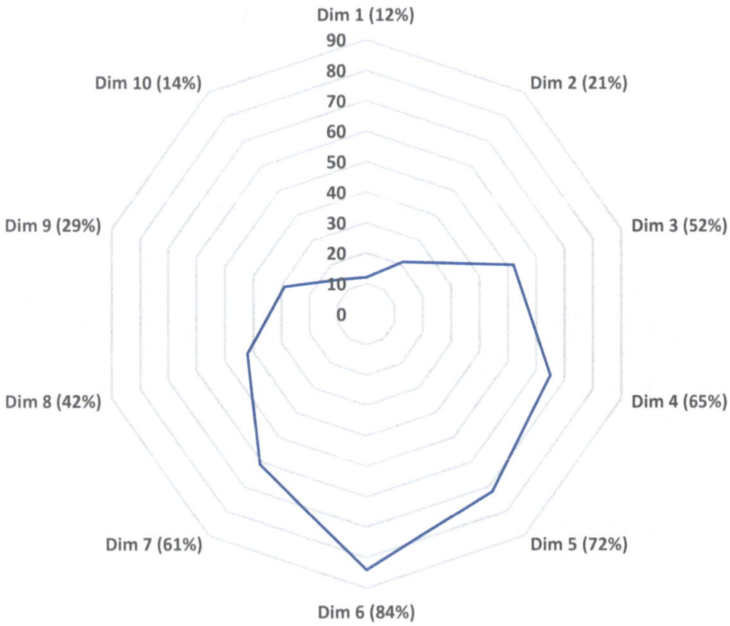

Fig. 4. Radar chart of I4.0 readiness for XYZ organization.

In the organization from this example, the highest level of readiness was accomplished for these dimensions: customers (4), suppliers (5) and products (6).

5 Conclusions

At the 4th International Conference – Industry 4.0, held from the 4th to the 7th of June 2019 in Belgrade, our country, as the 38th country in the world, defined its national Programme for Industry 4.0 [1–6]. After the conference, panels were held across Serbia to present this Programme. For this year's Conference, it is planned to appoint all the bodies of the Platform and to submit it to the Government of Serbia.

This paper is our contribution to the listed activities.

Acknowledgement. The article was prepared thanks to supporting by the Ministry of Education, Science and Technology Development of the Republic of Serbia through the grants TR35022 and TR35029.

References

1. Majstorovic, V., et al.: Industry 4.0 programs worldwide. In: Proceedings of the 4th International Conference on the Industry 4.0 Model for Advanced Manufacturing, AMP 2019, Belgrade, pp. 79–99 (2019). https://doi.org/10.1007/978-3-030-1818-2_7
2. Majstorovic, V., et al.: Industry 4.0 – a national program of Serbia. Faculty of Mechanical Engineering, Belgrade (2019)
3. Majstorovic, V., et al.: Program - Advanced Industrialization of Serbia and Industry Policy, Horizon 2020/2030. Mechanical Engineering Faculty (2016)
4. Majstorovic, V., et al.: Project - Advanced Industrialization of Serbia – Industry 4.0 for Serbia. Mechanical Engineering Faculty, Belgrade (2016)
5. Majstorovic, V., et al.: Cyber-physical manufacturing in context of Industry 4.0 model. In: Proceedings of 3rd International Conference on the Industry 4.0 model for Advanced Manufacturing, pp 227–238, Belgrade (2018). https://doi.org/10.1007/978-3-3-319-89563-5
6. Majstorovic, V., et al.: Cyber-physical manufacturing metrology model (CP3M) for sculptured surfaces – turbine blade application. Procedia CIRP **63**, 658–663 (2017). https://doi.org/10.1016/j.procir.2017.03.093
7. Recommendations for implementing the strategic initiative Industrie 4.0, Securing the future of German manufacturing industry Final report of the Industrie 4.0 Working Group, April 2013. https://www.din.de/blob/76902/e8cac883f42bf28536e7e8165993f1fd/recommendations-for-implementing-industry-4-0-data.pdf. Accessed Feb 2020
8. Nicka, G., Szallera, Á., Bergmanna, J., Várgedőa, T.: Industry 4.0 readiness in Hungary: model, and the first results in connection to data application. In: Proceedings of 9th IFAC Conference on Manufacturing Modelling, Management and Control Berlin, Germany, 28–30 August 2019. IFAC Papers on Line 52-13, pp. 289–294 (2019). https://doi.org/10.1016/j.ifacol.2019.11.185
9. Castelo-Branco, I., Cruz-Jesus, F.: Assessing Industry 4.0 readiness in manufacturing: evidence for the European Union. Comput. Ind. **107**, 22–32 (2019). https://doi.org/10.1016/j.compind. Accessed 7 Jan 2019

10. Basl, J., Doucek, P., et al.: A metamodel for evaluating enterprise readiness in the context of Industry 4.0. Information **10**, 89 (2019). https://doi.org/10.3390/info10030089. Accessed Feb 2020
11. Ganzarain, J., Errasti, N.: Three stage maturity model in SME's towards Industry 4.0. J. Ind. Eng. Manage. **9**(5), 1119–1128 (2016). https://doi.org/10.3926/jiem.2073
12. Masood, T., Egger, J.: Augmented reality in support of Industry 4.0—implementation challenges and success factors. Robot. Comput.-Integr. Manuf. **58**, 181–195 (2019). https://doi.org/10.1016/j.rcim.2019.02.003
13. Botlikova, M., Botlikov, J.: Local extremes of selected Industry 4.0 indicators in the european space—structure for autonomous systems. J. Risk Manage. **13**, 13 (2020). https://doi.org/10.3390/jrfm13010013
14. Paulo-Tadeu Pacchini, A., Lucato, W.C., Facchini, F., Mummolo, G.: The degree of readiness for the implementation of Industry 4.0. Comput. Ind. **113**, 103125 (2019). https://doi.org/10.1016/j.compind.2019.103125
15. Oesterreich, T.D., Teuteberg, F.: Understanding the implications of digitisation and automation in the context of Industry 4.0: a triangulation approach and elements of a research agenda for the construction industry. Comput. Ind. **83**, 121–139 (2016). https://doi.org/10.1016/j.compind.2016.09.006
16. Hofmann, E., Rüsch, M.: Industry 4.0 and the current status as well as future prospects on logistics. Comput. Ind. **89**, 23–34 (2017). https://doi.org/10.1016/j.compind.2017.04.002
17. Schumacher, A., Erolb, S., Sihn, W.: A maturity model for assessing Industry 4.0 readiness and maturity of manufacturing enterprises. Procedia CIRP **52**, 161–166 (2016). https://doi.org/10.1016/j.procir.2016.07.040
18. Machado, C.G., Winrotha, M., Carlsson, D., Almströma, P., Centerholtb, V., Hallin, M.: Industry 4.0 readiness in manufacturing companies: challenges and enablers towards increased digitalization. Procedia CIRP **81**, 1113–1118 (2019). https://doi.org/10.1016/j.procir.2019.03.262
19. Agca, O., Gibson, J., Godsell, J., Ignatius, J., Davies, C.W., Masons, P.: An Industry 4.0 readiness assessment tool, University of Warwick. www.warwick.ac.uk/scip. Accessed Feb 2020
20. Chonsawat, N., Sopadang, A.: The development of the maturity model to evaluate the smart SMEs 4.0 readiness. In: Proceedings of the International Conference on Industrial Engineering and Operations Management Bangkok, Thailand, 5–7 March 2019, pp: 354–363. http://www.ieomsociety.org/ieom2019/papers/97.pdf. Accessed Feb 2020
21. Basl, J.: Companies on the way to Industry 4.0 and their readiness. J. Syst. Integr. **9**(3), 3–6 (2018). https://doi.org/10.20470/jsi.v9i3.351
22. Stentoft, J., Jensen, K.W., Philipsen, K., Haug, A.: Drivers and barriers for Industry 4.0 readiness and practice: a SME perspective with empirical evidence. In: Proceedings of the 52nd Hawaii International Conference on System Sciences, pp. 5155–5164 (2019). https://scholarspace.manoa.hawaii.edu/handle/10125/59952
23. De Carolis, A., Macchi, M., Kulvatunyou, B., Brundage, M,P., Terzi, S.: Maturity models and tools for enabling smart manufacturing systems: comparison and reflections for future developments. In: Proceedings of IFIP 14th International Conference on Product Lifecycle Management (2017). https://doi.org/10.1007/978-3-319-7295-3_3
24. Batchkova, A., Popov, T., Ivanova, A.: Assessment of readiness for Industry 4.0. Int. J. Ind. **III**(6), 288–291 (2018). https://stumejournals.com/journals/i4/2018/6/288.full.pdf. Accessed Feb 2020

25. Viharos, Z.J., Soós, S., Nick, G., Várgedő, T., Beregi, R.: Non-comparative, Industry 4.0 readiness evaluation for manufacturing enterprises. In: 15th IMEKO TC10 Workshop on Technical Diagnostics. Technical Diagnostics in Cyber-Physical Era, Budapest, Hungary, 6–7 June 2017. https://www.imeko.org/publications/tc10-2017/IMEKO-TC10-2017-031.pdf. Accessed Feb 2020
26. Taurino, T., Villa, A.: A method for applying Industry 4.0 in small enterprises. In: Proceedings of 9th IFAC Conference on Manufacturing Modelling, Management and Control Berlin, Germany, 28–30 August 2019. IFAC Papers on Line 52-13, pp. 439–444 (2019). https://doi.org/10.1016/j.ifacol.2019.11.099
27. Kumar, R., Enose, N.: Making Industry 4.0 Real – Using the ACATECH I4.0 Maturity Index, White paper (2017). https://www.infosys.com/engineering-services/white-papers/Documents/industry-4.0-real.pdf. Accessed Feb 2020
28. Bibby, L., Dehe, B.: Defining and assessing Industry 4.0 maturity levels - case of the defence sector. J. Prod. Plann. Control **29**(12), 1030–1043 (2018). https://doi.org/10.1080/09537287.2018.1503355
29. Qina, J., Liua, Y., Grosvenora, R.: A categorical framework of manufacturing for Industry 4.0 and beyond. Procedia CIRP **52**, 173–178 (2016). https://doi.org/10.1016/j.procir.2016.08.005
30. Felch, V., Asdecker, B., Sucky, E.: Maturity models in the age of Industry 4.0 – do the available models correspond to the needs of business practice? In: Proceedings of the 52nd Hawaii International Conference on System Sciences, pp. 5165–5174 (2019). https://doi.hdl.handle.net/10125/59953
31. Essakly, A., Wichmann, M., Thomas, S.: A reference framework for the holistic evaluation of Industry 4.0 solutions for small-and medium-sized enterprises. In: Proceedings of 9th IFAC Conference on Manufacturing Modelling, Management and Control Berlin, Germany, 28–30 August 2019, IFAC Papers on Line 52-13, pp. 427–432 (2019). https://doi.org/10.1016/j.ifacol.2019.11.093
32. Schuh, G., Anderl, R., Gausemeier, J., ten Hompel, M., Wahlster, W.: Industrie 4.0 Maturity Index - Managing the Digital Transformation of Companies. www.acatech.de/publikationen. Accessed Feb 2020
33. Lichtblau, K., Stich, V., Bertenrath, R., Blum, M., Bleider, M., Millack, A., Schmitt, K., Schmitz, E., Schröter, M.: Industrie 4.0 Readiness, Aachen, Cologne, October 2015. https://industrie40.vdma.org/documents. Accessed Feb 2020

The Importance of CT in Industry 4.0 by Supplying Quality and GPS Standards for Several Production Methods Such as Additive Manufacturing

Cem Yurci[1(✉)] and Numan M. Durakbasa[2]

[1] Mechanical Engineering Department, Yildiz Technical University, Istanbul, Turkey
yurci_cem@yahoo.com

[2] Industrial Metrology and Adaptronic Systems, Institute of Production Engineering and Photonic Technologies, Vienna Technical University, Vienna, Austria
numan.durakbasa@tuwien.ac.at

Abstract. Today Industry 4.0 is spreading from developed countries into other developing economies even to small and medium sized enterprises. With a strong infrastructure and fast connection between components of the whole factory, Industry 4.0 will introduce innovations and a more robust, reliable and sustainable production. While Industry 4.0 is having several fundamental technology additive manufacturing has been playing an important role and takes part in today's Industry 4.0 technology, prominently for automotive vehicle producers. Additive manufacturing has many advantages and tasks such as in reverse engineering, by defect prediction or producing complex shaped parts. Its vitality becomes more important when its implementation is combined with the most recent measurement technology 'Computer Tomography'. CT can detect complex surfaces or inner structure with X-Ray usage which helps to additive manufacturing parts by inspection which can not be provided by a CMM in similar manner. That study presents a simple sample for the combination of these two technologies. Followingly, voids and GPS analyses after CT measurements of several parts will be introduced. Also, those CT inspections will give an example for a definite stage of Industry 4.0. In conclusion, a brief investigation of those innovative most recent technologies, which are developing in a continuous manner, will be made in the context of Industry 4.0 according to inspection results.

Keywords: Industry 4.0 · Computer Tomography · Additive manufacturing · Quality control

L. Wang et al. (Eds.): AMP 2020, LNME, pp. 80–91, 2020.
https://doi.org/10.1007/978-3-030-46212-3_5

1 Introduction

1.1 Industry 4.0

The third industrial revolution included automatization getting started of modern computers and the Internet [1] which led the need for another industrial sphere Industry 4.0 at a definite point as innovative technologies were being improved. After the first revolutions, the last industrial revolution 'Industry 4.0' has become the main vector in certain developed countries' several particular sectors which will cause a progressive modernization of other spheres of industry, in the future [2]. During their processes, Industry 4.0 employs the technological enablers such as 'Internet of Things', 'Big Data', 'Cloud Computing', 'Cyber Physical Systems' (CPS), 'Artificial Intelligence' (AI) and 'Augmented Reality' (AR) [1]. Besides crucial positive effects created with these superior technological aspects, the negative side effects of Industry 4.0 such as loss of jobs for non-qualified employees [3] can be reduced and transformed to having positive properties by aligning the education system to support digitalization and Industry 4.0 [4]. For instance, new courses in the curriculum in the Life Sciences Education should be centred on Data Science supported with other technological fundamentals of Industry 4.0 according to Catal [4] et al. The programme of that education will be organised according to the capability needs of Industry 4.0. For example, Benesova [5] et al. propose that a manufacturing engineer should have a secondary postgraduate education in electrical engineering with definite skills.

To meet the technologies' needs due to rapid developments (like achieving the increased wireless connection speed to satisfy big data through 5G) the emergence of Industry 5.0 will be inevitable by carrying especially data transfer properties similar to the previous industrial revolution. 5G allows these service-type categories: massive machine-type communication (mMTC) for connecting sensors [6], critical machine-type communication (cMTC) providing ultrareliable low-latency communication (URLLC) (or ultra-reliable and low-latency MTC (uMTC) [7]) for deterministic real-time applications and mobile broadband (MBB) for high data rate applications [8]. The rise of the rate and capabilities of Industry 4.0 technological tools will facilitate human-robot co-working which will be one of the main themes of Industry 5.0 through significant advancements in robotics and AI [9]. While the motivation will be transforming from mass production to sustainability the involved technologies and research areas will spread to Bionics, Sustainable Agriculture and Renewable Resources according to foresights of Demir [9] et al.

Production will always be one of the main concerns and research and application areas of the industrial innovations leading to revolutions. However, the applications can change according to the variety of the production. Thus, while tasks differ essentially in-between discrete manufacturing and process industry. they are also showing a changing characteristic within each specific industry sector [8]. As an exemplified case, a factory's mechanism operation is outlined by Gold [8] et al. illustrating the importance of supply management, operations control and robot motion control in a manufacturing workflow bottomed on assembly line promoted by sensors and automated guided vehicles. Ghobakhloo [10] et al. made a research about Small and Medium-sized Enterprises (SMEs) as well which integrate modern Smart Manufacturing with

Digitation via artificial intelligence applications within their business operations. At that point simulation plays a core functionality role via seamless assistance along the entire life cycle, for example supporting operation and service directed to operation data [11]. That concept is called Digital Twinning implemented as Digital Shadow and Digital Master [11].

1.2 Additive Manufacturing in Industry 4.0

As it is seen in Fig. 1, additive manufacturing takes an important role in Industry 4.0 applied as just-in time production [12]. That developing AM does not require a hammer like in forging or a lathe and miller like in machining. So, 3D and 4D local printers have began to be fed by teleported design and instruction sets [13]. Besides metal-matrix, polymeric and ceramic materials, biomaterials are being implemented for AM. Followingly, Li etc. [14] give 'biomedical beta Ti-24Nb4Zr-8Sn' and 'Ti-6Al-7Nb (wt %)' as biomaterial samples which are fully developed by SLA directly from CAD data. As the main utilization of AM is in aerospace, military and automotive sectors, automobile producers are also using more and more components manufactured by additive processes, for instance manifolds, interior trim, wheels with solid tyres, integrated motors and even the suspension elements [13].

As the main utilization of AM is in aerospace, military and automotive sectors, automobile producers are also using more and more components manufactured by additive processes [13]. Today, advanced printing can be employed for mass production as well. In addition, customers can design specific parts of automobiles according to their pleasure and let manufacture a unique car with some specific properties (an instance in Fig. 1 [15]).

Fig. 1. An example for customized parts of an automobile with AM: led door sills and side scuttles of BMW MINI 2018 [15]

1.3 A Brief Information for the Principles of Computer Tomography

Computer Tomography is based on the dispersion of X-Rays in a solid. 'X-ray tomography' means imaging by generating a superposition representation of sections or layers via X-rays [16].

The absorption of X-rays varies by different materials. Absorption increases with the density of the material, which increases with the atomic number [17]. According to the data from [18], plastic materials are inspected by CT easier and more efficiently then metallic materials. Besides the material composition, the geometry of the measured object plays a role as well. The arrangement of the X-Ray voltage by CT enables the inspection of metallic materials as well.

By a CT, an electrode beam formed with a voltage generation between two metal electrodes in an X-Ray tube hits on the target object [18]. That radiation is in fact a flow of photons having various frequencies and carrying different energy levels according to that [18]. 3D model of the inspected object can be created from 2D imaged slices by a reconstruction technique and a comparison with the CAD data can be implemented to realize the object such as failures. Such an application study was carried out by Gapinskia [17] et al. X-Ray Computed Tomography has the advantage to detect the defects in the volume with its as well when compared with non-destructive techniques such as ultrasonic testing, eddy current and non-interferometric optics. Such a property opens the door of micro and nano CT. A sample of that is given in the study of Bossa [19] et al. with a civil engineering application.

1.4 Computer Tomography in Industry 4.0 to Sustain Quality Standards

When mentioning of production is inevitable in Industry 4.0 and its efficiency is the crucial point, Production and Product Quality, Quality control, inspection and detection and the modifications during that have to be sustained persistently as well to achieve the main objects of Industry 4.0. Meanwhile the main tools of Industry 4.0 are being used. They are implemented to provide the efficiency; flexibility; productivity; cybersecurity; quality of products and services; and decision process intertwined with data analysis [20].

In quality control it is important to work and stay efficiently in the circle composed of the product design, control and verification processes. Today those processes are being supported with generative manufacturing including additive manufacturing. Modelling and simulation have been contributing to those periods [21]. In Industry 4.0 intelligent manufacturing is being utilized. That concept has been evolved from digital manufacturing with a transition phase called 'networked manufacturing' [22]. Testing and design are the main parts of digital manufacturing. So, the quality inspection and analysis methods will play the biggest role at that stage. Stojanovic [23] et al. propose a stronger, quality control process for modern Industry 4.0 and give for the modern industry system Industry 4.0 a microwave oven fan as a critical part sample. That way can help to solve more complex problems which can not be solved with traditional quality control applications. For instance, a method 'Quality Scorecard' is mentioned and proposed in the study of Shin etc. for the functioning of Industry 4.0 [24] to satisfy the qualitative and quantitative categories of performance measurement methodologies. Followingly, Colledani etc. investigated the production quality improvement during ramp-up-time (time-to-volume), defined as the period from the production of the first item after a manufacturing system reconfiguration of a manufacturing system until the accomplishment of the certain target output rate with the output aims 'increased production rate' and 'raised quality performance' [25]. Additionally, Rosa etc. exemplified the quality and productivity issue for the automotive industry in the assembly line with a steel wire-rope by applying Lean and PDCA methodologies [26].

PDCA cycle is one of the most efficient and important tools in lean manufacturing to fulfil the intent of Industry 4.0 at the same time. Quality tools are employed in the 'Check' phase [27]. They are prominently 5S, Failure Mode Analysis and Effects (FMEA), benchmarking, Statistical Process Control (SPC), Ishikawa diagrams and the

Pareto chart, Quality Function Deployment (QFD), the flowchart, histograms, Single Minute Exchange of Die (SMED), Poka Yoke, Servqual and Six Sigma [27]. Of course, there are quality actions in other PDCA periods as well such as 'Total Quality Management' and 'ISO 9001 Certification' in the Act and 'Quality Concepts and Objectives' in the Plan Phase [28].

During modern production it is obligatory to follow international ISO Standards based on GPS (Geometrical Product Specifications). Glukhow's work [29] is included in studies researching new alternative GPS methods and principles via tolerancing and datum works for several parts. Complying to tolerances is the main object of GPS. Software after necessary operations with measurement machines can give information about tolerance check when related alignment of created 3D model with CAD data is made. At that step, the measurement act can be implemented with CT having the advantage of the inspection of the whole object including the inner structure. Such a property is significant for quality control in the context of Industry 4.0's production objectives and Six Sigma's 0 defects mentality. Detailed GPS information can be used via Big data and Cloud properties of Industry 4.0. With the automatization of instant measurements via innovative developing innovative robotics applications of Industry 4.0 quality control can be utilized as early as possible. Such an implementation is introduced in the study of Bauza [30] et al. They showed the automized feeding operation for a CT to determine the external and internal features of manufactured parts' batches via a serial number for every part [30]. Also, that whole wide measurement information of all the parts can be used by Big Data step or transferred with fast wireless IoT connections and processed with data analysis techniques automatically to make modifications or corrections in the production phase. Furthermore, CT can take part in intelligent quality control systems supported by cyber-physical systems. In this way Albers [31] et al. give in their study a quality assurance mentality by machine applier and suppliers as well in the border of a company having internal quality control loop. These quality control processes can be promoted by AR applications supplying human-machine interactions for mobile information access and seamless navigation in the smart factory through run-time generated, context-sensitive user interfaces tablets [32] and AR glasses having even the function voice control for the operator. Also, all those mentioned steps have to be understood and realized through an integrative manner in the whole operation mechanism of a factory. AM and CT implementations are also included and significant in production and control phases supported by Industry 4.0's tools.

2 Experimental Procedure and Results

'Werth TomoScope® XS' model CT device (Fig. 2) is used for three samples in that study. These samples are plastic injected and additive manufactured parts. The used CT device is a has a power range from 160 kV to 80 W and CMM design. The maximum part diameter dimension can be 205 mm and has maximum length measurement deviation is 6.5 μm. Also the utilized device is medium-sized.

Fig. 2. The used CT device in the Precision Metrology and Nanotechnology Laboratory at Vienna University of Technology

After measurements, 'WinWerth®' measurement software has been applied to extract the 3D model of the inspected parts for the comparison with CAD models. So, form and position tolerance and voids analysis can be implemented.

2.1 CT Examination of a Turbine Blade (Specimen 1)

An additive manufactured turbine blade has been inspected via the CT device in Fig. 3. It has been placed on a support for easy detection.

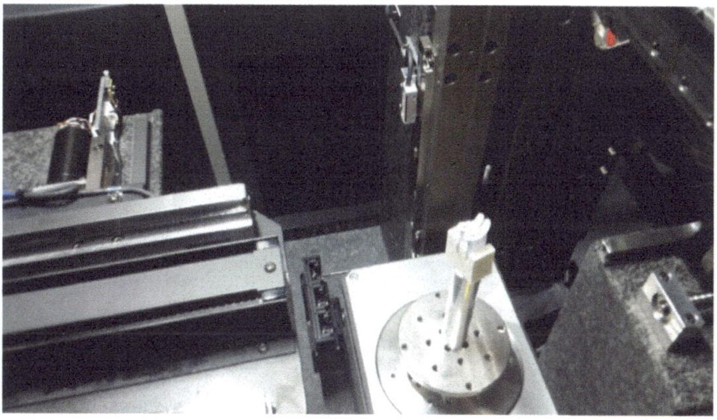

Fig. 3. CT inspection of the turbine blade

After the operations in Werth Software, the 3d model of the object (Fig. 4) has been obtained in STL format. Taht is an example for the CT application for AM parts.

Fig. 4. 3D STL model of the CT scanned turbine blade

2.2 CT Examination of a Plastic Injection Part (Specimen 2)

The inspected object is an off-road vehicle signal arm part.

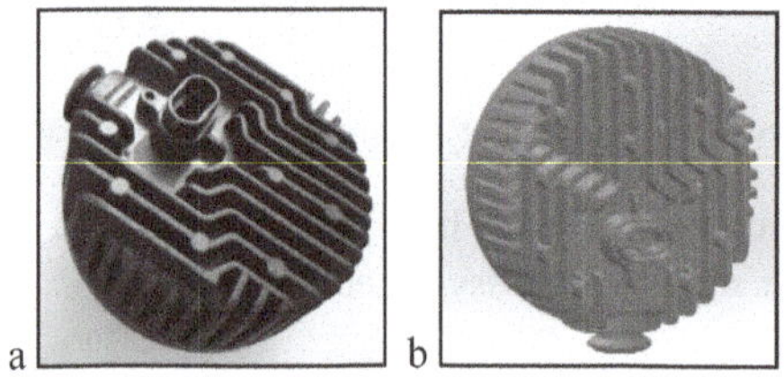

Fig. 5. a) Real part outlook b) its 3D data model in STL format after CT

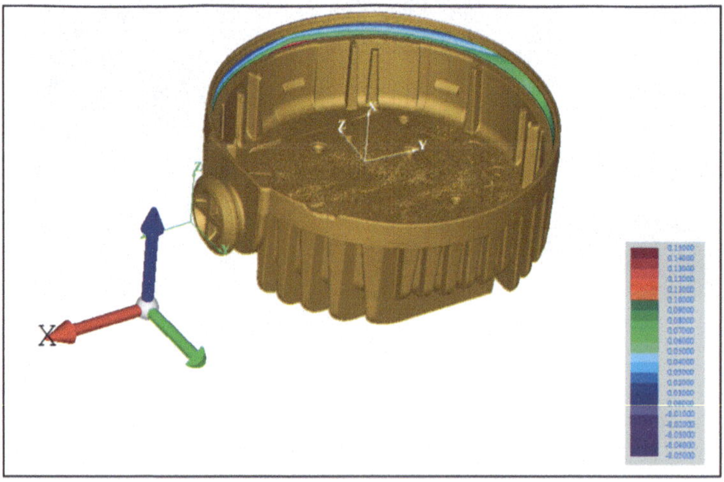

Fig. 6. 3D comparison in Werth Software with the aligning method ‘Reference Point System’ (RPS)

The investigation for the dimensional deviation of a local surface of the specimen 2 (Fig. 5) is investigated in Fig. 6. An important deviation is observed for a very small region despite the almost equal deviation distribution. Thus, an application based on GPS - tolerance information can be necessary to check the validity of the part.

2.2.1 Voids and GPS Analysis by CT Data

Table 1. Voids analysis data

Name	Symbol	Value	Value to be	+Tol	−Tol	Deviation
Volumen_Bauteil (Volume_ Component)	VOL	108800,610	108800,610	0,100	−0,100	0,000
Volumen_Lunker (Volume_Void)	VOL	44,965	0,000	0,800	−0,800	44,905

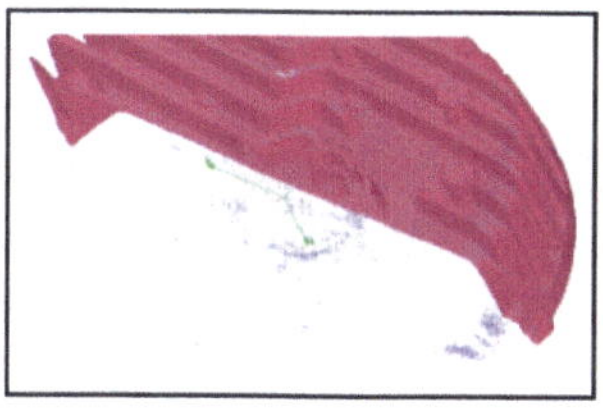

Fig. 7. Voids analysis illustration – specimen 2

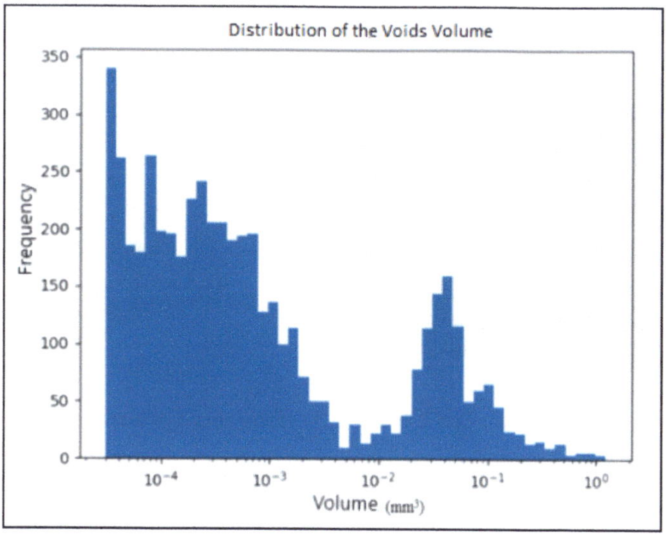

Fig. 8. Distribution of the voids volume the specimen 2

According to the void information in Fig. 7, Fig. 8 and Table 1, voids in the biggest amount with the dimension '0.1 mm^3' are distributed in Specimen 1. Other voids with different sizes are distributed among the part approximately equally. But especially small voids are observed in the part.

2.3 CT Examination of a Protective Cap (Specimen 3)

A protective cap placed on a basis (Fig. 9) has been CT scanned with a voltage 149 kV, a target current 336 μA and a target power 50.1 A. The voxel size by that measurement is 34.15 μm and the pixel size is 1000 μm. The measurement time is 45 min.

Fig. 9. CT inspection of the protective cap

In Fig. 10, the point cloud of the part and the its 3D STL result achieved with the reconstruction technique of the used software are shown.

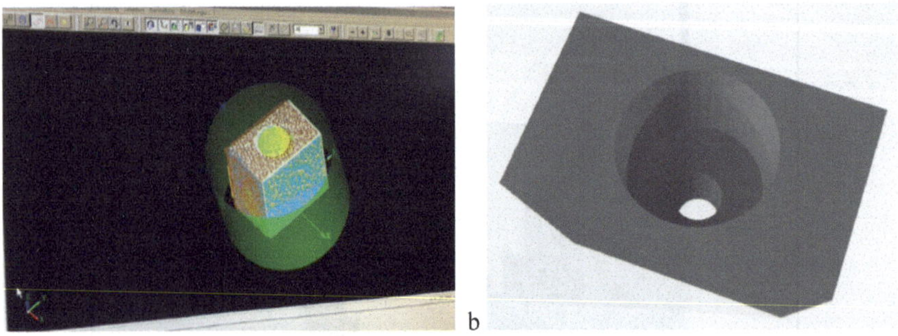

a b

Fig. 10. a. Point cloud, b. 3D STL model after CT measurement

3D-printed/CT		Actual	Nominal	Tol +	Tol -	Dev.
plane upper	Flatness/FT	1,452	0,000	0,300	0,000	1,452
plane downer	Flatness/FT	0,72	0,000	0,300	0,000	0,72
plane side view1	Flatness/FT	0,707	0,000	0,400	0,000	0,707
plane side view2	Flatness/FT	1,075	0,000	0,400	0,000	1,075
plane side view3	Flatness/FT	1,031	0,000	0,400	0,000	1,031
plane side view4	Flatness/FT	0,865	0,000	0,400	0,000	0,865
3//4	Parallelism/LT	1,154	0,000	0,500	0,000	1,154
1//2	Parallelism/LT	1,155	0,000	0,500	0,000	1,155
upper side1 perp	Perpendicularity/LT	0,792	0,000	0,500	0,000	0,792
upper side2 perp	Perpendicularity/LT	1,146	0,000	0,500	0,000	1,146
upper side3 perp	Perpendicularity/LT	1,218	0,000	0,500	0,000	1,218
upper_side4_perp	Perpendicularity/LT	0,864	0,000	0,500	0,000	0,864
cyl hole1 diameter	Diameter	10,736	11,000	0,500	0,500	0,264
cyl hole1 ft	FT	0,353	0,000	0,500	0,000	0,353
cyl hole2 diameter	Diameter	4,692	4,900	0,500	0,500	0,208
cyl_hole2 ft	FT	0,353	0,000	0,300	0,000	0,353
cone angle	Angle	89,093	90,500	1,000	1,000	1,407
cone ft	FT	0,182	0,000	0,300	0,000	0,182
length	Dst	23,45	23,500	0,500	0,500	-0,050
width	Dst	15,27	15,200	0,500	0,500	0,070
height	Dst	23,58	23,500	0,500	0,500	0,080
concentricity	LT	0,264	0,000	0,200	0,000	0,264

Fig. 11. GPS analysis data according to several specifications after CT

As it is seen in Fig. 11, the deviations from tolerances can be determined after CT based on GPS. Accordingly, by that measurement some deviations like flatness, parallelism and flatness have become in major magnitudes.

3 Conclusion

From the applied measurements for several parts it has been observed that CT can be used efficiently as a fast inspection method even for internal failures such as voids. Data obtained from point cloud can also be implemented for GPS analysis to detect deviations from tolerances.

As have been shown with a simple sample, CT detection can be adapted to AM parts. The usage of both methods for complex shaped parts can be combined so. Failures which can not be foreseen before AM method can be detected easily in a fast and detailed manner by CT. The same observations are valid for other production ways as well.

Additive manufacturing and such Quality Control methodologies like CT compose one of fundamental stages of Industry 4.0. It is certain that they become more powerful when they are supported by other main innovative and technological concepts of Industry 4.0 through an integrative mentality. Too fast wireless connections, a strong IoT infrastructure and efficient Big Data analysis will enable the just-in time communication between measurement devices and production machines and operators.

Even immediate operational corrections can be made in every subsection of a planned factory. That operational efficiency can be risen by applications via AR. In addition, according to Majstorovic et al. [33], the importance of Cyber-Physical Manufacturing Metrology Systems will be increasing in the subject 'Advanced Manufacturing Metrology'.

In future the superior rate of technological innovations will open the door of Industry 5.0. The importance of quality control will increase with pervading to every production sector. That will push of course the innovations in CT, AM as well which will cause new researches and findings for measurement and production techniques.

References

1. Bastbaz, M.: Industry 4.0 (i4.0): the hype, the reality, and the challenges ahead. In: Dastbaz, M., Cochrane, P. (eds.) Industry 4.0 and Engineering for a Sustainable Future, pp. 1–12. Springer, Switzerland (2019)
2. Sukhodolov, Y.A.: The notion, essence, and peculiarities of industry 4.0 as a sphere of industry. In: Popkova, E., Ragulina, Y., Bogoviz, A. (eds.) Industry 4.0: Industrial Revolution of the 21st Century, pp. 3–10. Springer, Switzerland (2019)
3. Feshina, S., Konovalova, O.V., Sinyavsky, N.G.: Industry 4.0—transition to new economic reality. In: Popkova, E., Ragulina, Y., Bogoviz, A. (eds.) Industry 4.0: Industrial Revolution of the 21st Century, pp. 3–10. Springer, Switzerland (2019)
4. Catal, C., Tekinerdogan, B.: Aligning education for the life sciences domain to support digitalization and industry 4.0. Procedia Comput. Sci. **158**, 99–106 (2019)
5. Benešová, A., Tupa, J.: Requirements for education and qualification of people in industry 4.0. Procedia Manuf. **11**, 2195–2202 (2017)
6. Dahlman, E., Mildh, G., Parkvall, S., Peisa, J., Sachs, J., Selén, Y., Sköld, J.: 5G wireless access: requirements and realization. IEEE Commun. Mag. **52**, 42–47 (2014). Communications Standards Supplement
7. Sachs, J., Popovski, P., Höglund, A., Gozalvez-Serrano, D., Fertl P.: Machine-type communications, pp. 77–106. UCL, Institute of Education (2016)
8. Gold, K., Wallstedt, K., Vikberg, J., Sachs, J.: Connectivity for industry 4.0. In: Dastbaz, M., Cochrane, P. (eds.) Industry 4.0 and Engineering for a Sustainable Future, pp. 23–48. Springer, Switzerland (2019)
9. Demir, K.A., Döven, G., Sezen, B.: Industry 5.0 and human-robot co-working. Procedia Comput. Sci. **158**, 688–695 (2019)
10. Ghobakhloo, M.: Adoption of digital technologies of smart manufacturing in SMEs. J. Ind. Inf. Integr. **16**, 100107 (2019)
11. Rodič, B.: Industry 4.0 and the new simulation modelling paradigm. De Gruyter Open Organizacija **50**, 193–207 (2017)
12. Arm, J., Zeulka, F., Bradac, Z.: Implementing industry 4.0 in discrete manufacturing: options and drawbacks. IFAC PapersOnLine **51**, 473–478 (2018)
13. Cochrane, P.: Why industry 4.0? In: Dastbaz, M., Cochrane, P. (eds.) Industry 4.0 and Engineering for a Sustainable Future, pp. 13–22. Springer, Switzerland (2019)
14. Li, N., Huang, S., Zhang, G., Qin, R., Liu, W., Xiong, H., Shi, G., Blackburn, J.: Progress in additive manufacturing on new materials. J. Mater. Sci. Technol. **35**, 242–269 (2019)

15. Ramezani, H., Luckow, A.: Big data, small data, and getting products right first time. In: Dastbaz, M., Cochrane, P. (eds.) Industry 4.0 and Engineering for a Sustainable Future, pp. 77–90. Springer, Switzerland (2019)
16. Villarraga-Gómez, H., Herazo, E.L., Smith, S.T.: X-ray computed tomography from medical imaging to current status in dimensional metrology. Precis. Eng. **60**, 544–569 (2019)
17. Gapinski, B., Janicki, P., Marciniak-Podsadnaa, L., Jakubowicz, M.: Application of the computed tomography to control parts made on additive manufacturing process. Procedia Eng. **149**, 105–121 (2016)
18. X-Ray Tomography in Industrial Metrology. Werth Messtechnik, Verlag Modene Industrie (Süddeutscher Verlag) (2011)
19. Bossa, N., Chaurand, P., Vicente, J., Borschneck, D., Levard, C., Aguerre-Chariol, O., Rose, J.: Micro- and nano-X-ray computed-tomography: a step forward in the characterization of the pore network of a leached cement paste. Cem. Concr. Res. **67**, 138–147 (2015)
20. Dalmarco, G., Ramalho, F.R., Barros, A.C., Soares, A.L.: Providing industry 4.0 technologies: the case of a production technology cluster. J. High Technol. Manag. Res. **30**, 1–9 (2019)
21. Balci, O.: Quality assessment, verification, and validation of modeling and simulation applications. In: Proceedings of the 2004 Winter Simulation Conference (2004)
22. Zhang, L., Zhou, L., Ren, L., Laili, Y.: Modeling and simulation in intelligent manufacturing. Comput. Ind. **112**, 1–11 (2019)
23. Stojanovic, L., Dinic, M., Stojanovic, N., Stojadinovic, A.: Big-data- driven anomaly detection in industry (4.0): an approach and a case study (2016)
24. Shin, W.S., Dahlgaard, J.J., Dahlgaard-Park, S.M., Kim, M.G.: A quality scorecard for the era of industry 4.0. Total Qual. Manag. Bus. Excell. **29**, 959–976 (2018)
25. Colledani, M., Tolio, T., Yemane, A.: Production quality improvement during manufacturing systems ramp-up. CIRP J. Manuf. Sci. Technol. **23**, 197–206 (2018)
26. Rosa, C., Silva, F.J.G., Ferreira, L.P.: Improving the quality and productivity of steel wire-rope assembly lines for the automotive industry. Procedia Manuf. **11**, 1035–1042 (2017)
27. Realyvásquez-Vargas, A., Arredondo-Soto, K.C., Carrillo-Gutiérrez, T., Ravelo, G.: Applying the plan-do-check-act (PDCA) cycle to reduce the defects in the manufacturing industry. A case study. MPDI Appl. Sci. **8**, 2181 (2018)
28. Sokovic, M., Pavletic, D., Pipan, K.K.: Quality improvement methodologies – PDCA cycle, RADAR matrix, DMAIC and FSS. J. Achiev. Mater. Manuf. Eng. **43**, 476–483 (2010)
29. Glukhov, V.I..: Geometrical product specifications: alternative standardization principles, coordinate systems, models, classification and verification (2014)
30. Bauza, M.B., Tenboer, J., Li, M., Lisovich, A., Zhou, J., Pratt, D., Edwards, J., Zhang, H., Turch, C., Knebel, R.: Realization of industry 4.0 with high speed CT in high volume production. CIRP J. Manuf. Sci. Technol. **22**, 121–125 (2018)
31. Albers, A., Gladysz, B., Pinner, T., Butenko, V., Stürmlinger, T.: Procedure for defining the system of objectives in the initial phase of an industry 4.0 project focusing on intelligent quality control systems. Procedia CIRP **52**, 262–267 (2016)
32. Gorecky, D., Schmitt, M., Loskyll, M., Zühlke, D.: Human-machine-interaction in the industry 4.0 era, pp. 289–294. IEEE (2014)
33. Majstorovic, V.D., Durakbasa, N., Takaya, Y., Stojadinovic, S.: Advanced manufacturing metrology in context of industry 4.0 model. In: IMEKOTC14, pp. 1–11, Springer, Cham (2019)

Challenges for Uncertainty Determination in Dimensional Metrology Put by Industry 4.0 Revolution

Adam Gąska(✉), Jerzy Sładek, and Piotr Gąska

Laboratory of Coordinate Metrology, Cracow University of Technology, al. Jana Pawla II 37, 31-864 Cracow, Poland
adam.gaska@pk.edu.pl

Abstract. The paper presents the most important challenges for uncertainty determination in dimensional metrology caused by the recent changes in manufacturing and production engineering related to fourth industrial revolution. Current trends in dimensional metrology are described and gaps in the state of art in measurement uncertainty determination are identified. Some propositions on how to fill this gaps are also given. The main finding of the paper is that simulation methods for uncertainty determination based on usage of numerical models of measurement and/or manufacturing processes seem to be the most promising in uncertainty determination for in-process and in-line/in-situ measurement systems and for modern measuring devices like industrial computed tomography systems.

Keywords: Uncertainty estimation · Dimensional metrology · CMM · Coordinate Metrology · Industry 4.0

1 Introduction

When analyzing changes in the field of metrology over the past decades, attention should be paid to two areas that are of key importance in the context of the fourth industrial revolution, i.e. the shift of the main focus from post-production metrology to control during the production of a part (in-process metrology, in-line metrology) and development of methods for assessing the accuracy of measurements in near real time.

The development of industry inevitably entails changes at individual stages of the production process. The discoveries in the field of theoretical physics and materials science result in the emergence of new manufacturing techniques pushing forward the existing possibilities of industrial production. Nowadays, we hear more and more about so-called nano-manufacturing that opens up completely new areas in fields such as medicine, aeronautics and molecular biology. One of the main limitations in this field is the problem of finding appropriate methods and tools for effective assessment of the quality of products. This issue illustrates well the importance of the quality control process for the entire development of industry. The changes that took place in metrology in the nineteenth and early twentieth centuries, primarily the normalization of the system of units of measurement and the development of measurement standards

L. Wang et al. (Eds.): AMP 2020, LNME, pp. 92–104, 2020.
https://doi.org/10.1007/978-3-030-46212-3_6

and tools were ones of the most important factors enabling the introduction of mass production, which marks the beginning of the second industrial revolution. However, it was only the application of computer techniques in manufacturing technologies that initiated a real revolution in the field of measuring systems. The possibility of conducting complex calculations in a short time allowed in the 50 s of the last century to introduce the first coordinate machines to industrial practice, which since then have gained importance in the quality control process, and today are primary tool in assessing the compliance of manufactured objects with specifications of their geometry [1–5].

In the coordinate technique, during the measurement, coordinates of the measuring points located on the surface of the measured workpiece are determined, which allows to recreate the geometry of the element being inspected and to assess the evaluated features. Coordinate systems offer high accuracy, automation of measurements and universality, which translates into shortening the measurement time. The most common tool of the coordinate technique is still a contact Coordinate Measuring Machine, which allows obtaining accuracy up to 0.1 μm. Contactless, optical systems cannot achieve such high accuracies, but they display other advantages, mainly shortening the time of the measurement. The trend observed in recent years clearly indicates that systems of this type, operating additionally in connection with e.g. industrial robots, in the future will be one of the most popular tools for quality assessment in advanced production processes [6].

2 Main Trends in Dimensional Metrology Related to Industry 4.0 Demands

2.1 In-Process Metrology

The most important advantage of this approach is that no additional time is required for inspection of processed parts, as the inspection process is parallel to the manufacturing process. Using this approach, it would also be relatively easier to implement 100% control of produced parts comparing to traditional post-production control. In-process metrology requires control of selected parameters of the production process, as well as measurements during the production of objects and automatic analysis of the results obtained so that it is possible to almost immediately prevent any dysregulation of production processes. Performing measurements during the manufacturing process is a difficult problem, which solution is mainly seen in the development of contactless techniques. Systems using laser triangulation or structured light scanners are already installed on production lines, enabling dimensional control (in-line metrology), but it should be noted that this type of measurement is made after processing of the object, and information obtained as a result may only be used to correct the process of manufacturing workpieces in the future.

The actual in-process metrology requires very detailed knowledge of the controlled process, especially information on the impact of various parameters on the properties of the manufactured element. Most often, this involves the need to develop a functional model of the system, which allows simulation of the impact of changes in selected parameters on the result of the machining process. Compliance of the simulation results with the operation of the real system is checked by conducting tests consisting in the production of a series of elements, using the values of process parameters included in the simulation, and then assessing the properties of the obtained elements, e.g. using Coordinate Measuring Machines. This approach also allows the observation of possible correlations between changes in the values of selected process parameters and the characteristics of the manufactured element. Figure 1 presents example of in-process system used in monitoring of rolling of toothed parts.

Fig. 1. In-process control of WPM rolling process [7]

Difficulty of efficient implementation of in-process metrology systems is clearly visible in low number of publications on that topic. Examples of their usage may be found in [7–11]. Analysis of case-studies presented in these publications shows that application of in-process control systems is based on the following steps:

1. Collecting of multi-physical process parameters.
2. Running a post-production measurement assessment of the controlled samples in order to create a "process fingerprint".
3. Finding correlation functions relating part quality with process parameters.
4. Validation of the quality control loop using predefined sets of manufacturing process parameters.

2.2 Challenges for Uncertainty Determination in In-Process Metrology

The challenges that metrologists dealing with in-process metrology systems have to face are numerous. The implementation of new systems of this type is always the new task because except from considerations regarding different manufacturing processes and their principles also machine-related influences must be included. But some challenges related to in-process metrology will be common for majority of usages [7]. They include:

a) functioning in shop-floor system environment,
b) requirement of near real-time data acquisition,

c) multisensor character of in-process measurement systems (for example integration of tactile and optical sensors),
d) data fusion of information given by sensors from several sources,
e) near real-time quality control feedback loop based on fused data from sensors and correlation functions used to control the manufacturing process stability.

Uncertainty determination in in-process control systems is an open problem and the simulation methods based on mathematical modeling of manufacturing and measurement processes and their parameters seems to be the best solution, as they are capable of uncertainty analysis in on-line mode so almost in real-time (which is consistent with the idea of in-process metrology that is also done in quasi real-time). However, development of such models is a difficult task because of complexity and high non-linearity of modelled processes.

The main tasks that has to be solved in the field of in-process metrology are:

- sensors and whole system calibration in order to achieve results traceable to meter unit,
- integration of in-process measurement infrastructure into machine tooling processes of different type,
- development of uncertainty determination methods based on numerical simulation models of manufacturing and measurement processes,
- assuring stability of quality control loops in regard of process fluctuations and influence of external factors,
- defining of models that accounts uncertainties coming from changes in environmental parameters, vibrations, etc.

2.3 In-Line/In-Situ Metrology

Second approach in bringing the inspection closer to manufacturing that may be observed last year and is related to Industry 4.0 is the in-line metrology. It is used especially in mass production and is based on locating the metrological systems on the production line. The desire to control 100% of the produced parts means that measuring systems with a high degree of automation and the shortest measuring time are installed on the production line, just after individual machining and assembly processes [12, 13]. As the inspection operations are performed in production cycle this approach gives significant time reduction comparing to post-production control. Among the systems of this type the most common are scanning systems based on the use of a structured light scanner mounted on an industrial robot. Thanks to this system configuration, the scanner's basic measuring range is significantly increased, and the high redundancy of the industrial robot enables the scanner to reach hard-to-reach places of manufactured parts, quite often large-volume ones. To achieve even greater reduction of working time, this type of systems are multiplied on one inspection cell (Fig. 2).

Fig. 2. Robotized measurements using structured light scanners and industrial robots on a production line in the automotive industry [14]

Another solution that is quite often used is robotization or automation of measurement stations with Coordinate Measuring Machines, in which the measured parts are delivered to the machine measuring space and positioned in it by industrial robots (Fig. 3) or by automatic feeders in the form of e.g. conveyors (Fig. 4).

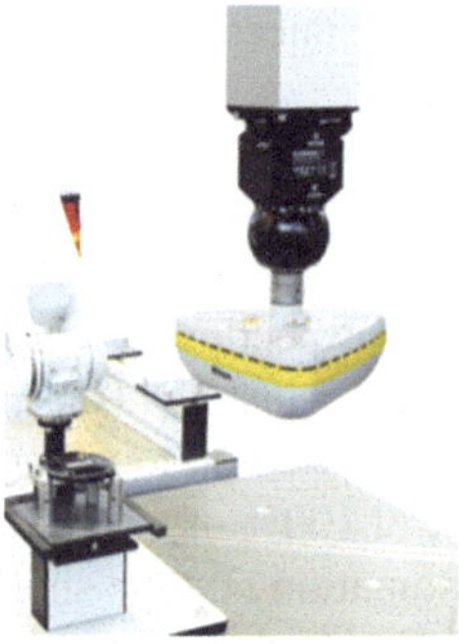

Fig. 3. Robotized CMM equipped with laser triangulation scanner cooperates with industrial robot [15]

Robotized measuring cells sometimes cooperate with vision systems that are able to recognize the type of part that is currently being picked up by the robot and, based on this, choose the right place and type of mounting of the measured part in the measuring volume of CMM. The cells of this type are very often controlled by PLCs.

Fig. 4. Automation of CMM measurements using automatic feeder [15]

It is also common nowadays to use special probe heads that are mounted on machine tools and so configured machine tools are used as CMMs. They are mainly utilized for checking the mounting of a blank in machine tool volume (tool offsets may be updated basing on results of such measurements), for measurement of machined part between different machining operations (key geometrical features are measured and in case of any divergences from specification the machining program may be updated) and measurement of part after machining, without necessity of dismounting the part and transporting it to CMM.

2.4 Challenges for Uncertainty Determination in In-Line/In-Situ Metrology

There are few options that may be chosen for uncertainty determination in in-line/in-situ metrology systems. Similarly as in case of in-process metrology the most promising method is simulation using virtual model of considered measuring system. For systems based on usage of CMMs with robotized or automated part mounting in CMMs measuring volume it is possible to use virtual models that are already in use, as industrial robots or automatic feeders are used only for part manipulations and do not affect strongly the measurement accuracy. Only some research should be undertaken on stability and repeatability of part mounting using automatic devices and its influence on the measurement uncertainty should be investigated. When it comes to virtual models of machine tools with probe heads also the existing models for CMMs may be used in that case with slight modifications, especially regarding models of probe head errors as they differ from models used for CMM's probes. However, models of this type were already developed within research presented in [16, 17], so modules for simulation of kinematic errors may be combined with mentioned probe head models constituting thus virtual models of machine tools with probe heads. Authors of this paper also developed virtual model of 5-axis CMM [18] and it may be used similarly for determination of uncertainty of measurements performed on 5-axis machine tools equipped with probe head.

The biggest challenge in this field is development of virtual model of optical systems, especially for structured-light scanners and laser triangulation scanners. Main limitation causing that such virtual models do not exist yet is the high number of different influences that has impact on optical measurement result and uncertainty. When models of this type will be developed it will be possible to use them with numerical models of industrial robots accuracy and create in that way the virtual model of measurement performed on industrial robot equipped with one of mentioned scanners.

Another possibility of uncertainty determination in in-line metrology systems is to use guidelines of ISO 15530-3 [19]. In-line measurements are in majority of cases performed in shop-floor conditions, for which it is specific that ambient conditions are changing and the range of this changes is much bigger than for laboratory conditions. This is why the most important challenge in uncertainty determination done using this method is development of material standards that are invariant for environmental changes. Some research has already been done in that scope and may be found in [20–22].

2.5 Simulation Systems for On-Line Uncertainty Determination

As was mentioned previously, the latest trends observed in the field of dimensional metrology clearly indicate great effort put on shortening of measurement time. This applies to the rapid growth of the importance of contactless techniques, as well as to the modifications introduced to classic, tactile coordinate machines. On the other hand, the measuring systems users awareness increase in terms of measurement accuracy assessment, especially in the context of the correct estimation of task specific uncertainty. The abovementioned considerations explain inextinguishable interest in so-called simulation methods of uncertainty estimation which is expressed by the most important manufacturers of measuring systems and by leading research centers in the world. The advantages of such approach are clearly visible when it is compared to the well-acclaimed calibrated workpiece method (based on [19]) which involves multiple repetition of measurements and usage of additional standards that should be precisely chosen in order to meet similarity requirements in relation to the measured object. In such case process of assessment of measurement results accuracy may take too much time and be difficult to automate, which hardly fits in to the requirements formulated by the fourth industrial revolution.

The simulation methods often utilize concept of so-called virtual machines, the metrological models of measuring systems. Such models should take into account main factors affecting measurement accuracy, and allow multiple simulation of measurement task. Then on the basis of obtained simulated results, the uncertainty can be estimated using simple statistics. The concept is rather straightforward, however its application for real measuring systems is challenging task. First of all the system for which such model would be applied have to be studied in order to know the main factors which influence its functioning. Next the impact of chosen factors on measurement accuracy need to be assessed through appropriate experiments. Finally the measurement process should be written in form of numerical model which accumulates influence of determined factors. These steps may be time-consuming, but after they are finished, the model may be used to estimate measurement uncertainty in almost real time (depending on computer performance).

Historically, first fully operational, virtual machine was developed in PTB (Physikalisch-Technische Bundesanstalt) for classic tactile CMMs. Two main factors have been identified as crucial for measurement accuracy: errors of machine kinematics and probing systems errors. Since then, many other solutions have been developed, some of which are also commercially available, delivered as separate software or additional functionality in existing metrological systems. Next paragraph presents examples of metrological models for different coordinate systems which were developed by scientists from Laboratory of Coordinate Metrology (LCM).

2.6 Examples of On-Line Uncertainty Determination Systems and Challenges for Their Future Development

Neuro CMM PK is a virtual machine developed for tactile CMMs. It utilizes artificial neural networks for modelling of main errors contributors that affects measurement accuracy. Two group of errors were indicated as dominant factors influencing the machine performance: the errors of probing system and errors connected with machine kinematic system. The virtual machine retain assumptions of the Matrix Method, mainly the concept of grid of reference points in which the machine errors are determined. Basing on implementation experiments separate artificial neural networks are prepared for modelling machine kinematics' related errors and for each probing system configuration including different stylus lengths, probe orientations or tip ball sizes. After preliminary research are done, the model can be used to simulate all points included in measurement task [4, 23].

Another virtual machine developed for classic tactile CMMs is called Virtual MCM PK. The name of model is related to the fact that it is based on Monte Carlo Method. As in previously described model, virtual machine consist of two main modules responsible for modelling of machine kinematics and probing systems errors. However, the model introduces fundamental innovations into the field of virtual machines development. The module for kinematics modelling utilize concept of residual errors distribution analysis. Residual errors are understood as errors that remains after application of correction methods. The method assumes that machines that would be modelled, utilize system of software correction of geometric errors, usually so-called CAA matrix (Computer Aided Accuracy). This assumption is fulfilled for almost all machines manufactured nowadays. In that case it is possible to experimentally determine the residual errors in points that was used during correction matrix determination. It is done using laser tracking system of appropriate accuracy. Full experimental procedure is given in [24]. Additionally model utilize spline and nearest neighbor interpolation methods to obtain information about residual errors distribution in points that weren't included in preliminary experiments directly (which is the vast majority of cases). For probing errors the idea of Probing Errors Function is used (for more details see [24]). The errors modelling is done using Monte Carlo Method and the two-dimensional interpolation method. The virtual machine is available for MODUS software as a script added to the program template.

One of the newest innovation introduced in the field of tactile CMMs are so-called five axis measuring systems that utilize articulated probing system with capability of continuous indexation. In such systems the probing process can be performed by rotational movements of probing system, what minimizes or even eliminates influence of kinematic structure on contact process. For some measurement tasks, especially measurements of rotational elements it can reduce time needed for inspection even by half comparing to classic three axes CMMs. The virtual machine for such five axis measuring systems was developed at LCM [25]. The concept is similar as in case of model that utilizes Monte Carlo method. The machine use two main modules responsible for modelling of kinematic errors and errors of probing system. The kinematic errors module is based on modelling of residual errors examined with laser tracking system (methodology described in previous paragraph). The second module was expanded in order to allow simulation of articulating probing system. The probe can rotate around two mutually perpendicular axes of rotation (vertical axis is called B, while horizontal axis is named A). The model of such solution uses three variables to link Probing Errors Function with appropriate probe orientation (determined by A and B angles) and the approach direction. Data needed for model functioning is obtained through series of measurements of material standard (reference ring) which is placed in machine volume in such a manner to cover whole working range of probing system (Fig. 5a). Then the results of such experiment is used as an input for Monte Carlo simulations. Three dimensional interpolation is used to calculate appropriate parameters for all orientations not included directly in experiment. Additionally the metrological model of five axis system uses Denavit–Hartenberg notation for description of kinematic of probing system. In order to know real geometric parameters of probe head the geometrical calibration is needed. Developed virtual machine was validated with methodology based on the concept of statistical consistency (for description of this concept see [26]). It was compared to normalized calibrated workpiece method. For validation the multi-feature standard was used which allows to check system performance for majority of typical measurement tasks (Fig. 5b). All results obtained during verification measurements proved correct functioning of the model. Moreover, it should be noted that presented methodology used for virtual machine development may also be used for modelling of five axis machine tools equipped with probe heads.

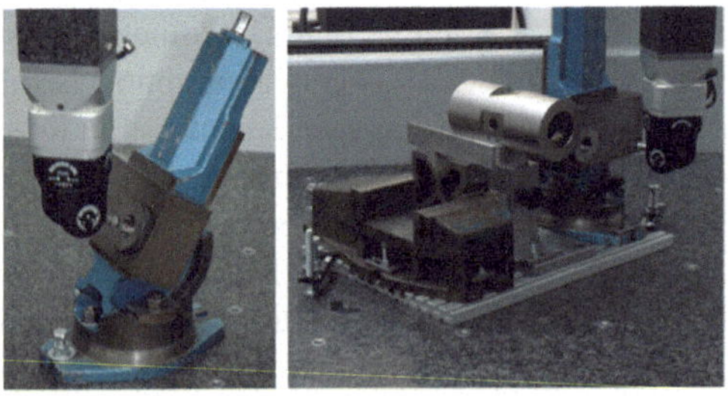

a) b)

Fig. 5. a) Preliminary tests for probing system module; b) Verification measurements

Virtual models are created not only for CMMs. Such solution was successfully developed for Articulated Arm Coordinate Measuring Machines (AACMM) [27]. These devices are very popular in industry due to their mobility and ease of use. The virtual model is based on the Denavit-Hartenberg notation and forward kinematics equations. Values obtained from angular encoders during real measurements are treated as model input. Multiple simulation of measurement is done using Monte Carlo Method. Model was successfully applied for different types of AACMMs, with absolute and incremental encoders and with different number of joints utilized in kinematic structure. It can also be used for industrial robots modelling. For now it is available as addon for PC-Dmis software.

As was shown in the previous sections virtual models can be successfully developed for different types of Coordinate Measuring Systems, utilizing different concepts. However there are still many challenges which must be overcome before virtual machines become a commonly used solution. In case of CMMs one of the biggest problems is large ambient conditions (mostly temperature) variation typically meet in industry. That is why virtual machines are usually used only by research or calibration laboratories which can ensure appropriate thermal stability. The solution of that problem was proposed in virtual machine based on residual errors modelling (Virtual MCM PK). The distribution of errors was checked for different temperature ranges below and above 20 °C. Then the new input data set were developed for kinematic errors module. Depending on the temperature in which points were measured, virtual machine change between data sets and adjust its behavior to changing ambient conditions. However, such approach requires more time for preliminary tests that need to be done prior virtual machine implementation. During these test machine is excluded from its normal functioning so it doesn't fulfil its basic tasks and generating losses for customer. Thus, another problem with virtual machines implementations is excessive length of pre-installation tests. By now the solution for this issue was proposed only by Virtual MCM PK. The series of experiments were conducted aimed at finding such number of points utilized by kinematic errors and probing system modules which guarantee good reality representation but also decrease experiments duration as much as possible. As an effect the whole installation process was limited to one day including virtual machine validation. The Virtual MCM PK model was implemented in one of the companies from automotive industry localized near Cracow [28].

New challenges for simulation methods of uncertainty estimation are also connected with development of new types of measuring systems. The rapidly growing number of multisensor machines which utilizes additionally contactless sensors demands virtual models that would take into account error sources typical to this kind of systems. Although growing interest in industrial CTs such systems hasn't been yet modelled. PTB developed virtual machine for laser tracking systems, however there is still problem of large influence of ambient conditions on laser light which is hard to assess and model for real-life conditions. Another issue that is barely studied is impact of operator influence on manually operated systems which can be one of major factors on resultant accuracy. As can be seen, there are many areas where additional research and work are required, however the advantages of the simulation approach seem to confirm that this subject will be very popular in the following years.

3 Summary

Changes in dimensional metrology that are related to fourth industrial revolution opens new area of research for metrologists. Development of novel measuring systems, possibility of usage of wide variety of sophisticated sensors and new tasks in theoretical metrology related to measurement errors and uncertainty cause that people professionally dealing with metrology will not complain on boredom for a long time.

The most important challenges for uncertainty determination in dimensional metrology put by Industry 4.0 revolution that were identified include:

1. Assuring traceability of in-process measurements through development of calibration and verification methods for in-process control systems. This task is strictly related to uncertainty determination in in-process metrology as there is not much sense in determination of uncertainty methods of measurements that are not traceable to primary standards of units.
2. Development of uncertainty determination methods for in-process metrology systems. Methods based on usage of numerical simulation models of manufacturing and measurement processes seems to be the most promising in solving this task.
3. Development of correct methodologies and relevant material standards for uncertainty determination based on usage of ISO 15530-3 standard guidelines in in-situ/in-line measuring systems.
4. Development of virtual model-based uncertainty determination methods for CTs, optical measurements using different working principles (structured-light, laser triangulation, digital image processing, confocal measurements, multi-focus technique, white light interferometry, etc.).

References

1. Hocken, R., Pereira, P.H.: Coordinate Measuring Machines and Systems. CRC Press, Boca Raton (2012)
2. Ratajczyk, E., Woźniak, A.: Coordinate Measuring Systems. Publishing House of Warsaw University of Technology (Oficyna Wydawnicza Politechniki Warszawskiej), Warsaw (2016)
3. Majstorovic, V.D., Durakbasa, N., Takaya, Y., Stojadinovic, S.: Advanced manufacturing metrology in context of industry 4.0 model. Lecture Notes in Mechanical Engineering, pp. 1–11 (2019)
4. Sładek, J.: Coordinate Metrology. Accuracy of Systems and Measurements. Springer, Heidelberg (2016)
5. Durakbasa, M.N., Bauer, J.M., Capuano, E., Acosta, J.M., Carlos Diaz, J.: The new paradigm in the task of direction-coordination of innovative technological projects in the environment of INDUSTRY 4.0. Lecture Notes in Mechanical Engineering, pp. 311–318 (2020)
6. Hamrol, A., Gawlik, J., Sładek, J.: Mechanical engineering in industry 4.0. Manag. Prod. Eng. Rev. **10**(3), 14–28 (2019)
7. Wieczorowski, M., Eichner, T., Lindner, I., Pereira, A.: A concept of in-process measurement system for spline forming. Manag. Prod. Eng. Rev. **6**(2), 73–81 (2015)

8. Gao, R.X., Tang, X., Gordon, G., Kazmer, D.O.: Online product quality monitoring through in-process measurement. CIRP Ann. – Manuf. Technol. **63**(1), 493–496 (2014)
9. Takaya, Y.: In-process and on-machine measurement of machining accuracy for process and product quality management: a review. Int. J. Autom. Technol. **8**(1), 4–19 (2013)
10. Lindner, I., Eichner, T., Sladek, J., Wieczorowski, M.: In-process quality control approach in metal forming of splined machine elements. In: Proceedings of 11th IMEKO TC14 International Symposium on Measurement and Quality Control, ISMQC 2013, IMEKO, Budapest, pp. 181–184 (2013)
11. Kazmer, D.O., Johnston, S.P., Gao, R.X., Fan, Z.: Feasibility analysis of an in-mold multivariate sensor. Int. Polym. Proc. **26**(1), 63–72 (2011)
12. Majstorovic, V.D., Mitrovic, R.: Industry 4.0 programs worldwide. Lecture Notes in Mechanical Engineering, pp. 78–99 (2019)
13. Majstorović, V.D., Velimirović, M., Glišić, M., Kostić, J., Đura, E., Rančić, M., Mitrović, R.: Cyber-physical manufacturing in context of industry 4.0 model. Lecture Notes in Mechanical Engineering, pp. 227–238 (2018)
14. Zeiss Industrial Metrology. https://www.zeiss.com/metrology/products/sens-ors/on-robot/aimax.html. Accessed 17 Jan 2020
15. Nikon Metrology. https://www.nikonmetrology.com/en-gb/product/in-line-cmm-automation. Accessed 17 Jan 2020
16. Jankowski, M., Wozniak, A.: Mechanical model of errors of probes for numerical controlled machine tools. Meas. J. Int. Meas. Confed. **77**, 317–326 (2016)
17. Wozniak, A., Jankowski, M.: Variable speed compensation method of errors of probes for CNC machine tools. Precis. Eng. **49**, 316–321 (2017)
18. Gąska, P., Gąska, A., Sładek, J., Jędrzejewski, J.: Simulation model for uncertainty estimation of measurements performed on five-axis measuring systems. Int. J. Adv. Manuf. Technol. **104**(9–12), 4685–4696 (2019)
19. ISO 15530-3:2011—Geometrical Product Specifications (GPS)—Coordinate Measuring Machines (CMM): Technique for Determining the Uncertainty of Measurement—Part 3: Use of Calibrated Workpieces or Measurement Standards. ISO Geneva, Switzerland (2011)
20. Viprey, F., Nouira, H., Lavernhe, S., Tournier, C.: Novel multi-feature bar design for machine tools geometric errors identification. Precis. Eng. **46**, 323–338 (2016)
21. Klobucar, R., Acko, B.: Experimental evaluation of ball bar standard thermal properties by simulating real shop floor conditions. Int. J. Simul. Model. **15**(3), 511–521 (2016)
22. Woodward, S., Brown, S., Dury, M., Mccarthy, M.: Traceable in-process measurement (TIM) - producing dimensional transfer artefacts for the assessment of workshop machine tool performance. In: Proceedings of the 16th International Conference of the European Society for Precision Engineering and Nanotechnology, EUSPEN 2016 (2016)
23. Sładek, J., Gąska, A.: Modelling of the coordinate measuring systems accuracy. J. Mach. Eng. **16**(3), 34–46 (2016)
24. Sładek, J., Gąska, A.: Evaluation of coordinate measurement uncertainty with use of virtual machine model based on Monte Carlo method. Meas. J. Int. Meas. Confed. **45**(6), 1564–1575 (2012)
25. Gąska, A., Gąska, P., Gruza, M.: Simulation model for correction and modeling of probe head errors in five-axis coordinate systems. Appl. Sci. **6**(5) (2016). Article no. 144, 12 pages

26. Gromczak, K., Gaska, A., Ostrowska, K., Sładek, J., Harmatys, W., Gąska, P., Gruza, M., Kowalski, M.: Validation model for coordinate measuring methods based on the concept of statistical consistency control. Precis. Eng. **45**, 414–422 (2016)
27. Ostrowska, K., Gąska, A., Sładek, J.: Determining the uncertainty of measurement with the use of a Virtual Coordinate Measuring Arm. Int. J. Adv. Manuf. Technol. **71**(1–4), 529–537 (2014)
28. Gąska, A., Harmatys, W., Gąska, P., Gruza, M., Gromczak, K., Ostrowska, K.: Virtual CMM-based model for uncertainty estimation of coordinate measurements performed in industrial conditions. Meas. J. Int. Meas. Confed. **98**, 361–371 (2017)

Cognitive Twins for Supporting Decision-Makings of Internet of Things Systems

Jinzhi Lu[1(✉)], Xiaochen Zheng[1], Ali Gharaei[1], Kostas Kalaboukas[2], and Dimitris Kiritsis[1]

[1] EPFL - École Polytechnique Fédérale de Lausanne, 1015 Lausanne, Switzerland
{jinzhi.lu, dimitris.kiritsis}@epfl.ch
[2] SingularLogic SA, Athens, Greece

Abstract. Cognitive Twins (CT) are proposed as Digital Twins (DT) with augmented semantic capabilities for identifying the dynamics of virtual model evolution, promoting the understanding of interrelationships between virtual models and enhancing the decision-making based on DT. The CT ensures that assets of Internet of Things (IoT) systems are well-managed and concerns beyond technical stake holders are addressed during IoT system development. In this paper, a Knowledge Graph (KG) centric framework is proposed to develop CT. Based on the framework, a future tool-chain is proposed to develop the CT for the initiatives of H2020 project FACTLOG. Based on the comparison between DT and CT, we infer the CT is a more comprehensive approach to support IoT-based systems development than DT.

Keywords: Cognitive Twins · Decision-making · Knowledge Graph · Internet of Things

1 Introduction

Internet of things (IoT) is a network of items embedded with sensors which are connected through the internet [1]. One IoT system consists of computing devices, physical plants and networks defined as a system-of-systems (SoS) [2]. During developing IoT systems, architectural dependencies across the entire SoS are challenged because of the massive compositions among them. During the lifecycle of IoT, virtual model assets for system, subsystems and components are needed to specify, detect and resolve dependencies across domains, such as interface definition. Compositions from different domains and hierarchies of IoT system are evolving fast. Well managed and predictable evolution dynamics reduce the risks brought by new compositions, such as new characteristics and interoperability. Moreover, the architecture of IoT systems should be permitted with easy connectivity, control and communication among domain-specific applications. Thus, understanding the interrelationships between systems, subsystems and components is very important.

The motivation of our work is to overcome the challenges identified in the above paragraph and provide a new concept and framework to support IoT system development as follows. First, during IoT development, the virtual model asset should be

L. Wang et al. (Eds.): AMP 2020, LNME, pp. 105–115, 2020.
https://doi.org/10.1007/978-3-030-46212-3_7

managed in a systematic way during initial phases. An integrated information infrastructure with virtual models should enable to describe the interrelationships of IoT compositions to promote the understanding of their dependency and traceability. Second, the dynamics of model evolution need to be identified in order to predict the evolution of IoT system, subsystem and compositions. Third, topologies between virtual model assets enable to represent interrelationships of IoT compositions. Thus, the topologies are required to be managed.

Our contribution is to illustrate a new concept called Cognitive Twins (CT) and a knowledge graph (KG)-centric framework supporting CT development. We first define the concept of CT and digital twins (DT) to distinguish the differences between them. Then based on the concept of CT, a KG-centric framework is proposed to develop CT. Using KG, the topologies of virtual model assets are identified and managed. Moreover, a tool-chain concept is designed to support the framework for developing future CT. The results will be used in the H2020 projects FACTLOG and QU4LITY.

The rest of the paper is organized as follows. We discuss the related work in Sect. 2 and introduce the definition of CT in Sect. 3. Moreover, the KG-centric framework is proposed in this section to create CT models. In Sect. 4, a future tool-chain concept is proposed for the related developments in the H2020 project FACTLOG[1]. Finally, we discuss about CT in Sect. 5 and offer the conclusions with a summary in Sect. 6.

2 Related Work

The concept of DT was fostered by the rapid development of various existing technologies such as 3D modeling, system simulation, digital prototyping etc. [3]. In the whitepaper [4] published in 2014, Grieves defined the concept of DT and proposed a three-dimension model of DT based on the previous conception of "a virtual, digital equivalent to a physical product". According to Grieves, a DT model should at least consist of three main parts including: physical products in Real Space; virtual products in Virtual Space; and the connections of data and information that tie the virtual and real products together [4]. Since then, DT and relevant technologies have been evolving rapidly, which reflects that the virtual world and the physical world are becoming increasingly linked to each other and integrated as a whole [5]. Tao F. et al. extended the existing three-dimension DT model by adding two more dimensions, DT data and services, and proposed a five-dimension model to promote the further applications of DT in more fields [6]. In a recent study, Qi et al. [5] reviewed the application fields, enabling technologies and tools for DT. Based on this study, it is concluded that universal design and development platforms and tools for DT are required to facilitate the integration of different technologies and tools which may have different formats, protocols and standards.

Data from different platforms and sources might be heterogeneous in syntax, schema, or semantics, which make data integration difficult. Semantic technologies provide solutions to achieve semantic interoperability in a heterogeneous system [7].

---
[1] H2020 Project FACTLOG: http://factlog.eu/.

Semantic models enable to capture complex systems in an intuitive fashion, which can be summarized in standardized ontology languages, and come with a wide range of off-the-shelf systems to design, maintain, query, and navigate semantic models [8]. This characteristic makes semantic modelling a promising paradigm to address the challenges that DT development is facing currently. The authors of [8] employed semantic technologies to design a system that supports semantics-based DT. Many of existing researches use ontologies as the knowledge base, but the manual construction of ontologies is a very time-consuming task [9]. To overcome this limitation, more advanced techniques such as KGs are being used. According to [10, 11] KGs acquire and integrate information into an ontology and utilize a reasoner to derive new knowledge and they can model information in the form of entities and relationships between them. KGs have been adopted in some studies to accelerate the implementation of DT. For example, in [12] the authors anticipated the paradigm of the next generation DT and KGs were considered as one of the main enabling technologies to link and retrieve all kinds of data, descriptive and simulation models etc. In [13], the authors analyzed the feasibility of backing DT with enterprise KGs based on the fact that DT could be strengthened by using semantic technologies to provide a formal representation of the DT domain. In [14] a graph-based query language was utilized to extract and infer knowledge from large scale production line data, to help generate DT models and therefore enhance manufacturing process management with reasoning capabilities.

Despite the importance of semantic technologies and KGs for the development of DT, there are still many gaps to be bridged, such as the lack of unified implementation architecture, integration of enabling technologies and tools etc. More research efforts are required for this topic.

3 Cognitive Twins

In this chapter, basic concepts of DT and CT are first introduced. Then the characteristics of IoT are introduced in order to formulate the problem of IoT systems. Then a KG- centric framework is proposed to construct CT for IoT systems.

3.1 Basic Concepts

In this chapter, concepts of DT and CT are introduced, as shown in Fig. 1, separately. Based on their respective concepts, the differences between them are summarized.

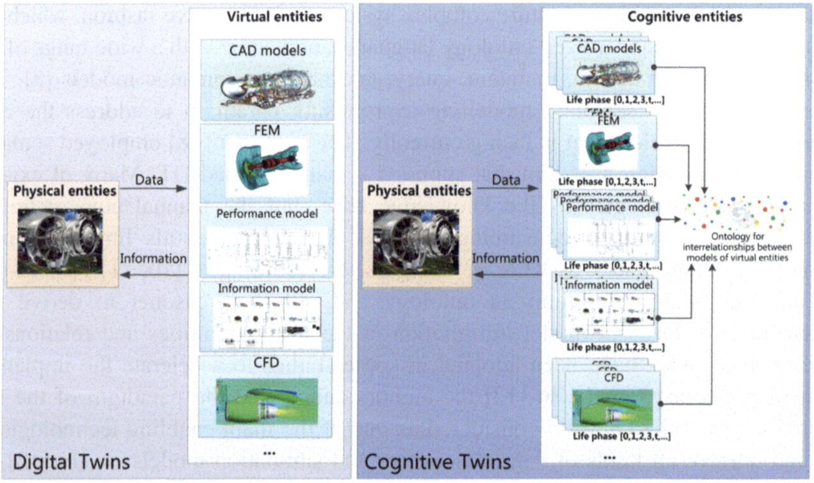

Fig. 1. Digital Twins vs Cognitive Twins

Digital Twins (DT). DT is a digital duplication of entities with real-time two-way communication enabled between the cyber and physical spaces [4]. It aims to support integration of IoT for connecting the physical and virtual spaces. As shown in Fig. 1, if the physical twin is defined as an areo-engine, the virtual entities of areo-engine include CAD models, FEM models etc. In this study, the concept of DT is formally defined as follows:

$$\begin{aligned} &\mathrm{DT_{Sys}} = \mathrm{PE}\{\mathrm{Sys}\} \cup \mathrm{VE}\{\Sigma\mathrm{Model}(\mathrm{Ms}, \mathrm{Mp}, \mathrm{Mt}, \mathrm{Ml}, \mathrm{Mt}, \mathrm{Mm})\} \cup \\ &\mathrm{Comm}\{\Sigma\mathrm{Data}(\mathrm{EntitySt}, \mathrm{EntityDe}, \mathrm{Dtype}, \mathrm{Datacontent})\} \end{aligned} \tag{1}$$

where DT_{Sys} refers to a DT of system *Sys*; *PE{Sys}* refers to the physical twin of *Sys*; *VE{ΣModel(Ms, Mp, Mt, Ml, Mt, Mm)}* refers to a collection of models related to *Sys*. Each model includes several items:

- *Ms* (Model Structure): topology of models, inputs, outputs and parameters.
- *Mp* (Model purpose): the views of modeling, "why is the model needed?"
- *Mt* (Modeling theory): the mathematical foundation of modeling, e.g. differential-algebraic system of equations.
- *Ml* (Modeling language): any language expressing information or knowledge or systems in a structure that is defined by a consistent set of rules.
- Mt (Modeling tool): tools implementing models.
- *Mm* (Modeling method): a set of concepts to explain "how to develop models using a given language in one modeling tool to represent the formalisms?", e.g. finite element modeling and structural equation modeling.

Comm{ΣData(EntitySt, EntityDe, Dtype, Datacontent)} refers to data and information flows between physical entities and virtual entities. Each flow includes several items:

- *EntitySt* (Entities of Start): start of the data and information flow.
- *EntityDe* (Entities of Destination): destination of the data and information flow.
- *Dtype* (Type of data): type of data, such as real-time data and off-line data.
- *Datacontent* (Content of data): the data used in this data flow.

Cognitive Twins (CT). DTs are expected to support the industrial area of design, production, prognostics, and health management, etc. [15]. Each DT has different models which are difficult to manage, because the model versions are updated across the lifecycle. Moreover, the virtual models in DT are across domains which are difficult to identify their interrelationships. The CT is proposed to solve this problem as shown in Fig. 1. One timestamp for each lifecycle spot is added to each virtual model. Moreover, topologies of models are required to be described.

$$\begin{aligned}&\mathrm{CT_{Sys}} = \mathrm{PE\{Sys\}} \cup \mathrm{VE\{\Sigma Model_t(Mt_t,\ Mp_t,\ Mt_t,\ Ml_t,\ Mt_t,\ Mm_t),}\\&\mathrm{Ontopology(entities,\ relationships)\}} \cup \mathrm{Comm\{\Sigma Data(EntitySt,\ EntityDe,\ Dtype,\ Datacontent)\},}\\&\mathrm{t} = 1, 2, 3, \ldots, \text{timespots in lifecycle}\end{aligned} \tag{2}$$

Where CT_{Sys} refers to a CT of system *Sys*; *PE{Sys}* refers to the physical twin of *Sys*; *VE{ΣModel*$_t$*(Ms*$_t$*, Mp*$_t$*, Mt*$_t$*, Ml*$_t$*, Mt*$_t$*, Mm*$_t$*), Ontopology(entities, relationships)}* refers to a collection of models related to *Sys*. Different from DTs, each model in the CT is added with a timestamp in the lifecycle. Except for the items in DTs, *Ontopology (entities, relationships)* refers to ontology to represent the *topology between Models*.

- The *entities* refer to all the information related to models, such as compositions.
- The *relationships* refer to all the interrelationships of entities.

3.2 Problem Formulation

Based on the basic concept of proposed CT, a KG-centric framework is proposed for supporting our EU Projects FACTLOG and QU4LITY and Swiss InnoSwiss IMPULSE project on DTs. These three projects are mainly focusing on IoT systems using DT. Based on the initiative definition [16], several technological and social aspects related to IoT are investigated to identify the industrial concerns for developing the framework in the next section.

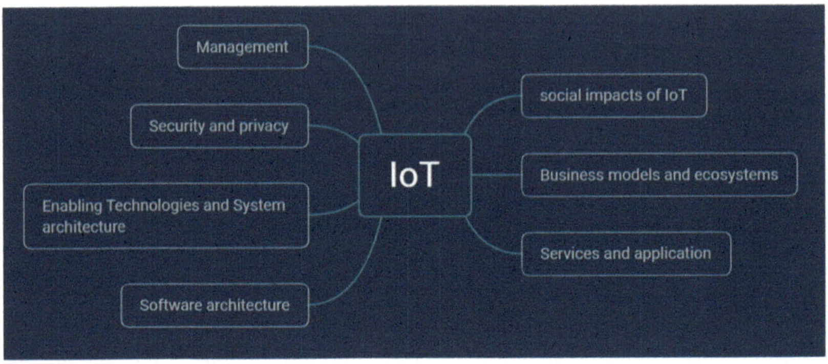

Fig. 2. IoT concerns

As shown in Fig. 2, seven aspects are considered during the entire lifecycle of IoT. The details are introduced as follows:

- Social impacts of IoT, such as impacts and acceptance of users.
- Business models and ecosystems, a new business model for IoT systems.
- Services and application, including domain specific services.
- Software architecture, such as operational systems, middleware.
- Enabling technologies and systems architecture, sensors, energy management
- Security and privacy, such as management of personal data.
- Management, such as autonomics and self-organization of large IoT systems.

3.3 A KG-Centric Framework for Cognitive Twins

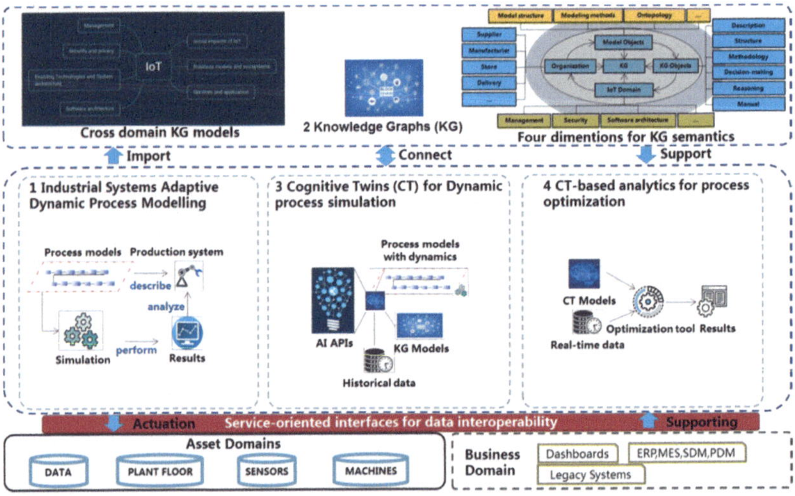

Fig. 3. A KG-centric framework for CT

In order to construct CT for IoT systems, a KG-centric framework is proposed as shown in Fig. 3. It requires inputs from business domains and performs outputs to asset domains. The five main patterns of the KG-centric framework are shown in details as follows:

Process Modeling and Simulation. IoT systems consist of computational compositions, sensors, networks and plants which are considered as hybrid systems including continuous systems and discrete systems [17]. DT is an integrated system consisting of mathematical models and data, which is closed to a real-time synchronization between real physical systems and their own virtual entities [18]. Such process can be represented as entire workflows where the computing composition and other plant nodes are linked together. In this pattern, a process modeling and simulation approach is used to formalize these workflows and to simulate the hybrid system behaviors.

Ontology-Based Knowledge Graph. KG models are at core to represent the topological interrelationships between physical entities and cognitive entities. Before developing KG models, ontologies for KG models are first designed in order to develop the semantics and syntax. Based on the basic concepts of CT and problem formulations in Sect. 3.3, the ontology includes:

- **IoT domains.** This part focuses on the contents related to IoT domains including physical entities and communications. Seven aspects in Sect. 3.2 are considered when defining the ontology.
- **Model objects.** This part mainly focuses on the contents related to CT, such as model structure and *Ontopology* (topology between models).
- **Organizations.** This part mainly focuses on the organizations related to IoT,
- such as suppliers and stores.
- **KG objects.** This part mainly focuses on the knowledge graphs including description, structure, methodology, decision-making, reasoning and manuals.

Cognitive Twins for Dynamic Process Simulation. Artificial Intelligence (AI) APIs, KG models, historical data and process models with dynamics are integrated to generate CT models. CT models aim to support decision-makings for dynamic processes of physical entities.

CT-Based Analytics for Process Optimization. Based on the CT models and real-time data, a tool is used to support process optimization. The optimization results are performed to make decisions for manipulating the physical entities.

Service-Oriented Interfaces for Data Interoperability. Aservice-oriented approach is proposed to develop interfaces for heterogeneous data. All the assets and business domain data are transformed to unified formats through the developed interfaces. Such unified data are used to support other patterns in the framework.

4 A Future Tool-Chain for Developing Cognitive Twins

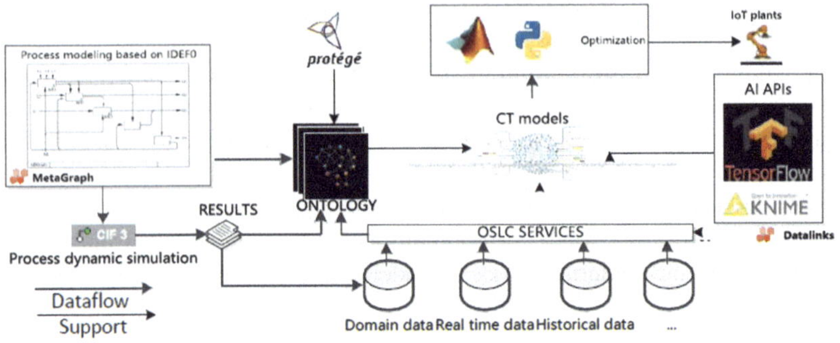

Fig. 4. Overview of the tool-chain

Based on the proposed KG-centric framework, a tool-chain is developed for developing the CT models in which several tools are adopted as compositions of the tool-chain, as shown in Fig. 4. The detailed tools are introduced in Table 1. *MetaGraph* is a DSM tool to develop process models based on *IDEF0* [19]. Moreover, *MetaGraph* generates CIF models for process dynamic simulation for process models. The domain data including process dynamic simulation results, real-time data and historical data of IoT systems are represented as Open Services for Lifecycle Collaboration (OSLC) services through datalinks (a tool for developing OSLC adapters). The OSLC services are RESTful services for linking data through defined URIs. Moreover, we make use of Protégé to formalize process models, process dynamic simulation results and OSLC services from domain data. AI APIs including TensorFlow and KNIME are used to generate CT models based on ontology and OSLC services. The CT models are used for supporting optimize the IoT plants which optimization algorithms are developed using Matlab and Python.

Table 1. The initial tool-chain for developing CT models

Tools	Descriptions
MetaGraph[a] & MetaEdit+ [20]	Process modelling
CIF simulator [21]	Process dynamic simulation
Protégé [22]	Ontology modeling
KNIME [23] & Tensorflow [24]	Develop CT models
Datalinks[b]	Developing OSLC services for domain data
Matlab [25] & Python [26]	Design optimization toolset for the dynamic processes

[a]A Domain-Specific Modeling tool of Z.K. Fengchao http://www.zkhoneycomb.com/.
[b]A tool for developing OSLC services [27] of Z.K. Fengchao http://www.zkhoneycomb.com/.

5 Discussion

Currently, DTs are proposed to support the entire lifecycle of IoT. Physical entities, virtual entities, data, service and connections between them are always concerned by industries, such as NASA [15]. From the literature review, traditional DTs at core focus on connections between the physical entities and virtual entities. The main difference between DT and the proposed CT are replacing the virtual entities by CTs. The CTs add timestamp for each model and provide topologies between all the models. Thus, the cognitive models are dynamically evolved rather than being static according to the physical entities. Several use cases are defined when CT is used (Table 2):

Table 2. The initial tool-chain for developing CT models

Use case	Description
Lifecycle dynamics	The added timestamp for each model is used to analyze the dynamics of virtual model evolutions
Decision-makings	The lifecycle dynamics provide clues for decision-makings for the system evolution
Data analysis across domains	The topology of virtual entities provides a unified description of across domain data which is the basis for data analysis at entire system level

Taking an example of aeroengines, DTs are used for constructing the prognostic health management system, which the physical engine is connected with the digital models in order to realize real-time aeroengine monitoring and fault detection. However, the lifecycle of aeroengine is very long leading to that there are various versions of models used before the aeroengine is finalized. Moreover, the aeroengine consists of different compositions which are used for different scenarios of production, operation and maintenances. The topologies between different virtual models with different versions, domains and hierarchies identify the lifecycle dynamics and domain interrelationships of each model which provide clues about dynamics of system lifecycle and a decision-making solution based on system-level data. Thus, several advantages are summarized:

- The time stamps for each model of CTs promote the dynamics of the virtual model evolution. Based on this dynamics, decision-making based on the CTs enable to predict not only the behaviors of physical entities, but also the model updates of the virtual entities (concepts in DTs).
- Ontology for representing interrelationships between models also provides more clues for analyzing the behaviors of physical entities.

This paper focuses on IoT system development, operation and maintenance. The IoT system developers expect to have a good dependency from requirement, function, behaviors and architecture when developing IoT systems. Moreover, the lifecycle of IoT systems is shorter than traditional equipment, such as areo-engine. The components are renewed quickly which means the entire IoT systems evolve fast. Furthermore, IoT requires flexible and standardized interfaces during they are developed because of such fast evolutions.

Based on the summarized advantages of CT, ontology promotes the understanding of dependencies between models, such as requirement models. In order to support fast evolution of IoT systems, dynamics of virtual models are useful to analyze the system changes and to identify the requirements for new system components. The flexible and standardized interfaces also require a good understanding of interrelationships between physical components or between models. Totally, CT has the better capabilities to support IoT system development compared with DTs.

6 Conclusion

This paper presents a conceptual definition of CTs supporting IoT system development and maintenance. Based on the definition, a knowledge graph based framework is proposed to develop CT models. Based on the framework, a future tool-chain concept is used to support an initiative solution for the H2020 project FACTLOG.

Acknowledgement. The work presented in this paper was supported by the EU H2020 project (869951) FACTLOG-Energy-aware Factory Analytics for Process Industries and EU H2020 project (825030) QU4LITY Digital Reality in Zero Defect Manufacturing and the InnoSwiss IMPULSE project on Digital Twins.

References

1. Chernyshev, M., Baig, Z., Bello, O., Zeadally, S.: Internet of Things (IoT): research, simulators, and testbeds. IEEE Internet Things J. **5**, 1637–1647 (2017). https://doi.org/10.1109/JIOT.2017.2786639
2. Jin, J., Gubbi, J., Marusic, S., Palaniswami, M.: An information framework for creating a smart city through Internet of Things. IEEE Internet Things J. **1**, 112–121 (2014). https://doi.org/10.1109/JIOT.2013.2296516
3. Bricogne, M., Le Duigou, J., Eynard, B.: Design Processes of Mechatronic Systems. In: Hehenberger, P., Bradley, D. (eds.) Mechatronic Futures, pp. 75–89. Springer, Cham (2016)
4. Grieves, M.: Digital Twin: Manufacturing Excellence Through Virtual Factory Replication (2014)
5. Qi, Q., Tao, F., Hu, T., et al.: Enabling technologies and tools for digital twin. J. Manuf. Syst. (2019). https://doi.org/10.1016/j.jmsy.2019.10.001
6. Tao, F., Zhang, M., Cheng, J., Qi, Q.: Digital twin workshop: a new paradigm for future workshop. Jisuanji Jicheng Zhizao Xitong/Comput. Integr. Manuf. Syst. CIMS (2017). https://doi.org/10.13196/j.cims.2017.01.001
7. Cho, S., May, G., Kiritsis, D.: A semantic-driven approach for industry 4.0. In: 2019 15th International Conference on Distributed Computing in Sensor Systems (DCOSS), pp. 347–354. IEEE (2019)
8. Kharlamov, E., Martin-Recuerda, F., Perry, B., et al.: Towards semantically enhanced digital twins. In: 2018 IEEE International Conference on Big Data (Big Data), pp. 4189–4193. IEEE (2018)

9. Ochoa, J.L., Valencia-García, R., Perez-Soltero, A., Barceló-Valenzuela, M.: A semantic role labelling-based framework for learning ontologies from Spanish documents. Expert Syst. Appl. **40**, 2058–2068 (2013). https://doi.org/10.1016/j.eswa.2012.10.017
10. Ehrlinger, L., Wöß, W.: Towards a definition of knowledge graphs. In: CEUR Workshop Proceedings (2016)
11. Nickel, M., Murphy, K., Tresp, V., Gabrilovich, E.: A review of relational machine learning for knowledge graphs. Proc. IEEE **104**, 11–33 (2016). https://doi.org/10.1109/JPROC.2015.2483592
12. Rosen, R., Boschert, S., Sohr, A.: Next generation digital twin. atp Mag **60**, 86 (2018). https://doi.org/10.17560/atp.v60i10.2371
13. Gómez-Berbís, J.M., de Amescua-Seco, A.: SEDIT: semantic digital twin based on industrial IoT data management and knowledge graphs, pp. 178–188 (2019)
14. Banerjee, A., Dalal, R., Mittal, S., Joshi, K.P.: Generating digital twin models using knowledge graphs for industrial production lines. In: Workshop on Industrial Knowledge Graphs, Co-located with the 9th International ACM Web Science Conference 2017 (2017)
15. Tao, F., Zhang, H., Liu, A., Nee, A.Y.C.: Digital twin in industry: state-of-the-art. IEEE Trans. Industr. Inf. **15**, 2405–2415 (2019). https://doi.org/10.1109/TII.2018.2873186
16. Minerva, R., Biru, A., Rotondi, D.: Towards a definition of the Internet of Things (IoT). IEEE Internet Initiat. (2015). https://doi.org/10.1111/j.1440-1819.2006.01473.x
17. Díaz, M., Martín, C., Rubio, B.: State-of-the-art, challenges, and open issues in the integration of Internet of things and cloud computing. J. Netw. Comput. Appl. **67**, 99–117 (2016). https://doi.org/10.1016/j.jnca.2016.01.010
18. Alaasam, A.B.A., Radchenko, G., Tchernykh, A., et al.: Scientific micro-workflows : where event-driven approach meets workflows to support digital twins. In: Proceedings of the International Conference on RuSCDays'18 - Russ Supercomput Days, Moscow, Russia, 24–25 September 2018, vol. 1, pp. 489–495. MSU (2018)
19. Director CSLNI of S and T: Integration Definition for Function Modeling (Idef0). Draft Federal Information Processing Standards Publication 183 (1993)
20. Smolander, K., Lyydnen, K., Tahvanalnen, V.-P., Marttiin, P.: MetaEdit - a flexible graphical environment for methodology modelling. In: Advanced Information Systems Engineering, pp. 168–193 (1991). https://doi.org/10.1007/3-540-54059-8_85
21. van Beek, D.A., Fokkink, W.J., Hendriks, D., et al.: CIF 3: model-based engineering of supervisory controllers. Lecture Notes in Computer Science (including subseries Lecture Notes in Artificial Intelligence and Lecture Notes in Bioinformatics), pp. 575–580 (2014)
22. Musen, M.A., Stevens, R.D.: The protege OWL experience. In: Proceedings of the OWLED, Workshop on OWL: Experiences and Directions (2005)
23. Berthold, M.R., Cebron, N., Dill, F., et al.: KNIME - the Konstanz information miner. ACM SIGKDD Explor. Newsl. **11**, 26 (2009). https://doi.org/10.1145/1656274.1656280
24. Ramsundar, B.: TensorFlow Tutorial. CS224d (2016)
25. Simulink, M., Natick, M.A.: The mathworks. MATHWORKS (1993)
26. Petnga, L., Austin, M.: Ontologies of time and time-based reasoning for MBSE of cyber-physical systems. Procedia Comput. Sci. **16**, 403–412 (2013). https://doi.org/10.1016/j.procs.2013.01.042
27. OASISOpenProject: Open Services for Lifecycle Collaboration Core Specification Version 3.0 (2018)

Similarity Based Methodology for Industrial Signal Recovery

Ramin Sabbagh[1], Alec Stothert[2], and Dragan Djurdjanovic[1(✉)]

[1] Department of Mechanical Engineering, University of Texas at Austin, Austin, TX 78712, USA
dragand@me.utexas.edu
[2] The Mathworks, Inc., Natick, MA 01760, USA

Abstract. The tremendous amount of data generated in the industry provides a massive opportunity to mine that data for decisions, such as prediction of outgoing product quality, process monitoring, etc. In addition, unlike computer and social networks, in the industrial data, the information is not directly observable and is embedded in the signals emitted during the corresponding processes, etc. However, in many cases and for many reasons these sensor signatures are not properly received at the very source causing missing segments in the signal sets. On the other hand, in many manufacturing facilities, large amounts of historical records of past sensor readings are available and can be used to enhance and reinforce the signal recovery process. In this paper, we propose the so-called *match matrix* methodology which uses signal similarity metrics to regenerate the missing segments in a signal from historical signal records. Three different incomplete signal set situations are simulated using a large dataset from a modern semiconductor manufacturing fab. The proposed method is validated utilizing the dataset and the results demonstrated a high fidelity in signal recovery in the all three cases.

Keywords: Signal recovery · Big data · Industry 4.0 · Match matrix · Prediction

1 Introduction

It is not widely known that industrial equipment is generating more data than computer and social networks, with almost double the growth rate, leading to tremendous amounts of pertinent data [1]. This provides a massive opportunity to mine that data for decisions, such as prediction of outgoing product quality, process monitoring or optimization of operations. In addition, this data can be used to extrapolate the knowledge learnt from past historical observations of the manufacturing process to facilitate better design and control of some new processes. However, unlike computer and social networks, in the realm of industrial data, events that are relevant to modeling and characterization of the underlying system are embedded in the data and are not directly visible. For instance, beginning and ending moments of a reaction in a chemical reactor, or moment and location of particle emission and trajectory of a particle in a semiconductor vacuum tool – all this information is not directly observable

L. Wang et al. (Eds.): AMP 2020, LNME, pp. 116–128, 2020.
https://doi.org/10.1007/978-3-030-46212-3_8

and is embedded in the signals emitted during the corresponding processes. Finding and characterizing such events in industrial data can leverage tremendous advancements in Artificial Intelligence (AI) and Machine Learning (ML) in the domains of computer and social networks and places great significance on one's ability to accurately and reliably perform process or quality control.

However, in many cases and for many reasons, the sensor readings are not properly received at the very source causing missing segment(s) in one or more signals. It can be expected that transmitted sensor data are lost or corrupted due to many reasons, such as power outage at the sensor's node, random occurrences of local interferences, or a higher bit error rate of the wireless radio transmissions as compared with wired communications. Simply re-querying data is a naïve and often unfeasible alternative, as it may induce long delays, quicken the power exhaustion of the node, and above all, it is still not guaranteed to recreate the original readings. Hence, enabling the use of missing or corrupted sensor readings, a much more plausible and feasible option is to estimate the missing stream values from the available historical sensor records [2].

A variety of techniques have been used in the past for time series modeling and prediction in order to recover the missing parts from the signals [3, 4]. Parametric linear prediction techniques, such as Auto-Regressive Moving Average (ARMA) [5, 6] or Kalman filtering [7], may work well only for short-term predictions and they could not capture the information from longer terms data. On the other hand, Recurrent Neural Networks (RNN) are used for short-term predictions of time-series in order to perform signal recovery [8]. In addition, variety of approaches has been studied in the literature, namely, fuzzy time series and clustering [9, 10], multi-resolution wavelet models [11–13] and neural networks [14]. However, without a priori knowledge about the signals under consideration, selecting an appropriate non-linear model and its structure is a challenging task.[1]

On the other hand, in many manufacturing facilities, large amounts of historical records of past sensor readings are available and can be used to enhance and reinforce the signal recovery process. In this paper, we propose the so-called *match matrix-based* method to regenerate the missing segments in the sensor readings. The employs inter-signal similarity metrics described in [16] to reconstruct the missing signal portions via weighted combination of corresponding portions of the signals available in the historical data records.

The rest of the paper is organized as follows. In Sect. 2, the new match matrix-based signal recovery method is introduced. The newly proposed approach has been tested on data generated during processing of 300 mm diameter wafers in a modern semiconductor manufacturing fab, with the results being shown in Sect. 3. Section 4 provides conclusions and future work.

[1] The commonly used gradient descent algorithms for RNN training exhibit certain problems during training, such as having difficulty dealing with long-term dependencies in the time series [15], which in turn limits their capability of achieving accurate long-term predictions. In addition, finding a suitable number of hidden neurons and appropriate RNN structure is another challenging problem [16].

2 Methodology

Let us assume that a manufacturing process is described by a multiple time series of sensor readings. In order to reliably recover the missing signals or their portions, we need an estimation method that effectively utilizes the existing historical records of sensor readings, which are connected to the corrupt sensors that are being reconstructed via common dynamics of the underlying process or machine. This task will involve comparison of sensory time series from a potentially large number of past production cycles and fusion of the results of those comparisons that enables estimation of the missing signals or signal portions of the current production cycle.

The Mahalanobis distance is commonly used to evaluate the distance between multidimensional feature vectors whose components are quantities that have different ranges and amounts of variations [17]. Following [16], in this paper, we will utilize a Mahalanobis distance-based similarity metric between two feature vectors, which is calculated as follows:

$$S_{i,j}^{p} = e^{-\sum_{k=1}^{n} \frac{\left[\left(f_k^i\right)_p - \left(f_k^j\right)_{Curent} \right]^2}{2\sigma_k^2}} \tag{1}$$

where vector

$$\left(\vec{f_i}\right)_p = \left[\left(f_1^i\right)_p \quad \left(f_2^i\right)_p \quad \cdots \quad \left(f_n^i\right)_p \right]^T$$

consists of n sensor readings observed at sample i of the past cycle p, vector

$$\left(\vec{f_j}\right)_{Curent} = \left[\left(f_1^j\right)_{Current} \quad \left(f_2^j\right)_{Current} \quad \cdots \left(f_n^j\right)_{Current} \right]^T$$

consist of the readings of the same n sensors, observed at sample j of the current production cycle, while σ_k^2 is the variance of the sensor reading k calculated from the signals corresponding to nominal system behavior. It is obvious that similarity $S_{i,j}^{p}$ between vectors of sensor readings at sample i of the past production cycle p and sample j of the current production cycle approaches 1 if those samples look alike (if the Mahalanobis distance between them approaches zero), and tends to 0 of those two samples happen to be very different (as Mahalanobis distance between them tends to infinity).

Let us also assume that a specific process in different manufacturing cycles takes the same time and sampling rate for all sensors are the same.[2] Thus, if each signal has n number of samples in the current and past runs $(i,j = 1, 2\ldots, n)$, the resulting match matrix will be in $n \times n$ dimensions. In addition, if one has historical records of P past production cycles, then one can observe P so-called match matrices $S_{i,j}^{p}, p = 1, 2, \ldots, p$ between the current cycle and all previously observed production cycles.

[2] In order to compare feature vectors, they need to be in the same length. If the sampling rates are different for different sensors, one may need to resample them to be in the same length.

Figure 1 illustrates the procedure of creating and updating the match matrices. Figure 1(A) shows a match matrix between the current cycle and the past cycles p. The red square in the Fig. 1(A) is a gray-scaled representation of $S^p_{i,j}$, the similarity between the feature vector $\left(\vec{f_j}\right)_{Curent}$ from the current cycle and the feature vector $\left(\vec{f_i}\right)_p$ from the past cycle p. Figure 1(B) demonstrates how a gray-scaled color represents the similarity between two feature vectors where white color shows identical similarity and black color shows no similarity. Figure 1(C) illustrates how we can update all the P past match matrices. Every time a new feature vector in the new cycle is extracted from sensor readings, its comparison with all the past feature vectors based on (1) incorporates a column of similarity measures to all P match matrices. This involves $\sum_{p=1}^{P} K_p$ comparisons, where K_p represents number of feature vectors in the pth past cycle.

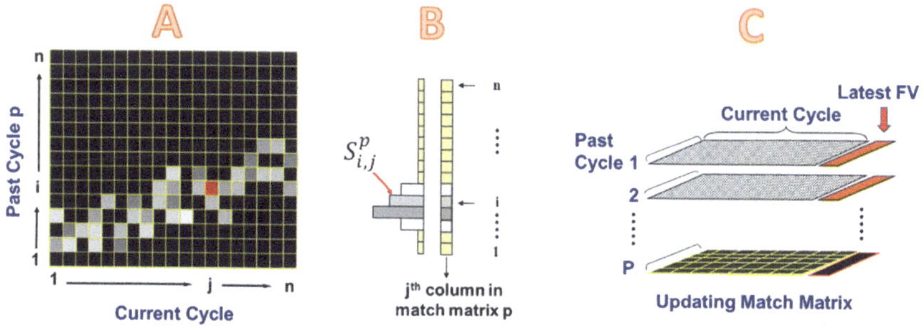

Fig. 1. Match Matrix based on two feature vectors (FVs) from the current cycle and a past cycle p. (A): X-axis and Y-axis are representing the sample numbers associated with the signals in the current cycle and past cycle p respectively. (B): A gray-scaled color representation of the similarity between two feature vectors. (C): The procedure of updating all the p number of past match matrices.

If all the feature vectors are collected from identical sensors and under identical conditions, one would expect similar sensor readings to permeate across multiple production cycles. Thus, in case of having incomplete signals, one could use the previous sensor readings to recover the missing segments in the signals. In this paper, we utilize the time series made of the best match indices in match matrices between the current production cycle and the past production cycles to fill the missing vectors of sensor readings with the most similar feature vectors in the past cycles. Figure 2 illustrates a typical index curve generated from a match matrix.

Nevertheless, previous experiences with industrial datasets demonstrate that very frequently, the feature vectors extracted out of real-life manufacturing processes contain a considerable level of noise. In order to reduce the variation of the best match indices, the expected index of similarities is calculated from each column j of any match matrix as

$$I_j^p = \frac{\sum_{i=1}^{n} iS_{i,j}^p}{\sum_{i=1}^{n} S_{i,j}^p} \quad (2)$$

where n denotes the number of feature vectors (i.e. length of signals) in the past cycle. The time series of mean best match indices makes it possible to apply the linear parametric prediction technique based on ARMA modeling [4] to predict the future values of the time series of mean best match indices between the current and the past cycles. Furthermore, ARMA prediction also allows one to analytically evaluate the uncertainty of prediction of the mean best match indices.

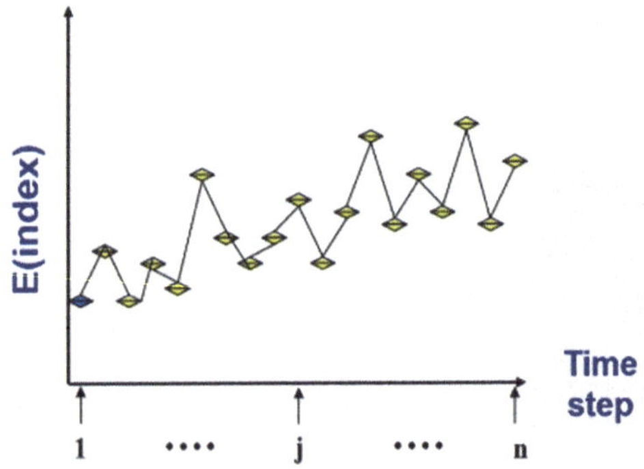

Fig. 2. Sample index curve generated from a match matrix

Let us assume that there are P past cycles, which means that there are P match matrices for a specific manufacturing cycle, each yielding a time series of expected indices of the best matches. Missing values of mean best match indices can be predicted using a linear utilizing as ARMA model-based prediction[3], with structure of the model is identified using e.g. Akaike Information Criterion (AIC) [4]. The predicted value of the best match index points to the index of the feature vector in the corresponding past cycle that is likely to be similar to the feature vector of the current cycle in the missing time steps. In order to emphasize the similarity of the past cycles, a linear combination of the P feature vectors (one from each match matrix, which correspond to each past production cycle) is evaluated to form a predicted feature vector. The weights are determined based on the average similarities W_p associated with match matrices $p = 1, 2, \ldots, P$ and calculated as

$$W_p = \frac{\sum \bar{S}_p}{\sum_k \bar{S}_K}, p = 1, 2, \ldots, P \quad (3)$$

[3] One advantage of utilizing an ARMA model is it allows analytical expression for the variance of prediction errors and can therefore yield uncertainty/confidence intervals for the prediction of the mean best match indices.

$\bar{S}_p$ denotes the average similarity between the latest several feature vectors in the current cycle with the feature vectors in the pth past cycle. Such weighting places stronger emphasis on the features observed in cycles that display higher similarity with the current cycle whose behavior we are trying to predict.

Three different situations of missing/corrup segments are studied in this paper.

i. In situation (i), one or more segments are missing from one signal (in the extreme case, the entire signal may be missing). In this case, one has a complete curve of expected best match indices from each match matrix, since the curves can be generated based on other elements of feature vectors along with the remaining elements from the missing feature vector. In other words, match matrix enables one to predict the incomplete signals based on the behavior all other signals, as well as that of the corrupt signal before and after the incompleteness occurs.
ii. In situation (ii), one or more segments is missing at exactly the same time portions from all the signals (i.e. there are no feature vectors available in specific time interval of the current cycle). In this situation, the expected best match index curves from all match matrices will be incomplete in those samples and one must predict the incomplete curves using adequate ARMA models. In order to increase the accuracy of the proposed method, the prediction is performed from both directions (forward and backward) in time, and two predicted results are being merged by weightings based on closeness (in time) to the nearest complete signal segments. Once the prediction is done, one could recover the missing feature vectors utilizing the estimated mean best match indices and the corresponding feature vectors from the past cycles.
iii. Finally, situation (iii) deals with the very extreme case where there are no feature vectors in the current production cycle (i.e. all the signals are missing), which means that there are no expected best match index curves. In this case, weighted expected valued of indices from a small portion of the most similar past cycles to the previous cycle will be used to regenerate the best match index curve and thus recover all the signals from scratch.

3 Results

In this section, we will demonstrate the results of applying the proposed methodology to a large industrial dataset. The match matrix created via the proposed method along with the expected best match index curve associated with that matrix will be used to reconstruct signals corrupt via one of the three mechanisms described at the end of the previous section. Capability to reconstruct the corrupt signals or signal portions will be demonstrated through comparisons of the recovered signals or signal portions with their corresponding actual signals or signal portions.

The dataset used in this research consists of sensor readings from a Plasma Enhanced Chemical Vapor Deposition (PECVD) process from a modern semiconductor manufacturing fab involving processing of 300 mm diameter wafers. PECVD tools are used for depositing thin films onto silicon wafer substrates, which is one of the key steps in manufacturing of logic and memory circuitry in semiconductor manufacturing. Inside a PECVD tool chamber, reactive gases pass over a silicon wafer and are absorbed onto its surface to form a thin film layer. The gases are excited into a plasma state using a strong radiofrequency (RF) electromagnetic field, which allows the film deposition to take place at lower temperatures, more suitable for integrated circuit fabrication on large silicon wafers [18]. This dataset contains sensor readings corresponding to three different recipes of thin film depositions involving the total of over 100,000 wafers, with readings from 11 different sensors collected using a 10 Hz sampling rate.

Initial set of sensor readings from 100,000 wafers was assumed to be available in the historical records, with sensor readings from the very next wafer being treated as the "current production cycle", which was corrupted in multiple ways and then reconstructed using the newly proposed methodology. Thus, we have 100,000 we deal with 100,000 match matrices and corresponding expected best match index curves (one for each past cycle). Each of the 11-simensional feature vectors (i.e. all sensor signals form all 100,001 wafers) nominally consisted of 125 sample points sampled at the same time-moment.[4]

The 100K match matrices are generated form the raw data in about 20 min using an ordinary Personal Computer.[5] Figure 3 illustrates one of the constructed match matrices, while Fig. 4 shows the 2-dimensional version of the match matrix shown in Fig. 3, with white color representing the similarity of 1 and black indicating that there is no similarity between the corresponding feature vectors.

Based on the index numbers on the Y-axis (past cycle) and similarity values associated with those indices, we create the expected best match index curve for each match matrix. Figure 5 shows the progression of the expected value of the best match indices associated with the match matrix shown in Figs. 3 and 4 (note that the ideal expected value index curve is a perfect diagonal line from (0,0) to (125,125) points; nevertheless, natural process noise and variations cause deviation from that ideal pattern).

For each match matrix, the average similarity of all its elements is used to ensure that past signals that show more similarities with the current one end up having more weighting in the signal reconstruction process. Weights are formed using the aforementioned average similarities and reconstruction is accomplished as a weighted average of missing signals or signal portions from the past, with all weights summing up to one.

---

[4] If the sampling rates are different for different sensors, one may need to resample them to be in same length.

[5] All evaluated on a PC with 32.0 GB RAM and Intel® Xeon® CPU E5-1650 v4 @ 3.60 GHz processor, 6 cores.

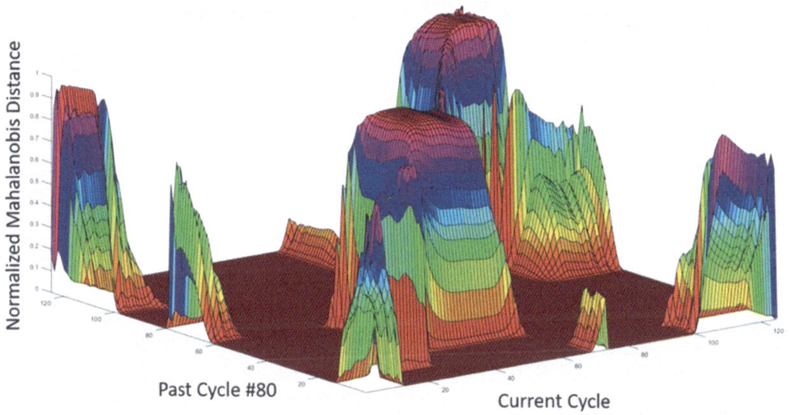

Fig. 3. Three-dimensional match matrix illustration. X-axis represents the current cycle and Y-axis represents one of the past cycles (cycle #80), and Z-axis shows the Mahalanobis-based similarity between these two cycles.

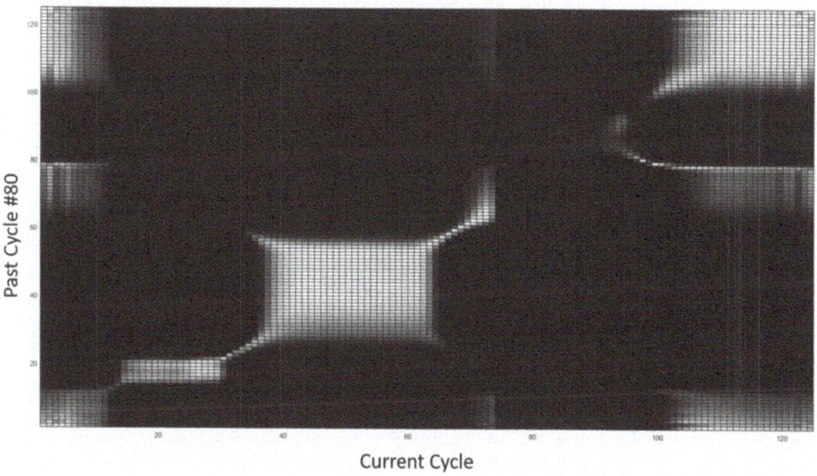

Fig. 4. Two-dimensional match matrix illustration. X-axis represents the current cycle and Y-axis represents one of the past cycles (cycle #80).

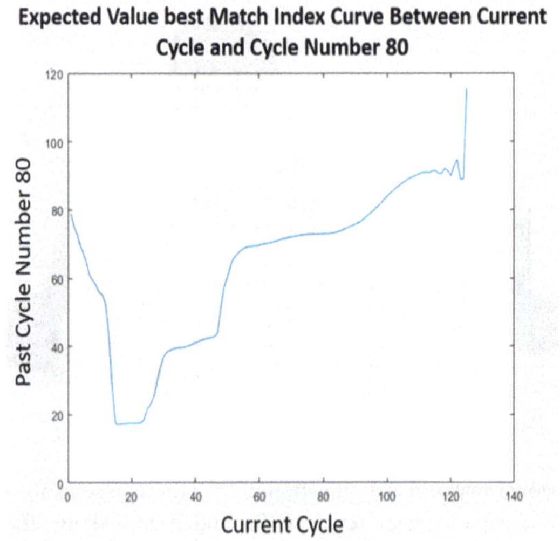

Fig. 5. Expected value best match index curve for match matrix number 80

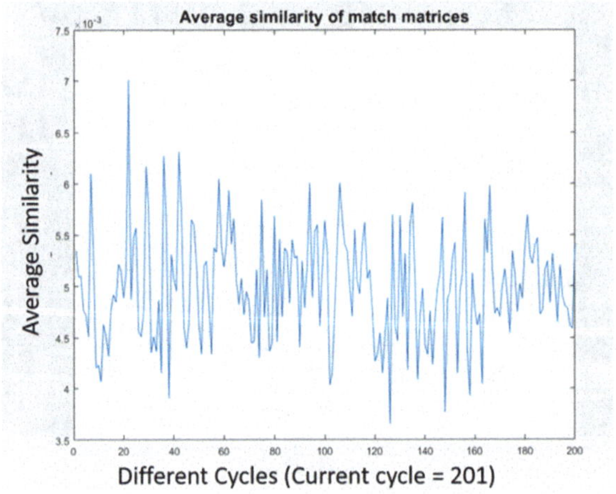

Fig. 6. Average similarity values associated with the first 200 match matrices.

Figure 6 shows the average similarity values associated with the first 200 match matrices. Note that these similarities seem to vary fairly randomly, which is understandable, given that we are dealing with a highly stable process in which dramatic signal changes and drifts are prevented through a very tight process control.

In order to validate the proposed methodology, the three scenarios discussed in the previous section are designed in the following manner. In the first case study, data points 60 to 90 of only one signal out of 11 signals are removed in the current cycle. We repeated this process for each of the 11 sensor readings and Fig. 7 shows the recovered and original readings for two different sensors. Average adjusted R^2 value over all 11 signals was 0.9723, ranging between 0.9316 and 0.9893.

Figure 8 illustrates the extreme situation in which the whole signal is missing from the current production cycle. In this case, the signal is reconstructed solely based on the past behavior of other signals. This experiment was also conducted for all 11 signals, resulting in average adjusted R^2 of 0.8902, ranging between 0.8643 and 0.9245.

Finally, let us consider the extreme situation that all signals are completely missing for the current production cycle. An experiment with 1,000 randomly chosen cycles in our dataset showed that the immediately preceding production cycle (wafer) seemed to be the most similar to the current one in more than 96.3% of times. Therefore, in the situations in which we do not have any information about the current cycle, we selected the top 0.5% most similar cycles to the previous cycle and estimated the missing sensor readings for the current cycle as the weighted average of signals that are reconstructed from the corresponding match matrices.

Consequently, the same production cycle considered in this section was corrupted in a way that all 11 signals disappeared completely. Figure 10 shows the result of reconstructing two of the 11 signals from the database. The average adjusted R^2 measure over all 11 signals was 0.7639, ranging between 0.7166 and 0.8274.

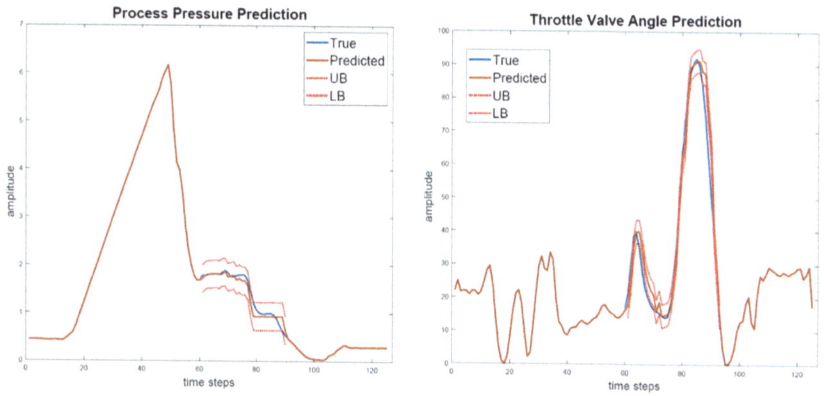

Fig. 7. Recovered and true signals along with the 95% confidence interval (UB - upper bound; LB - lower bound) associated with recovery of Process Pressure and Throttle Valve Angle signals. Data points 60 to 90 are missing.

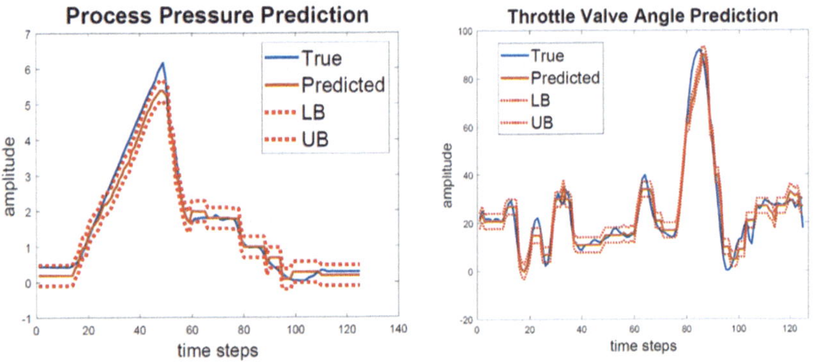

Fig. 8. Recovered and true signals along with the 95% confidence interval (UB - upper bound; LB - lower bound) associated with recovery of Process Pressure and Throttle Valve Angle signals. Note that the entire signal is missing in the current cycle.

Situation (ii) was simulated by corrupting (removing) samples 60 to 90 from all the signals in the current cycle. In this case, the best match index curves in each match matrix will miss indices 60 to 90 and those indices need to be reconstructed. First, the top 5% most similar match matrices are selected, and missing portions of the expected best match index curves are estimated using weighted predictions forward from the index series prior to the corrupted signal section, and backward from the signal portion following the corrupted section. The two predictions were combined into the estimated best expected match via weighted averaging, where weights associated with the backward and forward predictions were assessed based on how far each one of them needed to go to estimate the missing index. Once the curve of the expected best match

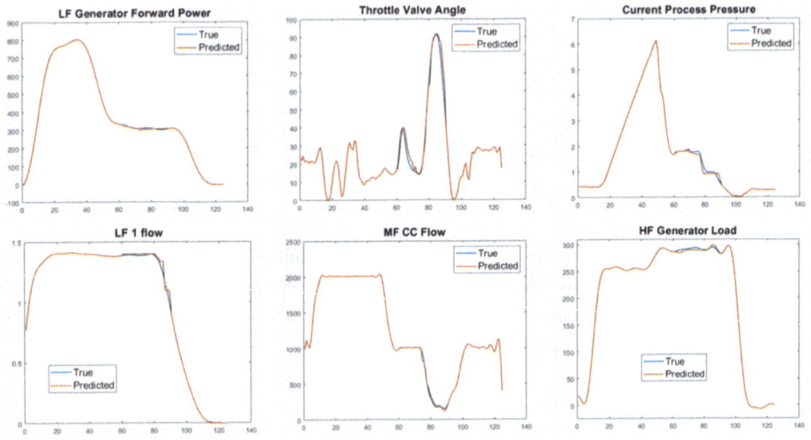

Fig. 9. Recovered and true sensor readings for 6 (out of 11) sensors, as obtained for situation (ii), when all 11 sensors were corrupted during the same time-interval (between samples 60 and 90).

indices was constructed, the corresponding past sensor readings could be pulled from the past to reconstruct the missing signal portions in all 11 sensors, with Fig. 9 showing the recovered and true sensor readings for six out of 11 of those signals. The average adjusted R^2 value for all 11 sensors was 0.9361, ranging between 0.9143 and 0.9766.

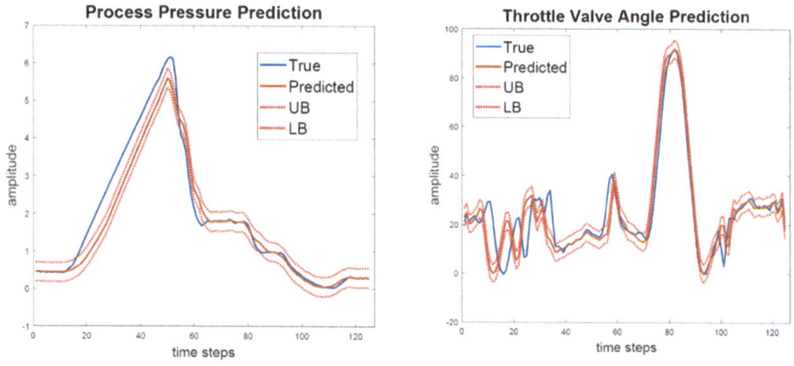

Fig. 10. Recovered and true signals along with the 95% confidence interval (upper bound – UB, and the lower bound – LB) associated with recovery of Process Pressure and Throttle Valve Angle signals – the whole feature vectors from the current cycle are missing in the current cycle.

4 Conclusions and Future Research

In this paper, we propose the so-called *match matrix-based method* to regenerate the missing segments in sensor readings using past historical records of sensor readings and signal similarity metrics between those signal records and the sensor readings that are being reconstructed. Multiple situations of signal corruption were simulated using a large dataset from a modern semiconductor manufacturing fab and comparison between reconstructed and actual sensor readings demonstrated a high fidelity in signal recovery in all considered cases.

In the future research, an unsupervised clustering method will be applied into match matrices to decrease the computational load of the method. Thus, instead of creating and utilizing entire number of match matrices and accordingly best match index curves, a selected proportion (such as cluster centroids) could be used for the signal recovery process.

Acknowledgements. This work is based upon work supported in part by the National Science Foundation under Cooperative Agreement No. EEC-1160494. Any opinions, findings and conclusions or recommendations expressed in this material are those of the author(s) and do not necessarily reflect the views of the National Science Foundation.

References

1. Kalyanaraman, S.: Industry 4.0 Meets Cognitive IoT: Internet of Things Blog (2016). https://www.ibm.com/blogs/internet-of-things/industry-4-0-meets-cognitive-iot/
2. Gruenwald, L., Chok, H., Aboukhamis, M.: Using data mining to estimate missing sensor data. In: Seventh IEEE International Conference on Data Mining Workshops (ICDMW 2007), pp. 207–212. IEEE, October 2007
3. Candes, E.J., Romberg, J.K., Tao, T.: Stable signal recovery from incomplete and inaccurate measurements. Commun. Pure Appl. Math. J. Issued Courant Inst. Math. Sci. **59**(8), 1207–1223 (2006)
4. Zhang, Z., Rao, B.D.: Sparse signal recovery with temporally correlated source vectors using sparse Bayesian learning. IEEE J. Sel. Top. Signal Process. **5**(5), 912–926 (2011)
5. Pandit, S.M., Wu, S.-M.: Time Series and System Analysis, with Applications. Wiley, New York (1983)
6. Box, G.E.P., Jenkins, G.M.: Time Series Analysis: Forecasting and Control, revised edn., Holden-Day, San Francisco (1976)
7. Kalman, R.E.: A new approach to linear filtering and prediction problems. Trans. ASME J. Basic Eng. **82**, 34–45 (1960)
8. Wang, Z.H., Horng, G.J., Hsu, T.H., Aripriharta, A., Jong, G.J.: Heart sound signal recovery based on time series signal prediction using a recurrent neural network in the long short-term memory model. J. Supercomput. 1–18 (2019)
9. Geva, A.B.: Non-stationary time-series prediction using fuzzy clustering. In: Presented at Proceedings of NAFIPS-99: 18th International Conference of the North American Fuzzy Information Processing Society, New York, NY, USA, 10–12 June 1999
10. Lee, C.-H.L., Liu, A., Chen, W.-S.: Pattern discovery of fuzzy time series for financial prediction. IEEE Trans. Knowl. Data Eng. **18**, 613–625 (2006)
11. Papagiannaki, K., Taft, N., Zhang, Z.-L., Diot, C.: Long-term forecasting of internet backbone traffic. IEEE Trans. Neural Netw. **16**, 1110–1124 (2005)
12. Renaud, O., Starck, J.-L., Murtagh, F.: Wavelet-based combined signal filtering and prediction. IEEE Trans. Syst. Man Cybern. Part B Cybern. **35**, 1241–1251 (2005)
13. Wei, H.L., Billings, S.A.: Long term prediction of non-linear time series using multiresolution wavelet models. Int. J. Control **79**, 569–580 (2006)
14. Chen, J.-L., Islam, S., Biswas, P.: Nonlinear dynamics of hourly ozone concentrations: nonparametric short term prediction. Atmos. Environ. **32**, 1839–1848 (1998)
15. Han, M., Xi, J., Xu, S., Yin, F.-L.: Prediction of chaotic time series based on the recurrent predictor neural network. IEEE Trans. Signal Process. **52**, 3409–3416 (2004)
16. Liu, J., Djurdjanovic, D., Ni, J., Casoetto, N., Lee, J.: Similarity based method for manufacturing process performance prediction and diagnosis. Comput. Ind. **58**(6), 558–566 (2007)
17. De Maesschalck, R., Jouan-Rimbaud, D., Massart, D.L.: The mahalanobis distance. Chemometr. Intell. Lab. Syst. **50**(1), 1–18 (2000)
18. Geng, H.: Semiconductor Manufacturing Handbook. McGraw-Hill, Inc. (2005)

Closed-Loop Control by Laser Power Modulation in Direct Energy Deposition Additive Manufacturing

Stefano Baraldo(✉), Ambra Vandone, Anna Valente, and Emanuele Carpanzano

SUPSI, Institute of Systems and Technologies for Sustainable Production, Department of Innovative Technologies, University of Applied Sciences and Arts of Southern Switzerland, Manno, Switzerland
stefano.baraldo@supsi.ch

Abstract. Direct Energy Deposition is a metal additive manufacturing technique that has raised great interest in industry thanks to its potential to realize complex parts or repairing damaged ones, but the complexity of this process still requires much effort from practitioners to achieve functionally sound parts. One of the recurring flaws of such parts is the phenomenon of over-deposition, which may occur due to unpredicted local increases of energy density.

The deposition of uniform metal tracks is critical in many practical cases, when parts are composed by a significant number of layers and/or when complex tool paths induce heat build-up, for example in thin structures. Therefore, detecting anomalies such as over-growth in real-time and dynamically correcting them is of paramount importance for achieving repeatable, first-time-right parts.

This work studies the use of a closed-loop control system for Direct Energy Deposition, proposing to adjust on-line the power of the laser beam according to the feedback provided by the analysis of melt pool images. The images are acquired by a camera, mounted coaxially into the optical chain of the deposition head, which records images at 100 fps while the process is running. The proposed approach is explored experimentally by comparing the over-deposition measured on sample test geometries obtained with a traditional feed-forward approach with the over-deposition obtained through the developed closed-loop control laser deposition system.

Keywords: Metal additive manufacturing · Vision systems · Process monitoring · Process control · Predictive model

1 Introduction

In the Additive Manufacturing (AM) process known as Direct Energy Deposition (DED), a mixture of carrier gas and metal powder particles is blown out from a set of nozzles [1]. The particles are projected toward the target deposition spot, where a laser beam fuses them along with the underlying substrate, forming a melt pool, i.e. a drop of molten metal. As the deposition head moves, the melt pool cools down, evolving into a solid metal track of the desired geometry (Fig. 1).

L. Wang et al. (Eds.): AMP 2020, LNME, pp. 129–143, 2020.
https://doi.org/10.1007/978-3-030-46212-3_9

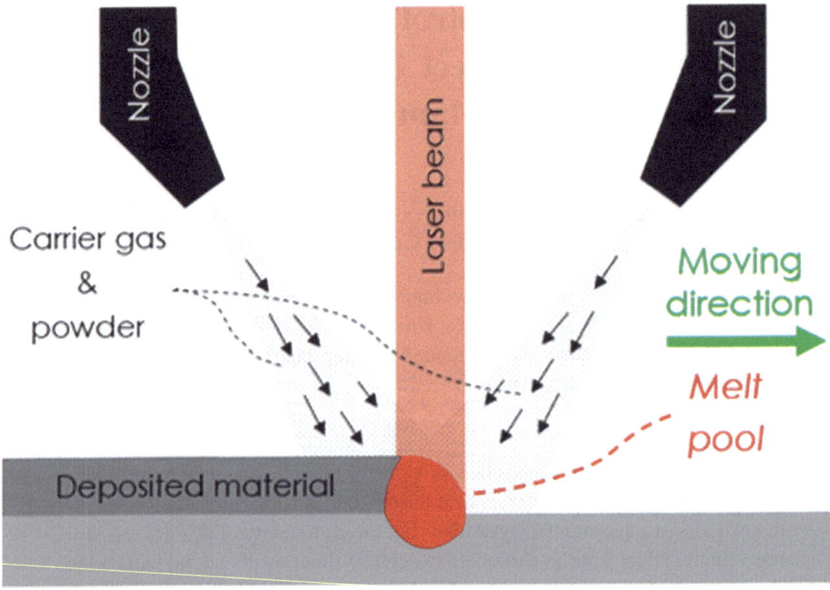

Fig. 1. Scheme of the DED functioning principle.

Despite the increasing diffusion of this technology and the multitude of foreseen benefits, improvements in process monitoring and control still have to be done for the stable adoption of DED in industry [2]. Compared to other metal AM techniques, like Powder Bed Fusion (PBF) and Electron Beam Melting (EBM), DED is less mature, and metal parts formed by this technique may present various defects: significant deviations from the reference geometry, low mechanical performances, presence of pores and residual stresses [3]. These poor qualities may arise when producing new, untested geometries and materials, because of the high complexity of this AM process. For these reasons, many researchers have been actively developing modelling and control of DED in recent years, trying to get the most out of such a promising technology.

In this work we propose a monitoring and control setup for DED, with the aim of improving the geometrical accuracy of deposited tracks when producing complex features. We propose a monitoring configuration based on beam-coaxial imaging, and show the results of controlling laser power during the deposition of reference geometries, called *V-tracks*, which are known to exhibit over-deposition phenomena.

The paper is structured as follows. Section 2 presents the state of the art of DED process monitoring and control. Section 3 presents the chosen setup and the experimental framework. Section 4 shows the results of the first tests of the proposed solution. Section 5 presents some concluding remarks about the presented control solution and future developments of on-line process control for DED processes.

2 State of the Art

This section describes the process monitoring methods proposed up to the present day in DED literature along with the control approaches, including the choices of command signals, feedback signals and control algorithms.

2.1 Process Monitoring

As far as process monitoring is concerned, various approaches have been examined in literature. In DED the occlusion caused to radiation-based sensors by the deposition head represents an additional difficulty to be overcome in the design of a monitoring system, with respect to PBF where there is normally a lot of empty space above the sample in production.

When monitoring is performed "recurrently", i.e. layer-wise, the deposition head can be moved away from the deposition plate, leaving space for inspection equipment to acquire information about the deposition status. In [4, 5] the use of fringe projection scanning during process interruptions is proposed, to acquire the actual geometry of the partially manufactured part and to evaluate its deviations from the reference geometry. In [6], this task is performed using a profilometer.

Various publications, as this paper, propose a configuration that includes sensors into the optical chain of the deposition head such as pyrometers, visible-light cameras, thermal or hyperspectral cameras and triangulation lasers [7]. These solutions allow acquiring information coming from the melt pool region following coaxially the laser path. The main advantage is a preferential top view of the region where the part is building up, even if signals might be heavily weakened by the passage through lenses or mirrors included into the optical chain of the deposition head. A hybrid solution has been proposed in [8] and is implemented on machines by DM3D Technology LLC: a set of three cameras and one pyrometer that are mounted onto the deposition head but outside the optical chain, and point towards the laser spot from an angled view. With this configuration great attention needs to be payed while the process is running to avoid collisions between sensors and machine or workpiece.

Among the aforementioned radiation-based monitoring technologies different trade-offs between signal quality, richness and frequency can be identified, as discussed in the following.

- Dual-color pyrometers can use Planck's radiation law evaluated in two different frequencies to measure temperature while eliminating the effect of material reflexivity. The traced temperature is an average over the measurement spot, and this single value can usually be streamed to the monitoring platform at very high frequencies (e.g. 100 kHz). This requires a calibration procedure to be performed in the exact optical configuration considering the used frequencies, as the optical path modifies the intensity of the signal, potentially in a different way at different frequencies. Also spurious information, like the temperature of the melt pool vapor plume, can affect the measured signal, although in a lighter way with respect to visible-light imaging, since the molten metal is normally the strongest source of radiation that is targeted by the measurement spot. The order of magnitude of the signal standard deviation at constant temperature while monitoring in-axis a DED deposition process by pyrometry is about 50–100 K, so although the signal frequency is very high, it must be filtered in a proper time window to be used reliably for temperature measurement. This variability is due to both the complexity of the phenomenon and to disturbance and attenuations introduced by the optical chain, in the case of in-axis pyrometers. Interestingly, [9] proposed an in-axis pyrometer as an indirect height measurement device, stating that when the laser spot is out of focus, the energy density on the target is lower, and the pyrometer reads a lower temperature.

- Visible-light cameras, on the contrary, can be heavily affected by phenomena that emit strong light, although they are possibly colder than the melt pool, e.g. sparks or vapor. Narrow band or NIR filters are often used, to exclude this spurious information and to avoid aberrations like ghosting and flares. For example, [10] shows how using an 850 nm filter improves the signal-to-noise ratio in the detection of intensity spikes due to over-deposition in track corners.

  The melt pool intensity is not directly interpretable as a temperature, but some authors performed ad hoc calibrations to try to interpret color or greyscale values as such [11]. The obvious advantage of using a camera is that the collected information is 2D and it may have a quite high resolution, e.g. 10 μm, at the expense of framerate. High-speed cameras used in literature can in some cases run at up to 1000 Hz, although the resolution can be limited and image processing can hardly run so fast, depending on its complexity, so usually sampling frequencies in the order of hundreds of fps are used.

  An interesting alternative is proposed by [12], which describes a system for Wire-DED that illuminates the welding area by a laser at 640 nm and collects only light at this frequency by a camera equipped with a proper narrow band filter. This solution could be effective for capturing more accurately the track geometry, although in this case the thermal information is completely lost.
- Thermal cameras are theoretically the best trade-off between pyrometers and visible-light cameras, but practically they are too limited in acquisition speed, and often in resolution, to be effective as matrix cameras, and on the other side they don't use dual-frequency, so their temperature measurements are valid only after a very specific calibration for very specific process conditions. Nonetheless, for low speed requirements and controlled experimental conditions, they can provide accurate and distributed information about the temperature map of the DED sample. Hyperspectral imaging has the potential to solve some of the limitations of thermal imaging, by using multiple frequencies to fit temperature on a pixel matrix. [13] worked on this approach on PBF, proposing also a method to evaluate and possibly account for the disturbance generated by the light radiation of the vapor plume.

All these methods offer, to some degree, information about cooling rates and/or thermal gradient, which are correlated to the resulting mechanical properties of the part [14]. Matrix sensors can provide information also about the geometry of the track in progress, from a 2D perspective.

Also some 3D measurement devices have been proposed by researchers. The two most notable may be the 3-cameras setup by [8], about which we have no information on performances, and the triangulation laser for track height proposed in [15]. This can be a promising tool to maintain the stand-off distance uniform during the whole deposition, even in presence of anomalies.

Other, less used monitoring methods have been proposed, mainly based on acoustic emissions. In [16], for example, a setup with two acoustic sensors placed at the sides of a deposition substrate has been proposed, with the aim of detecting cracks forming into the deposited part.

Remarkably, monitoring the part in production is not the only way to improve process quality. For example, [17] focuses on monitoring the powders flow, whose stability is undoubtedly important for the predictability of deposited features. The authors use a piezo-electric transducer to monitor acoustic emissions sampled in the powder transport tube, to estimate the actual flow of powders.

Considering also cost and ease of integration, beam-coaxial imaging with a visible-light camera remains a convenient and cost-effective solution, as many deposition heads for DED are equipped with a port for including a monitoring camera into the optical chain. For this reason, in this work we focus on a vision setup that is capable of detecting anomalies in the energy content of the melt pool, thus allowing the prediction and correction of potential deviations from the desired deposition result.

2.2 Process Control

The literature about on-line process control for DED is still limited, nonetheless some of the solutions proposed in the last years started providing promising results for the development of these techniques. Most researchers focus their attention on laser power modulation, based on the signals by one or more pyrometers and cameras, because it is a convenient solution for various reasons: first of all, it influences the deposition rate in an approximately linear way [18]; moreover, the system response time is quite low, compared to changing the powder feeding rate and waiting for the updated powder flow to reach the nozzles; finally, it does not affect motion dynamics, which may have unexpected consequences when modulating deposition speed.

Due to the difficulty of monitoring and interpreting feedback signals in real time, many authors focus on a feed-forward approach, possibly enriched with information updated recurrently by inspecting the partially deposited part during a process pause.

A basic approach to the estimation of the feed-forward component of the process control scheme is presented in [19], where a regression study that uses power, velocity, powder flow and layer number as regressors and melt pool width as response yields the optimal nominal parameters to obtain thin walls with uniform thickness. This approach can provide good results only in the same conditions used during the experimental campaign, i.e. with thin walls in this case, but it is a good starting point for determining the process parameters before closing the control loop.

[4, 5] propose a recurrent scheme where a 3D scan of the partially manufactured part is analyzed and an updated deposition strategy is generated for the remainder of the process. This kind of global approaches can be used complementarily to real-time control, which can act rapidly and locally but could hardly include dynamic tool path regeneration.

In [20] the process recipe is adapted after the deposition of each layer, based on the geometry acquired with a profilometer. The authors consider the layer-by-layer structure of a 2.5D deposition as a discrete time process, and after each layer scan they adapt the velocity profiles over the whole following layer to tune the resulting mass deposition rate. The result is a visible improvement on the regularity of thin-wall longitudinal profiles.

[6] proposes a hybrid approach in the middle of real-time and layer-wise, i.e. a "track-wise" approach. Basing on a profilometer, if the temperature near the end of the last deposited track is too high, they choose to start the next track from the opposite side, instead of continuing with the planned zig-zag path. According to their results, this approach yields more uniform microstructure, due to the reduced thermal gradients, and a slightly higher hardness.

[21] proposes an alternative approach to the regulation of deposition rate, by developing a system to modulate the powder flow. The modulation is performed in a feed-forward manner, basing on previous experiments on right-angled tracks, and shows improvements in the uniformity of track size. On the other side, this result could potentially be obtained also by modulating power and/or speed, in the same feed-forward way, with faster response times.

Since there are no available monitoring methods that can provide the full track geometry in real time, researchers who decided to pursue real-time control focused on stabilizing the available feedback signal on a reference value, establishing in advance the one yielding a "good quality" deposition. The control is performed with a variety of methods, ranging from proportional-integral-derivative (PID) controllers to generalized process control (GPC).

[22] use a PI controller with a pre-determined feed-forward component to stabilize the estimated melt pool width on a reference value, showing improvements in the uniformity of thin-wall cylinders.

[9] proposed the use of an off-axis pyrometer as feedback signal for a GPC that relies on a 4th order state space model. The controller aims at a temperature target that is considered normal at the right stand-off distance, and tunes laser power when the temperature appears lower due to lack of deposition, i.e. stand-off higher than expected. The experimentally induced lack of deposition (a 3 mm step) was compensated in a few layers. This approach has been extended in [8], where the system was coupled with a height controller based on three cameras mounted on the deposition head, pointing from the outside towards the melt pool. The solution is a proprietary system by DM3D, and uses a similar GPC approach with two feedbacks to improve the flatness of a part top.

[23] proposes a control loop involving both the laser power and spot movement speed, highlighting the potential of using both command variables to improve the deposition quality. Also [24] proposes an optimal controller that involves both power and velocity for controlling track height and shape, although such an approach has been tested only on simulations based on real process data. This work also proposes a general methodology for aligning acquired melt pool images to deposition geometry data, which is the basis for building systematically the datasets required for estimating a process model.

In this work, we chose to start from a simple PI controller, to validate the potential of using on-line image features to assess the status of the on-going deposition. Particularly, the image intensity will be used as a feedback to modulate laser power.

3 Experimental Setup and Proposed Control Methodology

This work assesses the potential of a feedback-based approach, which could dynamically adjust process fluctuations to achieve specific deposition quality targets, without needing to refer to a previously tested geometry/material/strategy and deposition history.

The proposed approach exploits on-line vision data, available in most DED machines, thus offering a simple and cost-effective monitoring solution. In this study, the objective has been focused on controlling track geometries: local variations of energy density and heat drift influence track size, and these phenomena may be revealed by the processing of melt pool images, thus allowing geometry control by power-based closed-loop control.

The proposed prototype of monitoring and control solution has been implemented in the framework of a proportional-integral process control: the considered monitoring signal is compared to a desired value, which can be considered a proxy of track height, and the residual is used to generate a new control action for laser power.

3.1 Machine and Materials

The DED machine used for the experiments is a 3-axis Laserdyne 430 laser cutting machine, retrofitted to DED by the integration of an Optomec 4-nozzle deposition head (Fig. 2). The Convergent CF1000 fiber laser source has 1 kW maximum power, and a spot diameter of 1 mm. The deposition head includes a slot for connecting a c-mount camera for a beam-coaxial view of the melt pool, thanks to a dichroic mirror that reflects the emitted radiation [2].

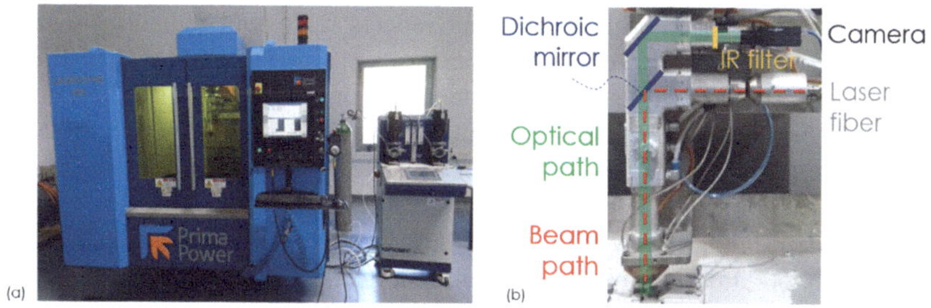

Fig. 2. Laserdyne machine (a) and deposition head (b).

The reference material used in the experiments is a high carbon content steel (C 0.85%), with grain size range 40–100 μm. This material presents good hardness, therefore it is an interesting material for the tooling and mold & die sectors, where DED is a strategic technology, thanks to its capability of performing repairs on any kind of geometry. The base plates used have the same chemical composition of the deposited material and a thickness of 12 mm.

3.2 Reference Experiment: V-Tracks

The baseline geometric features taken into account in this work are single tracks with corners, *V-tracks* in the following (Fig. 3a), as a test bench for the vision-based process control solution. Such features represent one of the main building blocks of almost any complex deposition, i.e. performing trajectories with changes of direction or curvature. Even simple rectangular geometries may present important deviations from the desired geometry, in regions where the heat concentration is higher (Fig. 3b–c).

Fig. 3. Example of V-track (a) and of over-deposition phenomena in tool path corners (b-c).

V-tracks represent an interesting type of basic features, since they present the typical presence of over-deposition, i.e. anomalous height and/or width of the track section, in the corner location, where both the high curvature and the acceleration ramps of the machine axis motion induce a local increment of energy density and a higher incident powder flow per surface unit. Moreover, the linear segments before and after the corner allow establishing a reference for track height.

3.3 Monitoring Setup

As anticipated, the monitoring camera is mounted on the deposition head, as represented in Fig. 2b. The dichroic mirror that reflects melt pool radiation toward the camera filters out the high-power laser frequencies, i.e. a narrow band around 1080 nm. Moreover, an 850 nm band-pass filter is added, to eliminate artifacts and improve the signal robustness. The used camera is equipped with a 2.3 Mpix CMOS SonyIMX174 sensor and connected via GigaEthernet to a dedicated PC that receives and stores the images acquired during the deposition process.

The post deposition quality inspection (track geometry) is executed by means of a Keyence VHX-6000 digital microscope with zoom lenses up to 2000X for a spatial resolution of 0.11 μm/pixel.

In this work, we focused our attention on a very simple image feature, the mean intensity I (in the following simply *intensity*) of images converted to greyscale. This value integrates all the light content of the image, so it should be highly correlated with the overall energy content of the melt pool. This insight is confirmed by plots like the one presented in Fig. 4a, which shows the intensity along two V-tracks. In this plot, the intensity has been normalized so that the mean and standard deviation within the first linear segment of the V-track were transformed respectively to 0 and 1. The clearly visible peak can be found in correspondence of the corner of the V-track, where the energy density increases and over-deposition occurs. Image intensity is also very fast to compute, so it is a good mean to validate the potential of retrieving information regarding deposition outcomes from melt pool images.

The correlation between image intensity and melt pool total energy implies indirectly correlation between intensity and profile height, as confirmed by the microscope measurements reported in Fig. 4b: the track presents a bulk in correspondence of the track inversion corner.

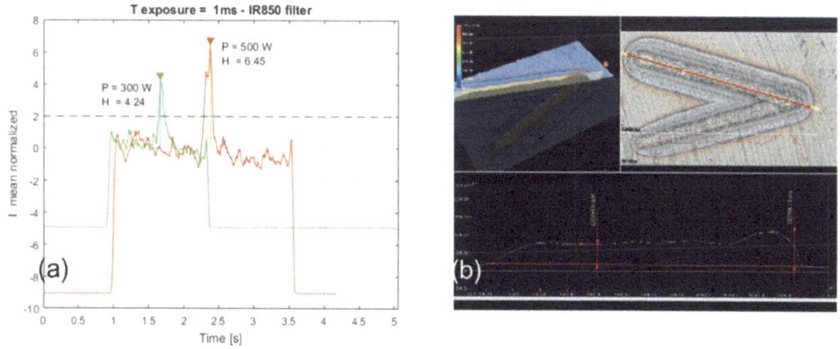

Fig. 4. (a) Normalized mean image intensity traced during the deposition of two different V-tracks. The peak height is, in both cases, well above the threshold of 2 standard deviations from the average value on the first linear segment. (b) V-track height profile.

3.4 Closed Loop PI Process Control

Focusing on track geometry, depositing a track of reasonably uniform height, independently from the performed trajectory, is an important target to achieve process stability and repeatability. This objective has been pursued by designing a control scheme through the following pipeline:

1. estimation of the feed-forward components, i.e. the process recipe and the relation between the latent, desired outcome (track height) and the observable feedback (image intensity);
2. choice of a control plant;
3. identification of the dynamic system generated by the process;
4. use of the identified system to tune controller gains in a virtual environment.

1. In detail: First of all, a feed-forward component has been estimated, based on a preliminary experimental campaign run on single tracks. The experimentation allowed determining the relations:

$$h = f(P, V), \quad I = g(P, V, \boldsymbol{M}),$$

where h represents track height, P and V represent laser power and cruise TCP velocity, I represents the image intensity and $\boldsymbol{M}$ represents an array of monitoring parameters (e.g. exposure time, gain and white balancing). Notice that $h = f(P, V)$ represents the process recipe, which is used also when the monitoring system is deactivated. When the optimal $\boldsymbol{M}$ is established as a function of P and V, we can obtain indirectly a relation $I = I(h)$ between image intensity and deposition height. The correlation between track height and image intensity allows adopting the following approach:

- Use the process map $h = f(P, V)$ to decide the nominal power P_{nom} and the cruise speed V_{nom} to be used during the whole deposition to obtain the desired track height $\overline{h}$. This yields also $\overline{I} = I(\overline{h})$.
- Stabilize I in time around the target value $\overline{I}$ Since I is a proxy of h, its regularity allows the achievement of a uniform track height.

2. The stabilization of image intensity is realized through the implementation of a PI (proportional-integral) controller (Fig. 5). A derivative component has not been included, as it yielded no significant benefits to the reduction of signal settling oscillations in the experiments considered for this work. In the future, a wider experimental campaign will reconsider the effects of the derivative component directly on the desired process outcome, i.e. track height.

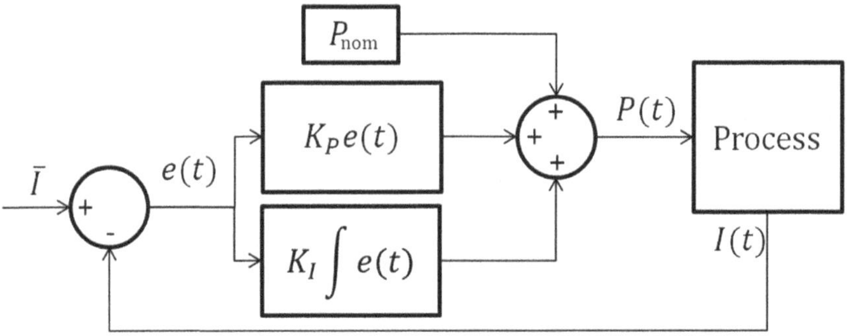

Fig. 5. PI control loop.

3. The tuning of the PI controller is performed first of all by commanding a random power signal to the dynamic system (the DED process) and acquiring the sensor feedback, i.e. image intensity. The unknown dynamic system is then identified as a state-space model, using the subspace method. The estimation yielded a stable fourth degree state-space model, coherently with the similar model identification approach used in [9].

   Figure 6 shows the comparison between a validation signal, generated with a new input, and its simulation by the identified model, which provides a good prediction of the model output.
4. The identified model has been then used to evaluate the step response and consequently to tune the K_P and K_I parameters. The results of the tuning are displayed in Table 1.

Table 1. PI tuning results.

K_P	2.5
K_I	0.1
Rise time	80 ms
Sample time	10 ms

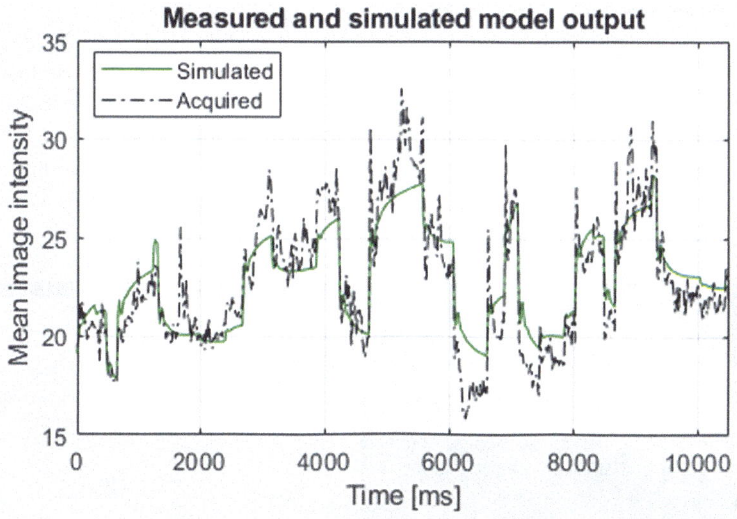

Fig. 6. Validation of identified dynamic model: acquired (dashed line) vs. simulated response (solid green) to random power input.

4 Approach Validation and Experimental Results

The preliminary experimental campaign, aimed at determining the feed-forward control component, has been performed by acquiring single tracks at nominal power levels P_{nom} from 100 to 1000 W with step 100 W, and by measuring track heights at the positions where images were taken, as described in [24]. The cruise velocity has been fixed to $V = 100$ mm/min, to simplify the experimentation and its further analysis. A linear regression analysis on collected data allowed to estimate the relation $h = f(p)$, which can provide a comprehensive set of process recipes to obtain the desired reference heights. The preliminary deposition runs have been exploited to capture the image intensities at 100 fps, while keeping V and the vision settings $\boldsymbol{M}$ fixed (in particular the exposure time of 10 ms). This provided the required function g that gives image intensities in relation to $P, V, \boldsymbol{M}$.

The effectiveness of the proposed control scheme has been tested by depositing multiple times V-tracks at P = 300 W, V = 100 mm/min and a powder flow rate of 30 mg/s. Five reference V-tracks have been deposited without melt pool control, while five samples with the same process parameters have been deposited with the melt pool control loop enabled, using the parameter values reported in Table 1.

For each track, the peak height in the corner region was measured, while track height in linear segments (before and after the corner) has been determined by averaging measurements taken in 80 equally-spaced positions along the track centerline (40 on the corner-entering branch and 40 on the corner-exiting branch). The resulting track heights in linear regions presented mean values in the range 562–779 μm and standard deviations in the range 42–63 μm. Compared to peak heights in the corners (931–1142 μm) the track heights resulted significantly different, as confirmed by one-sided Wilcoxon tests (p-values < 1e-3 in all cases).

The over-deposition ratios, i.e. the ratios between peak height and linear section height of each track, have been averaged across the repetitions with and without power control enabled. The results, resumed in Table 2, show that in tracks deposited with melt pool control enabled the over-deposition in the corner region (see Fig. 7) is reduced ($\sim$17% of the track height in linear regions). A Wilcoxon test confirms this difference, yielding a p-value lower than 5% for a one-sided two-samples test.

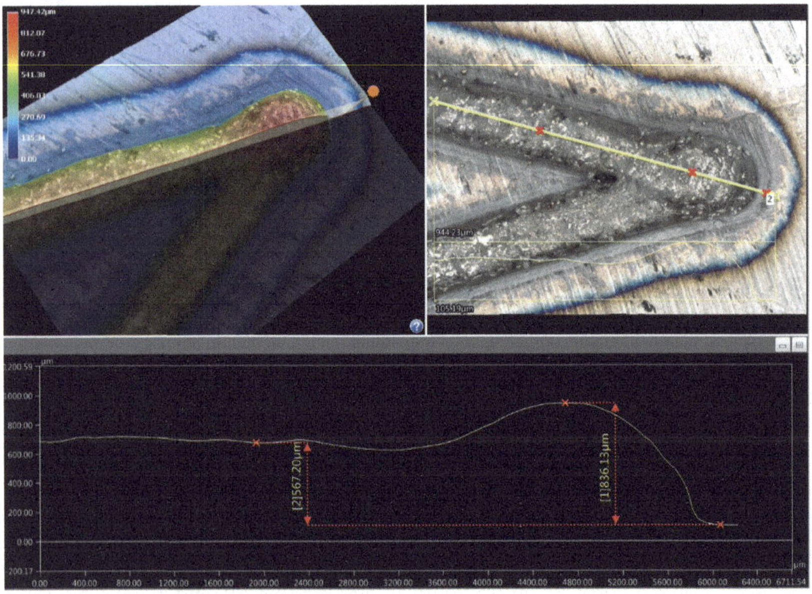

Fig. 7. V-track deposited with melt pool control enabled.

Table 2. Deposition tests summary

	Open loop		Closed loop	
	mean	st. dev.	mean	st. dev.
Over-deposition ratio (averaged on 5 tracks)	1.66	0.08	1.49	0.07

5 Conclusion and Future Works

The article presents a closed loop PI control approach for reducing the over-deposition rate in parts presenting sharp angles printed by DED; this AM process had not yet been studied in literature with a similar closed-loop approach. The proposed control system runs on a 3 axes machine and modulates the laser power in relation to the melt pool image intensity captured with an in-axis camera sensor. The implemented PI process control model shows the potential of this approach in treating height over-shootings, which have been reduced by 17% in the performed tests. Such improvement is to be appreciated especially in the tool path corners, where the risk of over-deposition is quite high as a result of increased local energy.

The proposed control model will be enriched by implementing the computation of different and more complex image features, and analyzing their correlation with process outcomes by more advanced regression methods, such as convolutional neural networks. Moreover, microstructural outcomes, like porosity, will be measured on track cuts and introduced into the control scheme. The approach will be tested on various reference geometries, starting from thin walls, where the heat build-up is faster, small 3D bulks, which require several and frequent motion inversions, and tensile samples, which will allow relating melt pool features, geometric and metallurgic quality also to the static mechanical behavior of a DED sample. Finally, the preliminary tests will scale up to more refined control approaches, such as GPC based on the identified dynamical model of the process, and to the integration of additional feedback and look-ahead signals, like the instantaneous laser spot velocity.

Acknowledgments. The research in this paper has been partially funded by EU H2020-CS2-CFP02-2015-01, AMATHO Additive MAnufacturing of Tiltrotor HOusing. Contract 717194.

References

1. Thompson, S.M., Bian, L., Shamsaei, N., Yadollahi, A.: An overview of direct laser deposition for additive manufacturing; part I: transport phenomena, modeling and diagnostics. Addit. Manuf. **8**, 36–62 (2015). https://doi.org/10.1016/j.addma.2015.07.001
2. Mazzucato, F., Avram, O., Valente, A., Carpanzano, E.: Recent advances toward the industrialization of metal additive manufacturing. In: Kenett, R.S., Swarz, R.S., Zonnenshain, A. (eds.) Systems Engineering in the Fourth Industrial Revolution: Big Data, Novel Technologies, and Modern Systems Engineering, pp. 273–319. John Wiley & Sons (2019)
3. Schmidt, M., et al.: Laser based additive manufacturing in industry and academia. CIRP Ann. **66**(2), 561–583 (2017). https://doi.org/10.1016/j.cirp.2017.05.011
4. Avram, O., Valente, A., Fellows, C.: Adaptive CAx chain for hybrid manufacturing. In: Fraunhofer Direct Digital Manufacturing Conference (DDMC 2018) (2018)
5. Garmendia, I., Leunda, J., Pujana, J., Lamikiz, A.: In-process height control during laser metal deposition based on structured light 3D scanning. Procedia CIRP **68**(April), 375–380 (2018). https://doi.org/10.1016/j.procir.2017.12.098
6. Nassar, A.R., Keist, J.S., Reutzel, E.W., Spurgeon, T.J.: Intra-layer closed-loop control of build plan during directed energy additive manufacturing of Ti-6Al-4V. Addit. Manuf. **6**, 39–52 (2015). https://doi.org/10.1016/j.addma.2015.03.005

7. Purtonen, T., Kalliosaari, A., Salminen, A.: Monitoring and adaptive control of laser processes. Phys. Procedia. **56**(C), 1218–1231 (2014). https://doi.org/10.1016/j.phpro.2014.08.038
8. Song, L., Bagavath-Singh, V., Dutta, B., Mazumder, J.: Control of melt pool temperature and deposition height during direct metal deposition process. Int. J. Adv. Manuf. Technol. **58**(1–4), 247–256 (2012). https://doi.org/10.1007/s00170-011-3395-2
9. Song, L., Mazumder, J.: Feedback control of melt pool temperature during laser cladding process. IEEE Trans. Control Syst. Technol. **19**(6), 1349–1356 (2011). https://doi.org/10.1109/TCST.2010.2093901
10. Vandone, A., Baraldo, S., Valente, A., Mazzucato, F.: Vision-based melt pool monitoring system setup for additive manufacturing. Procedia CIRP **81**, 747–752 (2019). https://doi.org/10.1016/j.procir.2019.03.188
11. Barua, S., Liou, F., Newkirk, J., Sparks, T.: Vision-based defect detection in laser metal deposition process. Rapid Prototyp. J. **20**(1), 77–86 (2014). https://doi.org/10.1108/RPJ-04-2012-0036
12. Motta, M., Demir, A.G., Previtali, B.: High-speed imaging and process characterization of coaxial laser metal wire deposition. Addit. Manuf. **22**(May), 497–507 (2018). https://doi.org/10.1016/j.addma.2018.05.043
13. Staudt, T., Eschner, E., Schmidt, M.: Temperature determination in laser welding based upon a hyperspectral imaging technique. CIRP Ann. **68**(1), 225–228 (2019). https://doi.org/10.1016/j.cirp.2019.04.117
14. Shamsaei, N., Yadollahi, A., Bian, L., Thompson, S.M.: An overview of direct laser deposition for additive manufacturing; part II: mechanical behavior, process parameter optimization and control. Addit. Manuf. **8**, 36–62 (2015). https://doi.org/10.1016/j.addma.2015.07.002
15. Donadello, S., Motta, M., Demir, A.G., Previtali, B.: Monitoring of laser metal deposition height by means of coaxial laser triangulation. Opt. Lasers Eng. **112**, 136–144 (2019). https://doi.org/10.1016/j.optlaseng.2018.09.012
16. Wang, F., Mao, H., Zhang, D., Zhao, X., Shen, Y.: Online study of cracks during laser cladding process based on acoustic emission technique and finite element analysis. Appl. Surf. Sci. **255**, 3267–3275 (2008). https://doi.org/10.1016/j.apsusc.2008.09.039
17. Whiting, J., Springer, A., Sciammarella, F.: Real-time acoustic emission monitoring of powder mass flow rate for directed energy deposition. Addit. Manuf. **23**, 312–318 (2018). https://doi.org/10.1016/j.addma.2018.08.015
18. Pinkerton, A.J.: Advances in the modeling of laser direct metal deposition. J. Laser Appl. **27**(2015), S15001 (2015). https://doi.org/10.2351/1.4815992
19. Ocylok, S., Alexeev, E., Mann, S., Weisheit, A., Wissenbach, K., Kelbassa, I.: Correlations of melt pool geometry and process parameters during laser metal deposition by coaxial process monitoring. Phys. Procedia **56**(C), 228–238 (2014). https://doi.org/10.1016/j.phpro.2014.08.167
20. Sammons, P.M., Gegel, M.L., Bristow, D.A., Landers, R.G.: Repetitive process control of additive manufacturing with application to laser metal deposition. IEEE Trans. Control Syst. Technol. **27**(2), 566–575 (2019). https://doi.org/10.1109/TCST.2017.2781653
21. Arrizubieta, J.I., Martínez, S., Lamikiz, A., Ukar, E., Arntz, K., Klocke, F.: Instantaneous powder flux regulation system for laser metal deposition. J. Manuf. Process. **29**, 242–251 (2017). https://doi.org/10.1016/j.jmapro.2017.07.018
22. Moralejo, S., et al.: A feedforward controller for tuning laser cladding melt pool geometry in real time. Int. J. Adv. Manuf. Technol. **89**(1–4), 821–831 (2017). https://doi.org/10.1007/s00170-016-9138-7

23. Seltzer, D.M., Wang, X., Nassar, A.R., Schiano, J.L., Reutzel, E.W.: System identification and feedback control for directed-energy, metal-based additive manufacturing. In: Proceedings of the solid Freeform Fabrication Symposium, pp. 592–601 (2015)
24. Vandone, A., Baraldo, S., Valente, A.: Multisensor data fusion for additive manufacturing process control. IEEE Robot. Autom. Lett. **3**(4), 3279–3284 (2018). https://doi.org/10.1109/LRA.2018.2851792

Manufacturing Process Monitoring and Control in Industry 4.0

Vinh Nguyen and Shreyes N. Melkote(✉)

George W. Woodruff School of Mechanical Engineering,
Georgia Institute of Technology, 813 Ferst Drive NW, Atlanta, GA 30332, USA
vnguyen43@gatech.edu, shreyes.melkote@me.gatech.edu

Abstract. The use of advanced process planning methodologies has enabled manufacturers to predict and optimize manufacturing processes in the planning stage. However, process faults and non-optimal conditions are always inherent to the manufacturing environment. The advent of Industry 4.0 has given rise to cyber-physical systems wherein online process monitoring and control can be performed autonomously. This paper discusses process monitoring and control in the context of Industry 4.0. With the focus on digital connectivity driving Industry 4.0, the advantages of cloud-based computing and knowledge inferred from a plethora of manufacturing processes can be leveraged for process monitoring and control to improve production speed, quality, and reliability. This paper presents a holistic framework for process monitoring and control in the context of Industry 4.0, where macro-level process control is conducted in the cloud and device-level process control occurs at the edge. A case study of tool life enhancement using such a framework is presented. Limitations of process monitoring and control in the context of Industry 4.0 are discussed along with proposals for new avenues of research.

Keywords: Process monitoring · Process control · Industry 4.0 · Edge computing

1 Introduction

No matter how well manufacturing processes and schedules are planned, they are subject to unforeseen disturbances that can result in scrap, rework, reduced process efficiency, or suboptimal operating conditions. Therefore, all production processes require process monitoring and control capabilities to ensure defect-free and optimal operation. The development of Industry 4.0 has resulted in increased digital connectivity at the machine and factory-floor levels [1]. A natural byproduct of the increased connectivity is the development of cyber-physical representations of manufacturing systems, which include digital twins of manufacturing processes. These cyber-physical representations can be used to provide near real-time and historical machine and/or process state information during production. The natural utility of cyber-physical systems is to use them for automated decision-making and control to ensure optimal operation in the presence of process disturbances.

L. Wang et al. (Eds.): AMP 2020, LNME, pp. 144–155, 2020.
https://doi.org/10.1007/978-3-030-46212-3_10

This paper discusses recent advances and future research opportunities in process monitoring and control in the context of Industry 4.0. Figure 1 shows an Industry 4.0 enabled communication paradigm for process monitoring and control. While the paradigm is generic, the focus of this paper is on application of this paradigm to process monitoring and control of discrete manufacturing processes. The paradigm consists of three major elements:

- **Machine:** The machine executes the manufacturing process(es) relevant to fabrication of the part of interest. With the advent of Internet of Things (IoT) technologies that have enabled easier digital connectivity, more legacy machines can be linked to a cloud-based computing infrastructure. In addition, manufacturing equipment vendors are selling more machines with network capability for cloud applications. Hence, a large number of machines in geographically distributed locations are able to participate in this architecture from a process monitoring viewpoint.
- **Edge Device:** The edge device consists of hardware/software that performs local data collection for real-time process monitoring and control of the machines and processes, while providing a communication interface with the cloud. While having a direct connection between the manufacturing machines and the cloud reduces complexity arising from multiple components and interfaces, edge devices reduce the need for high network data transmission rates, enable cloud connectivity for legacy machines, and provide an extra layer of security. Hence, edge devices have been proposed in smart cities [2] and in autonomous driving infrastructures [3]. Note that conducting edge processing does not necessarily require external sensors. Recently, manufacturing equipment vendors have enabled built-in sensor fusion and native network communication capabilities into their machines, which enable automatic edge computing and communication with the cloud [4].
- **Cloud:** While edge computing can be seen as a natural byproduct of Industry 4.0, the interconnectivity of geographically distributed cyber-physical systems through the cloud is where the maximum benefit of Industry 4.0 lies. The cloud is a software environment that hosts on-demand databases and computational resources without active management by the user [5]. This implies ease of connectivity from a variety of edge devices to the cloud in addition to vast computational and data storage capabilities. However, network transmission speeds between edge devices and the cloud are relatively low and inconsistent compared to edge device transmission speeds. For instance, network data rates have been shown to be on the order 100 ms compared to their 100 μs counterparts at the edge [6]. Hence, the cloud is suitable for storing large amounts of long-term trend data thereby relaxing edge computing requirements.

Note that a cloud architecture can have multiple levels of hierarchy. For instance, a local factory can have its data hosted on a factory level cloud while a higher cloud environment can manage data from several factory level clouds. This type of hierarchical architecture can give individual work environments the freedom to interface with their own cloud environment. In addition, note that measured process data does not always directly provide the state information of interest (e.g., acceleration data does not directly give information of part accuracy). Hence, process models that relate process data to process conditions of interest for process control applications are critical in both edge and cloud computing. Note that other architectures in Industry 4.0 have been proposed. For instance, Lucke et al. [7] have proposed an architecture involving direct communication between the cloud and machines without the use of edge devices. In addition, their cloud only hosts communication between offline planning and manufacturing machines, whereas the architecture in Fig. 1 directly hosts the planning and control scheme that monitors the factory.

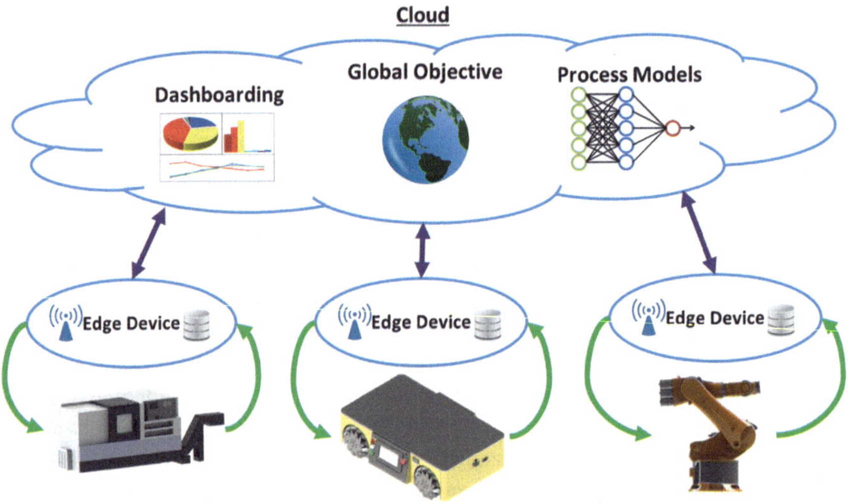

Fig. 1. Process monitoring and control paradigm.

2 Edge Device Monitoring and Control

Edge-based monitoring and control is performed at the machine level. Conducting these actions at the edge enables faster, machine-specific process control that cannot be conducted at the cloud level due to data transmission latencies associated with the Local Area Network (LAN) and/or the Wireless Local Area Network (WLAN). In addition, edge processing can condense high frequency data for transmission to the cloud to minimize data transmission volumes and the likelihood of security breaches. This section describes an overview of the sensing and localized control methods using edge processing.

2.1 Process Monitoring

The development of external sensors to obtain process data dates back at least 50 years [8, 9]. However, the scalability and decentralized requirements of Industry 4.0 have created a push for cheaper, flexible sensing technologies that can be useful at the edge processing level. Vibration signals using Micro-Electro-Mechanical Systems (MEMS) accelerometers have been demonstrated to be ideal due to their low cost and their ease of interfacing with embedded computers [10]. For systems with limited space for mounting external sensors, sound vibration measurements using low-cost microphones have been utilized for process monitoring [11]. Force measurements have been demonstrated since many process conditions are correlated with force. For instance, Suprock et al. [12] demonstrated the use of wireless strain sensors to monitor cutting forces in milling applications. In addition, low-cost temperature measurements generally include thermocouples for applications including monitoring defects, especially in additive manufacturing applications [13]. With the rise of more powerful embedded systems to process vision data, machine vision systems have increasingly been studied for in-situ process monitoring. For instance, vision systems have been used to measure die cracking in forging operations [14]. Generally, machine vision sensors generate extremely large amounts of data compared to time series measurements, and therefore machine vision data processing must be done at the edge for real-time control. Not all the aforementioned sensing techniques are generally used individually. Sensor fusion has been demonstrated to improve the fidelity of process state measurements and for decision making [15, 16].

2.2 Process Control

Because of the larger data throughput between edge devices and machines, the communication architecture shown in Fig. 1 requires that high bandwidth process control be implemented at the edge device level. To leverage the power of external sensors at the edge device level, researchers have attached external add-ons to machines for control. Generally, external process control methods coincide with external process monitoring capabilities, thus making them ideal for edge-based process control. For instance, Van de Wouw et al. [17] added an Active Magnetic Bearing (AMB) system in conjunction with eddy current sensors for chatter control. In addition, Farshidianfar et al. [18] added an infrared camera in conjunction with an external CNC feed drive controller to control temperature rates, and therefore microstructure, in additive manufacturing using PID control of the feed drives.

However, current industry trends show that manufacturers are still hesitant to allow control of their machines from external edge processing devices because of potential safety, security, and trade secret violations. A hypothetical IoT architecture could simply allow the machine to execute its native control strategy while the edge device just measures process data and the control outputs. However, machine vendors' planning and control algorithms are generally proprietary and difficult to obtain for edge processing. For instance, the remote sensor interface for KUKA industrial robots (KUKA RSI) includes knowledge of the robot joint encoders and motor torques [19]. However, the interface does not allow users to see important control details including robot trajectory interpolation strategies or end effector setpoints, and thus the information is unavailable for use at the edge device or cloud levels. This significantly limits the capabilities of machine-level process control in the context of Industry 4.0 as the lack of insight into the machine controller means that machines cannot leverage the benefits of external sensors and decision models that have been developed for Industry 4.0.

2.3 Process Models

In general, only gathering sensor data does not automatically provide knowledge of the state variable of interest (e.g., part quality, tool life). For instance, vibration measurements do not directly provide knowledge of the remaining useful life of a cutting tool. Hence, process models must be utilized to relate sensor data to the process condition or state variable of interest. Note that process models should be used in both cloud and edge-based process monitoring applications. With the recent rise of data-driven models and advancements in computational power, process models can easily be implemented in the cloud or in an edge device with minimal change in structure.

Physics-based models have been previously used to relate sensor data to the process condition. Physics-based models benefit from minimal calibration experiments, extrapolation ability, and model transparency. However, the addition of more sensors requires more complex physical models that relate sensor measurements to the process condition. Therefore, data-driven models have been developed to account for these flaws. Dating back at least 20 years, force sensors were used as an input to Neural Networks to predict tool wear [20] and tool breakage [21]. With the recent development of lower cost sensors and more sophisticated data-driven models, multiple low-cost sensors have been used in data-driven models to infer process condition [22, 23]. Implementation of data-driven models derived from multiple sensor inputs have been demonstrated to increase decision making fidelity over a single sensor [24]. However, combining sensors generally requires longer model calibration and prediction times, which has been mitigated somewhat by advances in computing hardware.

However, note that data-driven models suffer from drawbacks including increased computational complexity, the need for more data for model calibration, and greater uncertainty when predicting outside the calibration range. These limitations suggest the need for the development of hybrid process models that combine physics-based models with data-driven models to take advantage of both model types. A majority of hybrid models have been implemented for offline applications including surface finish prediction [25]. An example of the hybrid approach would be to use a physics-based model to create an initial model, and then utilize data-driven statistical techniques, such as Bayesian inference, to update the model. Alternatively, sensors can be used as inputs into a physics-based model, and the model's output is then fed into the data-driven model. This particular framework has been demonstrated in prediction of material removal rate in chemical mechanical planarization [26]. However, methods incorporating both physics-based models and data-driven models for process monitoring still pose unique challenges for researchers to study, such as quantifying improvements in processing time.

3 Cloud Process Monitoring and Control

Because communication speeds to cloud-based applications in Industry 4.0 are limited [27], process monitoring and control at the cloud level is often restricted to monitoring long term process trends and achieving an overarching goal such as optimize cost, develop equipment maintenance policies, etc. In addition, process models in the cloud have access to data gathered by edge devices and can therefore implement global actions that affect all edge devices.

3.1 Process Monitoring

Edge devices have the capability to send raw or processed sensor data to the cloud. Note that since sensor data are usually sampled at extremely high rates, raw data cannot be directly streamed to the cloud without suffering from internet bandwidth and latency limitations. However, segments of sensor data can be stored and then sent to the cloud at regular intervals e.g., after a process is completed. This enables the cloud to have access to all the data. However, transmission of all process data by edge devices increases cost, as cloud service providers charge based on data usage. In addition, recording all raw data in the cloud creates a high-risk environment where data hackers can access all the data related to a company's production facility.

Hence, processing the sensor data on the edge device before transmitting it to the cloud is a more desirable approach. This approach has been proposed by other researchers as a method to alleviate cloud computing constraints [28]. However, processing sensor data at the edge can lead to loss of information, which can result in the inability to diagnose faults. In addition, edge processing of sensor data imposes more computational burden on the edge device and can result in failure to achieve real-time process control speeds.

Note that use of multiple cloud hierarchies can result in a combination of both processed and raw sensor data being streamed to the multiple clouds in the hierarchy. For instance, a recently proposed cloud architecture recommended a local cloud for raw sensor data streaming, which would then stream only training datasets to a remote cloud [29]. However, an edge device that monitors a subset of the manufacturing machines can also replace a local cloud.

3.2 Process Control

For a production environment to enact corrective actions without minimizing machine downtime, the correction from the cloud must be automated. Research on automatic process control from the cloud is limited compared to cloud process monitoring and predictive data analytics. Previous research demonstrating cloud based process control involves sending digital notifications to equipment operators to make manual corrections to the process [30]. In addition, an architecture was proposed to utilize an agile approach to manually adjusting a pick and place robotic system using IoT data [31]. In the context of Industry 4.0, however, the cloud must automatically react to process model predictions and perform the corresponding corrections.

Note that automated cloud control actions are limited in the terms of cyber-physical systems, primarily owing to the latency constraints of LAN and WLAN protocols. For instance, Nguyen et al. [27] found that LAN and WLAN are slower than protocols that do not utilize networks owing to the larger overhead and longer distances. Hence, control action from the cloud can only involve less time constraint tasks including sending machine programming (G-code or PLC programs), general process models to edge devices, and on-off control. Machine programming from web applications is already in place in many production facilities, particularly through Systems

Applications and Products (SAP) systems [32]. However, an interesting use of cloud based feedback control is to send generic data-driven models from the cloud to edge devices that are updated for their machine specific configurations [33].

An interesting cloud based control concept is multi-input on-off (bang-bang) control [34]. A majority of residual thermostats are on-off controllers that attempt to achieve an analog setpoint only with on-off commands. On-off control refers to control performed by the cloud to activate/deactivate machines to achieve a global, long-term goal. This type of control architecture has been proposed by the autonomous driving research community where cars are relegated to the standby mode by the cloud to minimize energy consumption while minimizing customer wait times [35]. The cloud can utilize process monitoring data from all machines to create a global data-driven model that can be used to activate/deactivate particular machines to accomplish an objective. Research is being conducted to solve on-off controller issues including inertia effects and accelerated wear via fuzzification and machine learning control [36]. Note that research in autonomous cloud-based control will naturally grow as more cyber-physical representations of manufacturing systems are created.

4 Case Study

An example case study that utilizes the cloud-based process monitoring and control paradigm shown in Fig. 1 is summarized here. In this case study, edge and cloud-based process monitoring and control were implemented to enhance tool life in turning operations. The experimental setup utilized is shown in Fig. 2.

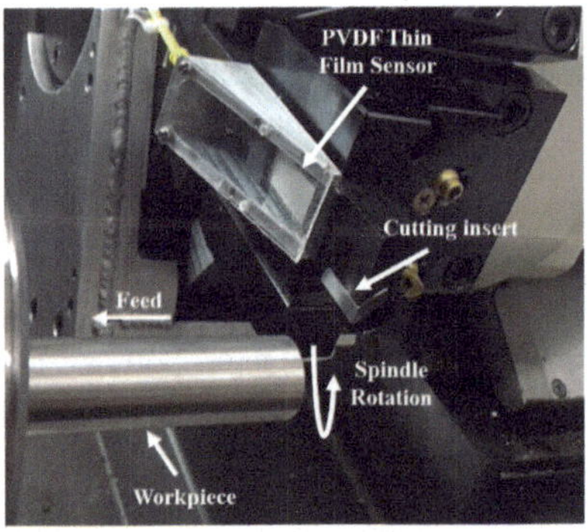

Fig. 2. Case study experimental setup [27].

In this work, a cloud-based process model forecasted the remaining useful tool life using real-time data from the machine tool controller and a low-cost thin-film polyvinyldene difluoride (PVDF) dynamic strain sensor mounted on the cutting tool. When the remaining useful tool life dropped below a threshold, spindle speed override control was executed from the cloud to the edge device connected to the machine to extend tool life. The architecture is shown in Fig. 3. The primary components consist of:

- **CNC Lathe:** An Okuma SpaceTurn LB2000 was used to conduct dry turning experiments of stainless steel 316L bars using an uncoated tungsten carbide insert mounted onto a 0° rake angle tool holder. The initial bar diameter (25.4 mm) was turned at a depth of cut of 1.27 mm, a length of cut of 38 mm, and a feed of 76 μm/rev. The average flank wear was measured intermittently using a Leica M125 digital optical microscope. The tool life was defined as when average flank wear exceeded 300 μm (ISO Standard 3685) [37]. The tests were used for both model calibration and system architecture validation.
- **Edge Devices:** The edge devices (denoted as Sensing Gateway and Correction Gateway in prior work) consisted of two Raspberry Pi 3B's [27]. The Sensing Gateway edge device collected machine state data (spindle load, axis load, and active machining time) obtained via the MTConnect protocol and analog voltage data from a PVDF piezoelectric polymer thin film sensor mounted on the tool. The Sensing Gateway sent condensed data (average and root mean square for the MTConnect and PVDF data, respectively) to the cloud. The Correction Gateway received spindle speed override commands from the cloud and transmitted them to a G-Code machine variable via RS232 communication.
- **Cloud Platform:** In this case study, the cloud platform environment (Siemens MindSphere) hosted the process models that received real-time data from the edge device and predicted the remaining useful tool life and the spindle speed override decisions required to extend tool life. The data-driven supervised learning model used to predict the average flank wear from process data was based on Gaussian Process Regression. If the average flank wear was larger than a predetermined threshold (225 μm), the cloud would send a control signal to alter the spindle speed of the machine to enhance tool life.

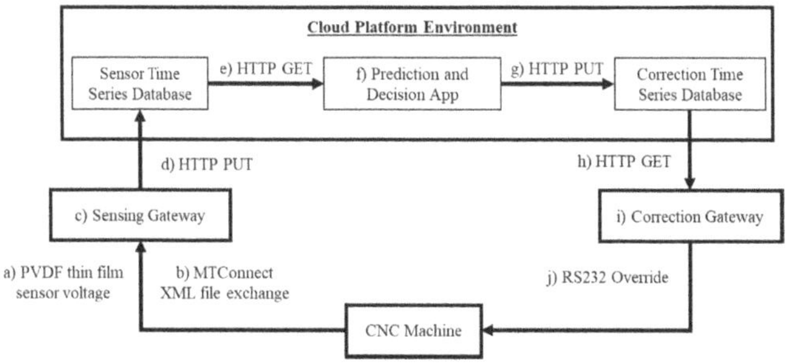

Fig. 3. Case study system architecture [27].

This case study was shown to improve tool life by 60% through spindle speed override control implemented via the cloud [27]. However, the process modelling and decision making was conducted in the cloud, and not at the edge devices as recommended in this paper. The authors of this case study found that the system had a maximum latency of 980 ms, and the majority of the latency (950 ms) was due to the interfacing of the Raspberry PI edge device with the cloud. Hence, process control from the cloud would not be suitable for applications requiring smaller latency, such as chatter detection and suppression, which should be performed at the edge. In addition, only one machine tool was used in this study and the capability of leveraging sensor data from multiple machines to make sophisticated macro-level decisions was not demonstrated.

5 Conclusions

In this paper, a framework for process monitoring and control in the context of Industry 4.0 technologies was discussed. However, the following fundamental limitations and challenges currently restrict the development of IoT based manufacturing process monitoring and control:

- **Security Concerns.** If any component of the network ranging from the individual sensors to the control infrastructure is breached, intentional flaws can be introduced into the manufacturing process. This is a broader concern in Industry 4.0 applications that extend beyond manufacturing environments. For instance, researchers in the medical field reported that medical devices including insulin pumps and pacemakers can be hacked through the network with the potential to alter patient health [38]. In addition, third party hosting of cloud services reduces transparency in the network infrastructure for manufacturers to discover hacking attempts. However, the security issue is difficult to address, as the nature of interconnectivity in Industry 4.0 naturally results in a less secure infrastructure. Innovative methods, such as leveraging blockchain technology in cloud manufacturing environments have been proposed [39], though actual physical implementation and testing of these solutions are limited.
- **Lack of Profit Analysis for Adoption.** For industry to adopt Industry 4.0 methodologies, an analysis of the cost versus benefit is required. However, owing to the variety of manufacturing machines and processes in a typical production environment, determining integration costs into the Industry 4.0 architecture is a complex task. The integration of edge computing for new manufacturing machines alleviates this problem, though cloud maintenance and development costs are still difficult to ascertain. A recent Cisco study showed that 39% of IoT adopters stated that the top unexpected benefit was profitability [40], meaning that profit was not considered during the adoption stage. For manufacturing environments, concrete and tangible use cases must be identified to demonstrate realistic profitability for adopting Industry 4.0 approaches in a production environment.
- **Lack of Combined Development Effort.** A large number of papers have proposed Industry 4.0 concepts. However, these papers document work by individual research groups with little physical experimental setup or testing. For instance, one research group will propose their own architecture for process monitoring while another will propose another cloud architecture for model-based decision-making. Thus, these research efforts appear disjointed and piecemeal. A comprehensive research effort into all aspects (e.g. process monitoring, decision-making, and control) simultaneously demonstrated in a laboratory scale must be performed for production environments to understand the potential cost versus benefit of Industry 4.0 technology implementation.

In light of these limitations, there is still a significant amount of work to be done before the advantages of Industry 4.0 are fully realized in manufacturing. However, overcoming these limitations provides opportunities for innovation and collaboration amongst all facets of manufacturing.

References

1. Lee, J., Bagheri, B., Kao, H.-A.: A cyber-physical systems architecture for industry 4.0-based manufacturing systems. Manuf. Lett. **3**, 18–23 (2015)
2. He, Y., Yu, F.R., Zhao, N., Leung, V.C., Yin, H.: Software-defined networks with mobile edge computing and caching for smart cities: a big data deep reinforcement learning approach. IEEE Commun. Mag. **55**(12), 31–37 (2017)
3. Sasaki, K., Suzuki, N., Makido, S., Nakao, A.: Vehicle control system coordinated between cloud and mobile edge computing. In: 2016 55th Annual Conference of the Society of Instrument and Control Engineers of Japan (SICE), pp. 1122–1127. IEEE (2016)
4. Zhong, R.Y., Xu, X., Klotz, E., Newman, S.T.: Intelligent manufacturing in the context of industry 4.0: a review. Engineering **3**(5), 616–630 (2017)
5. Wang, L., Von Laszewski, G., Younge, A., He, X., Kunze, M., Tao, J., Fu, C.: Cloud computing: a perspective study. New Gen. Comput. **28**(2), 137–146 (2010)
6. Singla, A., Chandrasekaran, B., Godfrey, P.B., Maggs, B.: The internet at the speed of light. In: proceedings of the 13th ACM Workshop on Hot Topics in Networks, pp. 1–7 (2014)
7. Lucke, D., Constantinescu, C., Westkämper, E.: Smart factory-a step towards the next generation of manufacturing. In: Manufacturing Systems and Technologies for the New Frontier, pp. 115–118 (2008)
8. Tlusty, J., Andrews, G.: A critical review of sensors for unmanned machining. CIRP Ann. **32**(2), 563–572 (1983)
9. Teti, R., Jemielniak, K., O'Donnell, G., Dornfeld, D.: Advanced monitoring of machining operations. CIRP Ann. **59**(2), 717–739 (2010)
10. Albarbar, A., Teay, S.: MEMS accelerometers: testing and practical approach for smart sensing and machinery diagnostics. In: Advanced Mechatronics and MEMS Devices II, pp. 19–40 (2017)
11. Kothuru, A., Nooka, S.P., Liu, R.: Application of audible sound signals for tool wear monitoring using machine learning techniques in end milling. Int. J. Adv. Manuf. Technol. **95**(9–12), 3797–3808 (2018)
12. Suprock, C.A., Nichols, J.S.: A low cost wireless high bandwidth transmitter for sensor-integrated metal cutting tools and process monitoring. Int. J. Mechatron. Manuf. Syst. **2**(4), 441–454 (2009)
13. Nie, Z., Wang, G., McGuffin-Cawley, J.D., Narayanan, B., Zhang, S., Schwam, D., Kottman, M., Rong, Y.K.: Experimental study and modeling of H13 steel deposition using laser hot-wire additive manufacturing. J. Mater. Process. Technol. **235**, 171–186 (2016)
14. Świłło, S., Perzyk, M.: Automatic inspection of surface defects in die castings after machining. Arch. Foundry Eng. **11**, 231–236 (2011)
15. Dornfeld, D.A., DeVries, M.: Neural network sensor fusion for tool condition monitoring. CIRP Ann. **39**(1), 101–105 (1990)
16. Cai, Y., Starly, B., Cohen, P., Lee, Y.-S.: Sensor data and information fusion to construct digital-twins virtual machine tools for cyber-physical manufacturing. Procedia Manuf. **10**, 1031–1042 (2017)

17. Van de Wouw, N., van Dijk, N., Schiffler, A., Nijmeijer, H., Abele, E.: Experimental validation of robust chatter control for high-speed milling processes. In: Time Delay Systems, pp. 315–331 (2017)
18. Farshidianfar, M.H., Khajepour, A., Gerlich, A.: Real-time control of microstructure in laser additive manufacturing. Int. J. Adv. Manuf. Technol. **82**(5–8), 1173–1186 (2016)
19. Schoepfer, M., Schmidt, F., Pardowitz, M., Ritter, H.: Open source real-time control software for the kuka light weight robot. In: 2010 8th World Congress on Intelligent Control and Automation, pp. 444–449. IEEE (2010)
20. Liu, Q., Altintas, Y.: On-line monitoring of flank wear in turning with multilayered feed-forward neural network. Int. J. Mach. Tools Manuf **39**(12), 1945–1959 (1999)
21. Tansel, I.N., Mekdeci, C., Mclaughlin, C.: Detection of tool failure in end milling with wavelet transformations and neural networks (WT-NN). Int. J. Mach. Tools Manuf **35**(8), 1137–1147 (1995)
22. Wang, J., Xie, J., Zhao, R., Zhang, L., Duan, L.: Multisensory fusion based virtual tool wear sensing for ubiquitous manufacturing. Robot. Comput. Integr. Manuf. **45**, 47–58 (2017)
23. Ghosh, N., Ravi, Y., Patra, A., Mukhopadhyay, S., Paul, S., Mohanty, A., Chattopadhyay, A.: Estimation of tool wear during CNC milling using neural network-based sensor fusion. Mech. Syst. Signal Process. **21**(1), 466–479 (2007)
24. Rao, P.K., Liu, J.P., Roberson, D., Kong, Z.J., Williams, C.: Online real-time quality monitoring in additive manufacturing processes using heterogeneous sensors. J. Manuf. Sci. Eng. **137**(6), 061007 (2015)
25. Joseph, V.R., Melkote, S.N.: Statistical adjustments to engineering models. J. Qual. Technol. **41**(4), 362–375 (2009)
26. Yu, T., Li, Z., Wu, D.: Predictive modeling of material removal rate in chemical mechanical planarization with physics-informed machine learning. Wear **426**, 1430–1438 (2019)
27. Nguyen, V., Malchodi, T., Dinar, M., Melkote, S.N., Mishra, A., Rajagopalan, S.: An IoT architecture for automated machining process control: a case study of tool life enhancement in turning operations. Smart Sustain. Manuf. Syst. **3**(2), 14–26 (2019)
28. Hu, L., Miao, Y., Wu, G., Hassan, M.M., Humar, I.: iRobot-factory: an intelligent robot factory based on cognitive manufacturing and edge computing. Future Gen. Comput. Syst. **90**, 569–577 (2019)
29. Wu, D., Liu, S., Zhang, L., Terpenny, J., Gao, R.X., Kurfess, T., Guzzo, J.A.: A fog computing-based framework for process monitoring and prognosis in cyber-manufacturing. J. Manuf. Syst. **43**, 25–34 (2017)
30. Mourtzis, D., Vlachou, E., Milas, N., Xanthopoulos, N.: A cloud-based approach for maintenance of machine tools and equipment based on shop-floor monitoring. Procedia Cirp **41**, 655–660 (2016)
31. Yusuf, Y.Y., Sarhadi, M., Gunasekaran, A.: Agile manufacturing: the drivers, concepts and attributes. Int. J. Prod. Econ. **62**(1–2), 33–43 (1999)
32. Knolmayer, G.F., Mertens, P., Zeier, A.: Supply Chain Management Based on SAP Systems: Order Management in Manufacturing Companies. Springer Science & Business Media (2002)
33. O'Donovan, P., Gallagher, C., Bruton, K., O'Sullivan, D.T.: A fog computing industrial cyber-physical system for embedded low-latency machine learning Industry 4.0 applications. Manuf. Lett. **15**, 139–142 (2018)
34. Bellman, R., Glicksberg, I., Gross, O.: On the "bang-bang" control problem. Q. Appl. Math. **14**(1), 11–18 (1956)
35. Lyu, X., Tian, H., Jiang, L., Vinel, A., Maharjan, S., Gjessing, S., Zhang, Y.: Selective offloading in mobile edge computing for the green Internet of Things. IEEE Netw. **32**(1), 54–60 (2018)

36. Duriez, T., Brunton, S.L., Noack, B.R.: Machine Learning Control-Taming Nonlinear Dynamics and Turbulence. Springer (2017)
37. Standard, I.: 3685. Tool-life Testing with Single Point Turning Tools (1993)
38. Baranchuk, A., Refaat, M.M., Patton, K.K., Chung, M.K., Krishnan, K., Kutyifa, V., Upadhyay, G., Fisher, J.D., Lakkireddy, D.R., Cardiology, A.C.O.: Cybersecurity for cardiac implantable electronic devices: what should you know? J. Am. Coll. Cardiol. **71**(11), 1284–1288 (2018)
39. Li, Z., Barenji, A.V., Huang, G.Q.: Toward a blockchain cloud manufacturing system as a peer to peer distributed network platform. Robot. Comput. Integr. Manuf. **54**, 133–144 (2018)
40. Cisco Survey Reveals Close to Three-Fourths of IoT Projects Are Failing. https://newsroom.cisco.com/press-release-content?articleId=1847422. Accessed 20 Dec 2019

On Standardization Efforts for Additive Manufacturing

Giovanni Moroni(✉), Stefano Petrò, and Huan Shao

Department of Mechanical Engineering, Politecnico di Milano, Via Giuseppe la Masa 1, 20156 Milan, Italy
giovanni.moroni@polimi.it

Abstract. Additive manufacturing is a set of technologies potentially covering the needs of many industrial sectors, some of which require the certification of the final product. This is the main motivation explaining why the International Organization for Standardization (ISO) and the American Society for Testing and Materials (ASTM) are putting significant efforts into defining standards covering different topics in this area. Efforts that very soon have become joint efforts to rapidly realize common standards under the name of ISO/ASTM standards. In this paper, the state of the art of these efforts is presented and discussed.

Keywords: Additive manufacturing · Standard · ISO/TC261 · ASTM F42

1 Introduction

Additive manufacturing is blooming and spreading with a variety of technologies and applications. Each of the seven process categories additive manufacturing technologies are classified in is a quite complex collection of methods, machines, and materials, potentially covering the needs of many industrial sectors, from medical to aerospace. Some of this industrial fields require for the certification of the final product, through the certification of the product design stage, of each manufacturing steps, and of the workers involved in.

This is the main motivation explaining why the International Organization for Standardization (ISO) and the American Society for Testing and Materials (ASTM) are putting significant efforts into defining standards covering different topics in this area. Those efforts were initially separated efforts, therefore bringing to separate ISO and ASTM standards, but very soon the two standardization bodies realized that they were doubling the efforts with no reasons and publishing slightly different standards, just creating confusion among the potential users. Therefore, the relevant decision to join the efforts to rapidly realize common standards under the name of ISO/ASTM standards.

In the following the additive manufacturing standards structure is presented and discussed.

L. Wang et al. (Eds.): AMP 2020, LNME, pp. 156–172, 2020.
https://doi.org/10.1007/978-3-030-46212-3_11

2 Additive Manufacturing Standards Structure

2.1 ISO AM Standards Structure

The ISO Technical Committee ISO/TC 261 Additive manufacturing is active since 2011, with the following scope [1]: "Standardization in the field of Additive Manufacturing (AM) concerning their processes, terms and definitions, process chains (Hard- and Software), test procedures, quality parameters, supply agreements and all kind of fundamentals." The ISO/TC 261 has 25 participating and 8 observing members, from all over the world.

The expected benefits achievable through AM standardization effort are:

1) systematic development, modification and use of AM processes resulting in innovative products;
2) guidelines to select the appropriate technology for the specified product demands;
3) specification of appropriate quality parameters and relative test procedures to assess the quality of products and processes;
4) standardization of AM process chains securing functionality and compatibility and of data formats and structures for AM models;
5) standardization of vocabulary required to define products and processes.

In order to achieve the considered benefits, ISO/TC 261 is organized in the following Working Group (WG) and Joint Working Group with other ISO/TCs (JWG):

- ISO/TC 261/WG 1 Terminology
- ISO/TC 261/WG 2 Processes, systems and materials
- ISO/TC 261/WG 3 Test methods and quality specifications
- ISO/TC 261/WG 4 Data and Design
- ISO/TC 261/WG 6 Environment, health and safety
- ISO/TC 261/JWG 10 Additive manufacturing in aerospace applications (Joint ISO/TC 261 - ISO/TC 44/SC 14)
- ISO/TC 261/JWG 11 Additive manufacturing for plastics (Joint ISO/TC 261 - ISO/TC 61/SC 9)

Table 1 shows the published standards by ISO, while in Table 2 the results of the joint cooperation of ISO and ASTM are presented.

2.2 ASTM AM Standards Structure

ASTM Committee F42 on Additive Manufacturing Technologies is active since 2009, with the following scope [2]: "The promotion of knowledge, stimulation of research and implementation of technology through the development of standards for additive manufacturing technologies". These standards are expected to play a preeminent role in all aspects of additive manufacturing technologies.

ASTM Committee F42 is composed of subcommittees addressing the following specific segments:

- F42.01 Test Methods
- F42.04 Design
- F42.05 Materials and Processes
- F42.06 Environment, Health, and Safety
- F42.07 Applications
  - F42.07.01 Aviation
  - F42.07.02 Spaceflight
  - F42.07.03 Medical/Biological
  - F42.07.04 Transportation/Heavy Machinery
  - F42.07.05 Maritime
  - F42.07.06 Electronics
  - F42.07.07 Construction
  - F42.07.08 Oil/Gas
  - F42.07.09 Consumer
- F42.91 Terminology
- F42.95 US TAG to ISO TC 261

Table 1 shows the published standards by ASTM, while in Table 2 the results of the joint cooperation of ISO and ASTM are presented.

2.3 ISO/ASTM AM Standards Structure

In order to eliminate duplication of efforts, in September 2011 ISO and ASTM have signed a cooperative agreement to govern the ongoing collaborative efforts between the two Organisations to adopt and jointly develop International Standards that serve the global marketplace in the field of additive manufacturing.

The active Joint Group are the following:

- ISO/ASTM JG51 - Terminology
- ISO/ASTM JG52 - Standard test artifacts
- ISO/ASTM JG53 - Requirements for purchased AM parts
- ISO/ASTM JG54 - Fundamentals of design
- ISO/ASTM JG55 - Standards specification for Extrusion Based AM of Plastic Materials
- ISO/ASTM JG56 - Standard practice for Metal Powder Bed Fusion to meet rigid quality requirements
- ISO/ASTM JG57 - Process-specific design guidelines and standards
- ISO/ASTM JG58 - Qualification, quality assurance and post processing of powder bed fusion metallic parts
- ISO/ASTM JG59 - Non-destructive testing for AM parts
- ISO/ASTM JG60 - Guide for intended seeding flaws in AM parts

- ISO/ASTM JG61 - Guide for anisotropy effects in mechanical properties of AM parts
- ISO/ASTM JG62 - Guide for conducting round robin studies for AM
- ISO/ASTM JG63 - Test methods for characterization of powder flow properties for AM applications
- ISO/ASTM JG64 - Additive Manufacturing File Format (AMF)
- ISO/ASTM JG66 - Technical specification on metal powders
- ISO/ASTM JG67 - Technical specification for the design of functionally graded AM parts
- ISO/ASTM JG68 - EH&S for 3D printers
- ISO/ASTM JG69 - EH&S for use of metallic materials
- ISO/ASTM JG70 - Optimized medical image data
- ISO/ASTM JG71 - Powder quality assurance
- ISO/ASTM JG72 - Machine – Production process qualification
- ISO/ASTM JG73 - Digital product definition and data management
- ISO/ASTM JG74 - Personnel qualifications
- ISO/ASTM JG75 - Industrial conformity assessment at AM centres
- ISO/ASTM JG76 - Revision of ISO 17296-3& ASTM F3122-14
- ISO/ASTM JG77 - Test method of sand mold for metalcasting
- ISO/ASTM JG78 - Safety regarding AM-machines

The main structure of the joint efforts in standardization is shown in Fig. 1.

The first level is the general top-level AM standards concerning topics like: Terminology, Design guides, Data formats, Qualification guidance, Inspection Method, Test methods, Test artifacts, System performance and reliability, Round robin test protocols, Safety.

The second level is related to specific category, and is divided with respect to materials, process and equipment, and finished parts. Then, the following levels refer to specific materials or processes, and to applications, at present mainly related to aerospace, medical, and automotive sectors.

The results of the ISO/ASTM joint efforts are shown in Table 2 as published standards, in Table 3 as standards under development, and in Table 4 as preliminary work items.

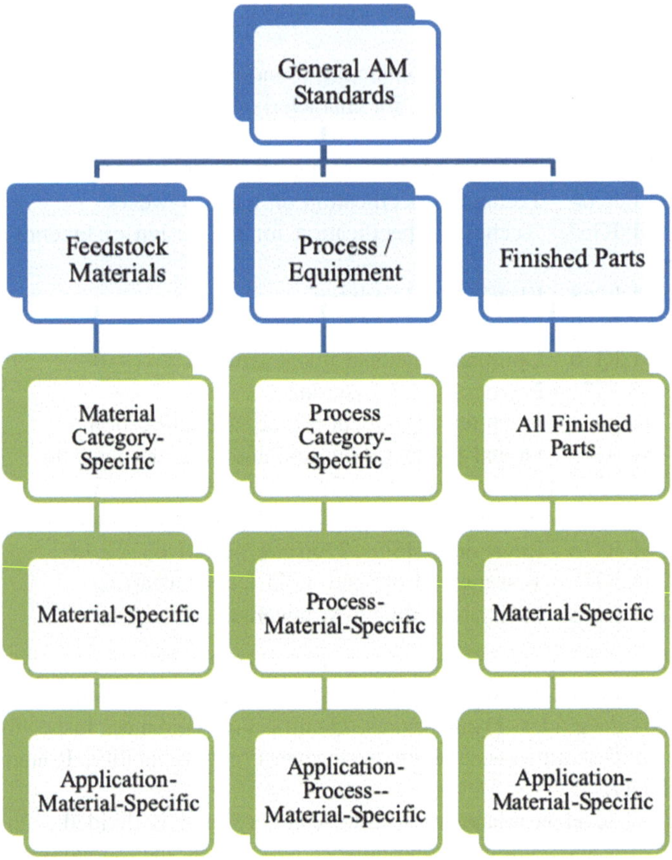

Fig. 1. ISO/ASTM additive manufacturing standards structure

Table 1. ISO and ASTM published standards.

Number	Title
ISO 17296-2:2015	Additive manufacturing - General principles - Part 2: Overview of process categories and feedstock
ISO 17296-3:2014	Additive manufacturing - General principles - Part 3: Main characteristics and corresponding test methods
ISO 17296-4:2014	Additive manufacturing - General principles - Part 4: Overview of data processing
ISO 27547-1:2010 (confirmed in 2015)	Plastics - Preparation of test specimens of thermoplastic materials using mouldless technologies - Part 1: General principles, and laser sintering of test specimens
ASTM F2924-14	Standard Specification for Additive Manufacturing Titanium-6 Aluminum-4 Vanadium with Powder Bed Fusion

(*continued*)

Table 1. (*continued*)

Number	Title
ASTM F2971-13	Standard Practice for Reporting Data for Test Specimens Prepared by Additive Manufacturing
ASTM F3001-14	Standard Specification for Additive Manufacturing Titanium-6 Aluminum-4 Vanadium ELI (Extra Low Interstitial) with Powder Bed Fusion
ASTM F3049-14	Standard Guide for Characterizing Properties of Metal Powders Used for Additive Manufacturing Processes
ASTM F3055-14a	Standard Specification for Additive Manufacturing Nickel Alloy (UNS N07718) with Powder Bed Fusion
ASTM F3056-14e1	Standard Specification for Additive Manufacturing Nickel Alloy (UNS N06625) with Powder Bed Fusion
ASTM F3091/F3091M-14	Standard Specification for Powder Bed Fusion of Plastic Materials
ASTM F3122-14	Standard Guide for Evaluating Mechanical Properties of Metal Materials Made via Additive Manufacturing Processes
ASTM F3184-16	Standard Specification for Additive Manufacturing Stainless Steel Alloy (UNS S31603) with Powder Bed Fusion
ASTM F3187-16	Standard Guide for Directed Energy Deposition of Metals
ASTM F3213-17	Standard for Additive Manufacturing - Finished Part Properties - Standard Specification for Cobalt-28 Chromium-6 Molybdenum via Powder Bed Fusion
ASTM F3301-18a	Standard for Additive Manufacturing - Post Processing Methods - Standard Specification for Thermal Post-Processing Metal Parts Made Via Powder Bed Fusion 1, 2
ASTM F3302-18	Standard for Additive Manufacturing - Finished Part Properties - Standard Specification for Titanium Alloys via Powder Bed Fusion
ASTM F3318-18	Standard for Additive Manufacturing - Finished Part Properties - Specification for AlSi10Mg with Powder Bed Fusion - Laser Beam

Table 2. ISO/ASTM published standards.

Number	Title	Note
ISO/ASTM 52900:2015	Additive manufacturing - General principles - Terminology	Will be replaced by ISO/ASTM DIS 52900
ISO/ASTM 52901:2017 (JG 53)	Additive manufacturing - General principles - Requirements for purchased AM parts	
ISO/ASTM 52902:2019	Additive manufacturing - Test artifacts - Geometric capability assessment of additive manufacturing systems	Will be replaced by ISO/ASTM AWI 52902

(*continued*)

Table 2. (*continued*)

Number	Title	Note
ISO/ASTM 52904:2019 (JG 56)	Additive manufacturing - Process characteristics and performance - Practice for metal powder bed fusion process to meet critical applications	
ISO/ASTM 52907:2019 (JG 66)	Additive manufacturing - Feedstock materials - Methods to characterize metal powders	
ISO/ASTM 52910:2018 (JG 54)	Additive manufacturing - Design - Requirements, guidelines and recommendations	
ISO/ASTM 52911-1:2019 (JG 57)	Additive manufacturing - Design - Part 1: Laser-based powder bed fusion of metals	
ISO/ASTM 52911-2:2019 (JG 57)	Additive manufacturing - Design - Part 2: Laser-based powder bed fusion of polymers	
ISO/ASTM 52915:2016	Specification for additive manufacturing file format (AMF) Version 1.2	Will be replaced by ISO/ASTM FDIS 52915
ISO/ASTM 52921:2013	Standard terminology for additive manufacturing - Coordinate systems and test methodologies	Will be replaced by ISO/ASTM DIS 52921

Table 3. ISO/ASTM standards under development.

Number	Title	Note
ISO/ASTM DIS 52900 (JG 51)	Additive manufacturing - General principles - Fundamentals and vocabulary	Approved as FDIS
ISO/ASTM AWI 52902 (JG 52)	Additive manufacturing - Test artifacts - Geometric capability assessment of additive manufacturing systems	New project approved
ISO/ASTM FDIS 52903-1 (JG 55)	Additive manufacturing - Material extrusion-based additive manufacturing of plastic materials - Part 1: Feedstock materials	Close Voting
ISO/ASTM DIS 52903-2 (JG 55)	Additive manufacturing - Standard specification for material extrusion based additive manufacturing of plastic materials - Part 2: Process - Equipment	Ballot
ISO/ASTM CD 52903-3 (JG 55)	Additive manufacturing - Standard specification for material extrusion based additive manufacturing of plastic materials - Part 3: Final parts	Project Deleted
ISO/ASTM DTR 52905 (JG 59)	Additive manufacturing - General principles - Non-destructive testing of additive manufactured products	Approved as DIS

(*continued*)

Table 3. (*continued*)

Number	Title	Note
ISO/ASTM CD TR 52906 (JG 60)	Additive manufacturing - Non-destructive testing and evaluation - Standard guideline for intentionally seeding flaws in parts	Committee Draft
ISO/ASTM AWI 52908 (JG 58)	Additive manufacturing - Post-processing methods - Standard specification for quality assurance and post processing of powder bed fusion metallic parts	Project Started
ISO/ASTM AWI 52909 (JG 61)	Additive manufacturing - Finished part properties - Orientation and location dependence of mechanical properties for metal powder bed fusion	Project Started
ISO/ASTM CD TR 52912 (JG 67)	Additive manufacturing - Design - Functionally graded additive manufacturing	Approved as DIS
ISO/ASTM 52915:2013	Standard specification for additive manufacturing file format (AMF) Version 1.1	Withdrawn
ISO/ASTM FDIS 52915 (JG 64)	Specification for additive manufacturing file format (AMF) Version 1.2	Close Voting
ISO/ASTM WD 52916 (JG 70)	Additive manufacturing - Data formats - Standard specification for optimized medical image data	Working Draft
ISO/ASTM WD 52917 (JG 62)	Additive manufacturing - Round Robin Testing - Guidance for conducting Round Robin studies	Approved as CD
ISO/ASTM CD TR 52918 (JG 64)	Additive manufacturing - Data formats - File format support, ecosystem and evolutions	Committee Draft
ISO/ASTM WD 52919-1 (JG 77)	Additive manufacturing - Test method of sand mold for metalcasting - Part 1: Mechanical properties	Approved as CD
ISO/ASTM WD 52919-2 (JG 77)	Additive manufacturing - Test method of sand mold for metalcasting - Part 2: Physical properties	Approved as CD
ISO/ASTM DIS 52921 (JG 51)	Additive manufacturing - General principles - Standard practice for part positioning, coordinates and orientation	Approved as FDIS
ISO/ASTM DIS 52924 (JWG 11)	Additive manufacturing - Qualification principles - Classification of part properties for additive manufacturing of polymer parts	Close Voting
ISO/ASTM DIS 52925 (JWG 11)	Additive manufacturing - Qualification principles - Qualification of polymer materials for powder bed fusion using a laser	Close Voting
ISO/ASTM AWI 52931 (JG 69)	Additive manufacturing - Environmental health and safety - Standard guideline for use of metallic materials	New project approved
ISO/ASTM WD 52932 (JG 68)	Additive manufacturing - Environmental health and safety - Standard test method for determination of particle emission rates from desktop 3D printers using material extrusion	Working Draft
ISO/ASTM WG 52938-1 (JG 78)	Additive manufacturing - Environmental health and safety - Part 1: Safety requirements for laser beam powder bed fusion machine using metallic feedstock	Working Draft

(*continued*)

Table 3. (*continued*)

Number	Title	Note
ISO/ASTM FDIS 52941 (JWG 10)	Additive manufacturing - System performance and reliability - Standard test method for acceptance of powder-bed fusion machines for metallic materials for aerospace application	Approved as FDIS
ISO/ASTM FDIS 52942 (JWG 10)	Additive manufacturing - Qualification principles - Qualifying machine operators of laser metal powder bed fusion machines and equipment used in aerospace applications	Approved as FDIS
ISO/ASTM DIS 52950 (JG 67)	Additive manufacturing - General principles - Overview of data processing	Approved as FDIS

Table 4. ISO/ASTM preliminary work items (PWI).

Number	Title
ISO/ASTM PWI 52911-3 (JG 57)	Additive manufacturing - Technical design guideline for powder bed fusion - Part 3: Standard guidcline for electron-based powder bed fusion of metals
ISO/ASTM PWI 52913 (JG 63)	Additive manufacturing - Process characteristics and performance - Standard test methods for characterization of powder flow properties
ISO/ASTM PWI 52914 (JG 54)	Additive manufacturing - Design - Standard guide for material extrusion processes
ISO/ASTM PWI 52920-1 (JG 75)	Additive manufacturing - Qualification principles - Part 1: Conformity assessment for AM System in industrial use
ISO/ASTM PWI 52920-2 (JG 75)	Additive manufacturing - Qualification principles - Part 2: Conformity assessment at Industrial additive manufacturing centers
ISO/ASTM PWI 52922 (JG 54)	Additive manufacturing - Design - Directed energy deposition
ISO/ASTM PWI 52923 (JG 54)	Additive manufacturing - Design decision support
ISO/ASTM PWI 52926-1 (JG 74)	Additive manufacturing - Qualification principles - Part 1: Qualification of machine operators for metallic parts production
ISO/ASTM PWI 52926-2 (JG 74)	Additive manufacturing - Qualification principles - Part 2: Qualification of machine operators for metallic parts production for PBF-LB
ISO/ASTM PWI 52926-3 (JG 74)	Additive manufacturing - Qualification principles - Part 3: Qualification of machine operators for metallic parts production for PBF-EB

(*continued*)

Table 4. (*continued*)

Number	Title
ISO/ASTM PWI 52926-4 (JG 74)	Additive manufacturing - Qualification principles - Part 4: Qualification of machine operators for metallic parts production for DED-LB
ISO/ASTM PWI 52926-5 (JG 74)	Additive manufacturing - Qualification principles - Part 5: Qualification of machine operators for metallic parts production for DED-Arc
ISO/ASTM PWI 52927 (JG 76)	Additive manufacturing - Process characteristics and performance - Test methods
ISO/ASTM PWI 52928 (JG 71)	Powder life cycle management
ISO/ASTM PWI 52930 (JG 72)	Guideline for Installation - Operation - Performance Qualification (IQ/OQ/PQ) of laser-beam powder bed fusion equipment for production manufacturing
ISO/ASTM PWI 52933 (JG 68)	Additive manufacturing - Environment, health and safety - Consideration for the reduction of hazardous substances emitted during the operation of the non-industrial ME type 3D printer in workplaces, and corresponding test method
ISO/ASTM PWI 52934 (JG 69)	Additive manufacturing - Environmental health and safety - Standard guideline for hazard risk ranking and safety defence
ISO/ASTM PWI 52935 (JG 74)	Additive manufacturing - Qualification principles - Qualification of coordinators for metallic parts production
ISO/ASTM PWI 52936-1 (JWG 11)	Additive manufacturing - Qualification principles - Laser-based powder bed fusion of polymers - Part 1: General principles, preparation of test specimens
ISO/ASTM PWI 52937 (JG 74)	Additive manufacturing - Qualification principles - Qualification of designers for metallic parts production
ISO/ASTM PWI 52938-1 (JG 78)	Additive manufacturing - Environmental health and safety - Part 1: Safety requirements for laser beam powder bed fusion machine using metallic feedstock
ISO/ASTM PWI 52943-1 (JWG 10)	Additive manufacturing - Process characteristics and performance - Part 1: Standard specification for directed energy deposition using wire and beam in aerospace applications
ISO/ASTM PWI 52943-2 (JWG 10)	Additive manufacturing - Process characteristics and performance - Part 2: Standard specification for directed energy deposition using wire and arc in aerospace applications
ISO/ASTM PWI 52943-3 (JWG 10)	Additive manufacturing - Process characteristics and performance - Part 3: Standard specification for directed energy deposition using laser blown powder in aerospace applications
ISO/ASTM PWI 52944 (JWG 10)	Additive manufacturing - Process characteristics and performance - Standard specification for powder bed processes in aerospace applications
ISO/ASTM PWI 52951 (JG 73)	Additive manufacturing - Data packages for AM parts

3 AM Published Standards Overview

In the following a short overview of the published standards is presented, according to structure presented in Fig. 1.

3.1 General AM Standards

3.1.1 ISO 17296 Series of Standards [3–5]

This series of standards is devoted to general principles related to additive manufacturing. In particular, ISO 17296-2:2015 gives an overview of existing process categories and describes the process fundamentals. It also explains how different process categories make use of different types of materials to shape a product's geometry. ISO 17296-3:2014 specifies the main quality characteristics of parts and appropriate test procedures. It is aimed at machine manufacturers, feedstock suppliers, machine users, part providers, and customers to facilitate the communication on main quality characteristics, whatever the process category is. ISO 17296-4:2014 specifies terms and definitions which enable exchanging information on geometries or parts. It is aimed at users and producers of additive manufacturing processes and associated software systems.

3.1.2 ASTM F2971 [6]

It describes a method for reporting results by testing or evaluation of specimens produced by AM. This practice provides a common format for presenting data for two purposes: to establish further data reporting requirements, and to provide information for the design of material property databases.

3.1.3 ISO/ASTM 52900 [7]

It establishes and defines terms used in AM technology. The terms have been classified into specific fields of application. New terms emerging from the future work within ISO/TC 261 and ASTM F42 will be included in upcoming amendments and overviews of this International Standard.

3.1.4 ISO/ASTM 52901 [8]

It defines and specifies requirements for purchased parts made by additive manufacturing. It gives guidelines for the elements to be exchanged between customers and the part providers at the time of the order, including the customer order information, part definition data, feedstock requirements, final part characteristics and properties, inspection requirements and part acceptance methods.

3.1.5 ISO/ASTM 52902 [9]

This standard covers the general description of benchmarking test piece geometries along with quantitative and qualitative measurements to be taken on the benchmarking test piece to assess the performance of AM systems. This performance assessment may serve to evaluate capability and to calibrate AM systems. The benchmarking test piece is primarily used to quantitatively assess the geometric performance of an AM system. It describes a set of geometries, each designed to investigate one or more specific performance metrics and several example configurations of these geometries into test piece. It prescribes quantities and qualities of the test geometries to be measured but does not dictate specific measurement methods. This document does not discuss a specific procedure or machine settings for manufacturing a test piece, which are covered by other standards.

3.1.6 ISO/ASTM 52910 [10]

It gives requirements, guidelines and recommendations for using additive manufacturing (AM) in product design. It is applicable during the design of all types of products, devices, systems, components or parts that are fabricated by any type of AM system.

3.1.7 ISO/ASTM 52915 [11]

It provides the specification for the Additive Manufacturing File Format (AMF), an interchange format to address the current and future needs of additive manufacturing technology. It does not specify any explicit mechanisms for ensuring data integrity, electronic signatures and encryptions.

3.1.8 ISO/ASTM 52921 [12]

It includes terms, definitions of terms, descriptions of terms, nomenclature, and acronyms associated with coordinate systems and testing methodologies for additive manufacturing (AM) technologies.

3.2 Materials Category-Specific

3.2.1 ASTM F3049 [13]

This guide introduces the techniques for metal powder characterization useful for powder-based AM processes including binder jetting, directed energy deposition, and powder bed fusion. It refers to other existing standards that may be applicable for the characterization of new and used metal powders processed in AM systems.

3.2.2 ASTM F3122 [14]

This standard serves as a guide to original or variations of existing standards that may be applicable to determine specific mechanical properties of metal materials made with an AM process.

3.2.3 ISO/ASTM 52907 [15]

It provides technical specifications for metallic powders intended to be used in additive manufacturing and covers the following aspects: documentation and traceability, sampling, particle size distribution, chemical composition, characteristic densities, morphology, flowability, contamination, packaging and storage. It also gives specific requirements for used metallic powders in additive manufacturing. It does not deal with safety aspects.

3.3 Process Category-Specific

3.3.1 ASTM F3187 [16]

This standard is intended to serve as a guide for defining the direct energy deposition applicability, system set-up considerations, machine operation, process documentation, work practices, and available system and process monitoring technologies.

3.3.2 ISO/ASTM 52904 [17]

It describes the operation and production control of metal powder bed fusion machines and processes to meet critical applications such as aerospace components and medical implants. The requirements contained herein are applicable for production components and mechanical test specimens using powder bed fusion with both laser and electron beams.

3.3.3 ISO/ASTM 52911 [18, 19]

This series of standards specifies the features of laser-based powder bed fusion of metals (PBF-LB/M) and provides detailed design recommendations. It also provides a state-of-the-art review of design guidelines associated with the use of powder bed fusion. In particular, ISO/ASTM 52911-1:2019 refers to metals and ISO/ASTM 52911-2:2019 refers to polymers.

3.4 Process Material-Specific

3.4.1 ISO 27547 Series of Standards [20]

This series of standards is devoted to general principles related on testing specimen. In particular, the ISO 27547-1:2010 specifies the general principles to be followed when test specimens of thermoplastic materials are prepared by laser sintering. It provides a basis for establishing reproducible sintering conditions. Its purpose is to promote uniformity in describing the main parameters of the sintering process and also to establish uniform practice in reporting sintering conditions.

3.4.2 ASTM F2924 [21]

It covers additively manufactured Titanium-6Aluminum-4Vanadium (Ti-6Al-4V) parts using powder bed fusion such as electron beam melting and laser melting. The parts produced by these processes are used in applications that typically require mechanical properties similar to machined forgings and wrought products. Parts manufactured to this specification are often, but not necessarily, post processed via machining, grinding, electrical discharge machining, to meet necessary surface finish and dimensional and geometrical requirements.

3.4.3 ASTM F3001 [22]

It covers additively manufactured Titanium-6Aluminum-4Vanadium with extra low interstitials (Ti-6Al-4V ELI) parts using powder bed fusion. The parts produced by these processes are used in applications that typically require mechanical properties similar to machined forgings and wrought products. Parts manufactured to this specification are often, but not necessarily, post processed via machining, grinding, electrical discharge machining, polishing, and other finishing processes, to meet necessary surface finish and dimensional and geometrical requirements.

3.4.4 ASTM F3055 [23]

It covers additively manufactured UNS N07718 alloy parts using powder bed fusion. The parts produced by these processes are used in applications that typically require mechanical properties similar to machined forgings and wrought products. Parts manufactured to this specification are often, but not necessarily, post processed via machining, grinding, electrical discharge machining, polishing, and other finishing processes, to meet necessary surface finish and dimensional and geometrical requirements.

3.4.5 ASTM F3056 [24]

It covers additively manufactured UNS N06625 alloy parts using full-melt powder bed fusion such as electron beam melting and laser melting. The parts produced by these processes are used in applications that typically require mechanical properties similar to machined forgings and wrought products. Parts manufactured to this specification are often, but not necessarily, post processed via machining, grinding, electrical discharge machining, polishing, and other finishing processes, to meet necessary surface finish and dimensional and geometrical requirements.

3.4.6 ASTM F3091/F3091M [25]

This specification describes a method for defining requirements and ensuring integrity for plastic parts created using powder bed fusion processes. Materials include unfilled formulations and formulations containing fillers, functional additives, and reinforcements or combinations thereof.

3.4.7 ASTM F3184 [26]

It covers additive manufacturing of UNS S31603 alloy parts by means of powder bed fusion processes. The parts produced by these processes are used in applications that typically require mechanical properties similar to machined forgings and wrought products. Parts manufactured to this specification are often, but not necessarily, post processed via machining, grinding, electrical discharge machining, polishing, and other finishing processes, to meet necessary surface finish and dimensional and geometrical requirements.

3.5 Finished Parts Material-Specific

3.5.1 ASTM F3213 [27]

It covers additively manufactured cobalt-28 chromium-6 molybdenum alloy parts with similar chemical composition to UNS R30075 by means of powder bed fusion processes. The parts produced by these processes are used typically in applications that require mechanical properties similar to cast or wrought products. Parts manufactured to this specification are often, but not necessarily, post processed via machining, grinding, electrical discharge machining, polishing, and other finishing processes, to meet necessary surface finish and dimensional and geometrical requirements.

3.5.2 ASTM F3301 [28]

It specifies the requirements for thermal post-processing of parts produced via metal powder bed fusion to achieve the material properties and microstructure required to meet engineering requirements.

3.5.3 ASTM F3302 [29]

It covers additive manufacturing of parts by means of powder bed fusion processing of titanium alloys. The parts produced by these processes are used typically in applications that require mechanical properties similar to wrought products. Parts manufactured to this specification are often, but not necessarily, post processed via machining, grinding, electrical discharge machining, polishing, and other finishing processes, to meet necessary surface finish and dimensional and geometrical requirements.

3.5.4 ASTM F3318 [30]

It covers additively manufactured AlSi10Mg (similar to DIN EN 1706:2013-12 EN AC-43000) parts using powder bed fusion. The parts produced by these processes are used in applications that typically require mechanical properties similar to or exceeding those of cast aluminum products of equivalent alloys. Parts manufactured to this specification are often, but not necessarily, post processed via machining, grinding, electrical discharge machining, polishing, and other finishing processes, to meet necessary surface finish and dimensional and geometrical requirements.

4 Conclusions

Standardization is essential for the use of AM in critical applications such as energy saving applications in aerospace or implants fabrication for medical applications. Standards will enable the certification and approval for medical and aerospace applications. Without standards such certifications and approvals are very complicated if not impossible. For Jörg Lenz, Former Chair of ISO/TC 261 on Additive Manufacturing, "*the industry really needs International Standards to provide clarity and dispel concerns, to provide reliability, acceptance and safety, and to further push the technology in the market*" [1].

The International Organization for Standardization (ISO) and the American Society for Testing and Materials (ASTM) are putting significant joint efforts into defining standards covering different topics in this area: from terminology and data formats, to design guidelines, parts and processes qualification assessment and health and safety issues, considering both general and specific aspects and applications.

Nevertheless, all these efforts are missing a relevant point as suggested by [31–33]: the geometric dimensioning and tolerancing of additively manufactured parts. As a matter of fact, the AM processes enabled "complexity for free" requires new design approaches and the appropriate methods to define the geometrical product specifications, so that the uncertainty may be properly managed along the lifecycle of the AM products.

Acknowledgements. This work was supported by European Union, Repubblica Italiana, Regione Lombardia and FESR for the project MADE4LO under the call "POR FESR 2014-2020 ASSE I - AZIONE I.1.B.1.3".

References

1. ISO/TC261 Additive Manufacturing Homepage. https://committee.iso.org/home/tc261. Accessed 10 Feb 2020
2. ASTM Committee F42 on Additive Manufacturing Technologies Homepage. https://www.astm.org/COMMITTEE/F42.htm. Accessed 20 Feb 2020
3. International Organization for Standardization: ISO 17296-2 Additive manufacturing - General principles - Part 2: Overview of process categories and feedstock (2015)
4. International Organization for Standardization: ISO 17296-3 Additive manufacturing - General principles - Part 3: Main characteristics and corresponding test methods (2014)
5. International Organization for Standardization: ISO 17296-4 Additive manufacturing - General principles - Part 4: Overview of data processing (2014)
6. American Society for Testing and Materials: ASTM F2971-13 Standard Practice for Reporting Data for Test Specimens Prepared by Additive Manufacturing (2013)
7. International Organization for Standardization: ISO/ASTM 52900 Additive manufacturing - General principles - Terminology (2015)
8. International Organization for Standardization: ISO/ASTM 52901 Additive manufacturing - General principles - Requirements for purchased AM parts (2017)
9. International Organization for Standardization: ISO/ASTM 52902 Additive manufacturing - Test artifacts - Geometric capability assessment of additive manufacturing systems (2019)
10. International Organization for Standardization: ISO/ASTM 52910 Additive manufacturing - Design - Requirements, guidelines and recommendations (2018)
11. International Organization for Standardization: ISO/ASTM 52915 Specification for additive manufacturing file format (AMF) Version 1.2 (2016)
12. International Organization for Standardization: ISO/ASTM 52921 Standard terminology for additive manufacturing - Coordinate systems and test methodologies (2013)
13. American Society for Testing and Materials: ASTM F3049-14 Standard Guide for Characterizing Properties of Metal Powders Used for Additive Manufacturing Processes 1 (2014)
14. American Society for Testing and Materials: ASTM F3122-14 Standard Guide for Evaluating Mechanical Properties of Metal Materials Made via Additive Manufacturing Processes (2014)
15. International Organization for Standardization: ISO/ASTM 52907 Additive manufacturing - Feedstock materials - Methods to characterize metal powders (2019)
16. American Society for Testing and Materials: ASTM F3187-16 Standard Guide for Directed Energy Deposition of Metals (2016)
17. International Organization for Standardization: ISO/ASTM 52904 Additive manufacturing - Process characteristics and performance - Practice for metal powder bed fusion process to meet critical applications (2019)
18. International Organization for Standardization: ISO/ASTM 52911-1 Additive manufacturing - Design - Part 1: Laser-based powder bed fusion of metals (2019)
19. International Organization for Standardization: ISO/ASTM 52911-2 Additive manufacturing - Design - Part 2: Laser-based powder bed fusion of polymers (2019)

20. International Organization for Standardization: ISO 27547-1 Plastics - Preparation of test specimens of thermoplastic materials using mouldless technologies - Part 1: General principles, and laser sintering of test specimens (2010)
21. American Society for Testing and Materials: ASTM F2924-14 Standard Specification for Additive Manufacturing Titanium-6 Aluminum-4 Vanadium with Powder Bed Fusion (2014)
22. American Society for Testing and Materials: ASTM F3001-14 Standard Specification for Additive Manufacturing Titanium-6 Aluminum-4 Vanadium ELI (Extra Low Interstitial) with Powder Bed Fusion (2014)
23. American Society for Testing and Materials: ASTM F3055-14a Standard Specification for Additive Manufacturing Nickel Alloy (UNS N07718) with Powder Bed Fusion (2014)
24. American Society for Testing and Materials: ASTM F3056-14e1 Standard Specification for Additive Manufacturing Nickel Alloy (UNS N06625) with Powder Bed Fusion (2014)
25. American Society for Testing and Materials: ASTM F3091/F3091M-14 Standard Specification for Powder Bed Fusion of Plastic Materials (2014)
26. American Society for Testing and Materials: ASTM F3184-16 Standard Specification for Additive Manufacturing Stainless Steel Alloy (UNS S31603) with Powder Bed Fusion (2016)
27. American Society for Testing and Materials: ASTM F3213-17 Standard for Additive Manufacturing - Finished Part Properties - Standard Specification for Cobalt-28 Chromium-6 Molybdenum via Powder Bed Fusion (2017)
28. American Society for Testing and Materials: ASTM F3301-18a Standard for Additive Manufacturing - Post Processing Methods - Standard Specification for Thermal Post-Processing Metal Parts Made Via Powder Bed (2018)
29. American Society for Testing and Materials: ASTM F3302-18 Standard for Additive Manufacturing - Finished Part Properties - Standard Specification for Titanium Alloys via Powder Bed Fusion (2018)
30. American Society for Testing and Materials: ASTM F3318-18 Standard for Additive Manufacturing - Finished Part Properties - Specification for AlSi10Mg with Powder Bed Fusion - Laser Beam (2018)
31. Thompson, M.K., Moroni, G., Vaneker, T., Fadel, G., Campbell, R.I., Gibson, I., Bernard, A., Schulz, J., Graf, P., Ahuja, B., Martina, F.: Design for additive manufacturing: trends, opportunities, considerations, and constraints. CIRP Ann. Manuf. Technol. **65**(2), 737–760 (2016)
32. Moroni, G., Petrò, S., Polini, W.: Geometrical product specification and verification in additive manufacturing. CIRP Ann. Manuf. Technol. **66**(1), 157–160 (2017)
33. Morse, E., Dantan, J.Y., Anwer, N., Soderberg, R., Moroni, G., Qureshi, A.J., Jiang, X., Luc Mathieu, L.: Tolerancing: managing uncertainty from conceptual design to final product. CIRP Ann. Manuf. Technol. **67**(2), 695–717 (2018)

Artefacts Used for Testing 3D Optical-Based Scanners

Maria Grazia Guerra, Fulvio Lavecchia, and Luigi Maria Galantucci(✉)

Dipartimento di Meccanica Matematica e Management, Politecnico di Bari, 70123 Bari, Italy
Luigimaria.galantucci@poliba.it

Abstract. Performance verification of 3D optical based scanners is currently a topic of great discussion, since they are ever more used in the industrial manufacturing field for the dimensional verification of components. Among their advantages, there is the capability of acquire large amounts of points in a very short time, regardless the geometrical complexity of the object under measurement. Although, traceability is still critical and the uncertainty assessment conducted using artefacts calibrated through the more traceable Coordinate Measuring Machines (CMMs), is still the most implemented method. In this paper, some of the most interesting geometries used for the performance verification of optical based scanners have been reported, considering the most widespread measuring tasks which require the use of these instruments, both prismatic geometries and freeform shapes.

Keywords: 3D scanners · Calibrated artefacts · Traceability · Uncertainty · Freeform · Performance verification

1 Introduction

Optical scanners are ever more used thanks to their many advantages and the variety of geometries and materials, which can be easily acquired and measured. Those systems fulfil most of the requirements of the new paradigm of Industry 4.0 in terms of fast, accurate and holistic measurement systems. Among their advantages, with respect to the more traditional Coordinate Measuring Machines (CMMs), there is the capability of acquire large amount of points in short time, regardless the geometrical complexity of the part to be measured, thus, the time needed for the measurement of wide and complex shapes is strongly reduced. It is a really great advantage, since the widespread adoption of the additive manufacturing techniques, whose main advantage is the freedom in choosing the part geometry, creates the need for measuring instruments capable of acquire freeform parts [1]. However, traceability of those systems is still an issue. The main difference with respect to CMMs, is the dependency of the measurement results from the object material, intended as optical properties of its surface. This aspect is further emphasized by the wide range of materials and geometries usually acquired via optical systems, among which, there are polymers. This means that considerations regarding accuracy and reliability of an optical based scanner are not

L. Wang et al. (Eds.): AMP 2020, LNME, pp. 173–189, 2020.
https://doi.org/10.1007/978-3-030-46212-3_12

generalizable and they have to be evaluated for each specific case, determined by the combination of *geometry-material-measuring principle*, which is identifiable with the term *measuring task*. One way to deal with the calibration of those instruments is the use of properly designed and calibrated artefacts.

Due to these limitations, in all cases where accuracy is of paramount importance, the use of measuring techniques capable of acquire the real shape of the test surface with measurement capabilities an order of magnitude better than the desired, is still necessary for understanding the behavior of the optical based scanners. This leads to the necessity of calibrating the object under measurement through a more traceable instrument. Several geometries have been implemented and tested in literature. Those geometries have some common characteristics and are specifically designed for the identification of inaccuracies and errors affecting the measuring instruments.

Usually, artefacts suggested by the currently available standards are simple geometries, such as gauge blocks and spheres made of well known materials characterized by very low thermal expansion coefficient (steel and ceramics). However, in most cases these artefacts are not well representative of the complexity of the measuring tasks actually required by the industrial reality, in terms of both geometry and material.

Indeed, due to the increase of polymers used many industrial fields, measuring polymers components with 3D optical scanners became a widely spread measuring task considering both, conventional manufacturing processes and the newer additive technologies. Although, they exhibit optical properties, which interact with the measuring system in a way dependent on the specific scanner and material and this aspect has to be properly considered [2].

In this paper, some of the most interesting geometries, both freeform and prismatic, used for the performance verification of optical scanners are described with their applications and their main results achieved. The focus is on those artefacts related to the most interesting measuring tasks characterizing the manufacturing field. The optical scanners considered are 3D scanners adopted for the dimensional verification of components in the close range, with measuring ranges below one meter and structural resolution on the order of few cents of millimeters or below. The accuracy and uncertainty of those scanners, considering the measuring tasks analysed in literature and reported in this paper, is comprised within 0.020 mm.

The term physical standard will be used within the paper referring to calibrated artefacts used for assessing the measuring performance of 3D optical scanners.

2 State of the Art of the International Standards

International standards represent the main tool for communicating in a global economy. In the context of measuring systems, among other things, they regulate the relation between manufacturers of those systems and customers especially in a global commercial context. The measurement limits and characteristics, such as accuracy, precision, as well as, resolution of a system should be obtained using standardized methods so that the manufacturer can provide specifications meaning something to the customer.

Even though, 3D optical scanning systems have been developed and used for a few decades, they are still considered to be an emerging technology [3].

2.1 Performance Verification and Testing of Optical Instruments

Performance verification tests are widely used as acceptance testing for optical 3D imaging systems [3]. Acceptance and reverification tests are fundamental for manufacturers (acceptance) and users (reverification) for:

- proving the applicability of a given system to the task (fitness-for purpose);
- comparing different instruments using proper methodologies and metrics;
- managing instrument warranty issues;
- reducing costs through effective use of 3D imaging systems;

The quality parameters for the acceptance tests are clearly defined in terms of recommended artefacts, the procedure, the method to calculate the results, and their interpretation. The reverification of the optical 3D measuring systems ensures long-term compliance with limits specified by the user and it allows to detect trends for preventive maintenance.

In this context and with these purposes, physical standards, along with the different tests methods, are used to characterize 3D optical imaging systems.

The ISO standards dealing with the performance verification of optical measuring instruments are quite recent: ISO 10360-7 [4], ISO 10360-8 [5]. Both these standards refer to the CMMs equipped with optical based systems: in particular, the ISO 10360-7 deals with CMM equipped with imaging probing systems, while the part 8 deals with CMM equipped with optical distance sensors.

The German VDI/VDE, as well, has been very active in defining standards for coordinate metrology, in particular, for coordinate measuring machines (CMMs) equipped with optical probing and, more importantly, for stand-alone 3D optical measuring systems.

- *VDI/VDE 2617 Part 6.2 (2005)* – Guideline for the application of ISO 10360 to coordinate measuring machines with optical distance sensors;
- *VDI/VDE 2634 Part 2 (2012)* – Optical 3-D measuring systems: Optical systems based on area scanning;
- *VDI/VDE 2634 Part 3 (2008)* – Optical 3-D measuring systems: Optical systems based on area scanning in several single images.

The VDI/VDE 2617-Part 6.2 proposes a revision to the ISO 10360-2 tests specific for coordinate measuring machines equipped with an Optical Distance Sensor (ODS), which could be both triangulation and interferometry-based sensors. The VDI/VDE 2634 series closely follows the recommendations of the VDI/VDE 2617 but the optical measuring systems can be mobile and considered as standing alone.

Part 2 and Part 3 of the VDI/VDE 2634 [6, 7] are important to manufacturers and users for verifying systems compliance with required performance specifications. This is realized through acceptance tests performed by the manufacturers and verification tests performed by the users. Part 2 include single-view optical systems based on area scanning and Part 3, multiple-view systems. Area scanning is based on triangulation methods, which include fringe projection, moiré techniques, and photogrammetry or scanning systems with area-based measuring capabilities. According to these standards, all the involved parameters (e.g. probing error, length error, flatness error, etc..) are usually computed on a single point cloud or multi-view registered point clouds; however, not all 3D imaging systems provide data in that format. For this reason, the possibility to use polygonised or triangulated data files is discussed in the VDI\VDE 2634. Filtering and pre-processing of the measured values are allowed only if they are part of the boundary conditions or it is a routine operation of the system's software procedure.

2.2 Uncertainty Assessment of a 3D Optical-Based System

With the aim to fully characterize a measurement result obtained with an optical based system, as for every other measuring instrument, the uncertainty assessment has to be reported together with the measured value. Although, quantifying the uncertainty associated to an optical-based system is not a trivial task due to the complexity of those systems. If, on one hand, they allow to conduct very tough measuring tasks, on the other hand, there are many sources of error affecting these systems. For these reasons, a specific procedure for the uncertainty computation of a 3D optical scanner does not exist and some approaches are usually implemented. Generally, the base guideline for the uncertainty assessment is the GUM approach (ISO/IEC Guide 98-3 – "Guide to the estimation of Uncertainty in Measurement") [8], however, this approach often results to be hard and not easily implementable in a production environment. Thus, starting from this fundamental guideline, other method for the uncertainty assessment have been developed. In particular, the ISO 14253-2 [9] is a guidance for the estimation of the uncertainty in the Geometrical Product Specification field (GPS) and comprises methods for calibration of measuring equipment and for product verification. The ISO 14253-2 introduces a guideline for the uncertainty assessment, which encompasses the GUM [8] and it is an iterative procedure with a value of target uncertainty (PUMA method [9]). It is well suited for the industrial environment, to reduce time, risks and costs. The method for the uncertainty computation, proposed in this standard, comprises all the possible sources of error of a measuring system. Each source of error, which produces an uncertainty component, is considered in the uncertainty budget and computed according to the GUM approach. The latter [5] comprises type A and type B uncertainty components, depending on the method used for their computation, statistical method, type A, or non-statistical method, type B. The approach described in the ISO 14253-2 is defined as an upper-bound model, due to the natural overestimation of the resulting uncertainty and it is a precautionary measure in order to avoid wrong decisions based on measuring results.

The use of artefacts for calibrating optical base scanners, takes inspiration by the standard ISO 15530-3 [10], which is related to Coordinate Measuring Machines implemented in industrial environment. Indeed, with some adaptations, it has been extensively adopted for optical and x-ray-based scanners [11, 12].

Generally, in the field of dimensional verification, most of these standards were developed for CMMs and, in particular, a well-known method, widely used in the production environment is the substitution method explained in the ISO 15530-3 [10].

It transfers traceability from a calibrated reference object to an actual part and it is considered also a good method for assessing the uncertainty of an optical based system related to a specific task and within specific conditions. The ISO 15530-3 is addressed to CMMs and describes a simplified method for the computation of the uncertainty. It is thought for the application in the industrial environment and it implements the substitution approach. The basic concept is the transfer of the traceability from an artefact with known uncertainty to an actual artefact with unknown uncertainty, under some specific conditions and assumptions, which are identified as similarity conditions. Similarity includes: the identical measuring equipment, procedure and environmental conditions, as well as, closeness of the two artefacts (calibrated and actual) in terms of material, mechanical properties and thermal expansion coefficient. All the variations respect to the similarity requirements must be taken into account in the total uncertainty. Indeed, due to the great flexibility of optical systems, it is almost impossible to calibrate them for all measuring tasks and a correct approach could be to calibrate it for each measuring task, under specified measuring conditions. Although, even if the ISO 15530-3 seems to fit the needs of optical based systems, the uncertainty assessment, currently implemented in that standard and used for contact probing systems, cannot be directly applied to optical-based instruments, due to the necessity to add or neglect some specific error sources. In particular, for the ISO 15530-3, the similarity requirements must be satisfied not just as similar dimensions or form error or material, but also as surface finishing, colour and in terms of the same optical characteristics. In Fig. 1, typical error sources for a 3D optical-based instrument are reported in the form of an Ishikawa diagram [13].

Differently from contact instruments, there are other factors affecting the measurement:

- Lens distortion and aberration;
- Algorithms for reconstruction;
- Algorithms for point clouds and mesh managing;
- Algorithms for analysis;
- Optical interaction with the object surface;
- Interaction with the environmental conditions.

Among them, the interaction between the measuring system and the object is of paramount importance. The optical properties of the object, transparency, translucency and reflectivity play a fundamental role and, according to their suitability to be acquired via optical systems, there is a distinction between cooperative and non-cooperative surfaces. Generally, the latter produce large measurement errors undermining the application of those systems.

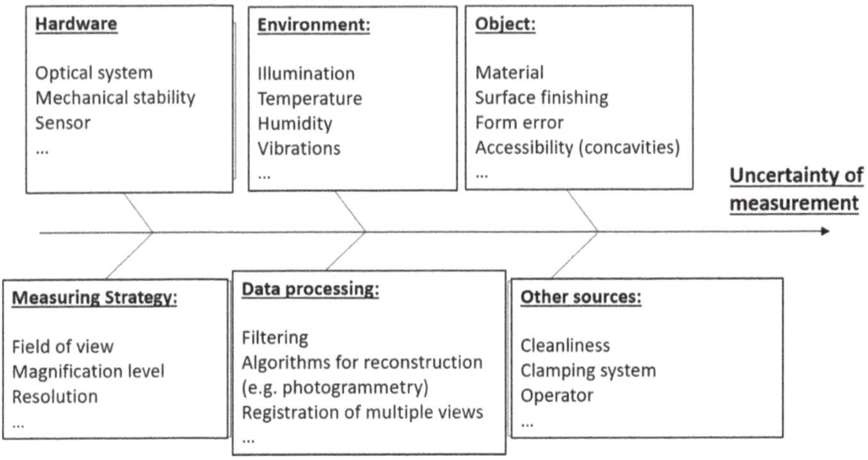

Fig. 1. Typical error sources for a 3D optical-based instrument [13].

2.3 Physical Standards

As described, the international standards ISO 10360 series and the German guideline VDI/VDE 2634 require the use of physical standards for the performance verification of optical based scanners. However, the required reference objects used for acceptance and reverification tests are basic geometries, such as spheres with certified form and diameter and flat planes with certified flatness (Fig. 2).

These simple geometries are sufficient for characterizing some of the test parameters reported in the VDI/VDE which can be grouped in *Form Error Measurement* parameters and *Length Measurement Error* parameters [14, 15].

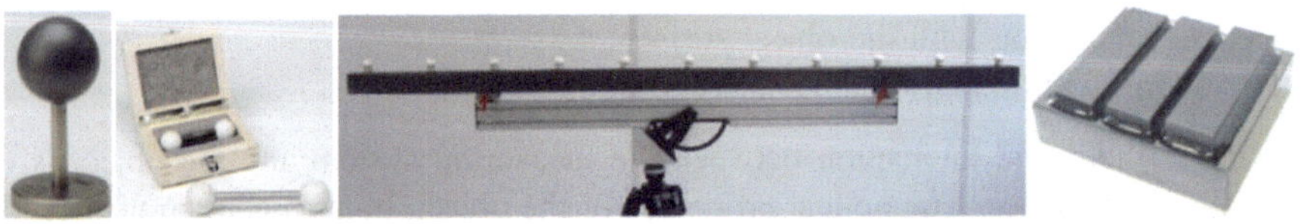

Fig. 2. Standard reference objects for testing optical based scanners [16].

The physical standards must be measurable and they are chosen for their physical and optical qualities, since optical instruments can be more or less sensitive to a specific surface characteristic depending on their measurement principles. For example, interferometry-based and confocal-based are capable of measuring opaque, transparent and translucent materials, while triangulation-based techniques require reference artefacts with cooperative optical surface characteristics (the optical and mechanical surface should coincide and the colour should be compatible with the light source). For these reasons, metallic surfaces are sometimes treated to make them diffusely reflecting.

Some surface treatments, like vapour blasting, light particle blasting, or spray particle coating, are able to change a specular surface into a diffusely-reflecting surface (Fig. 3).

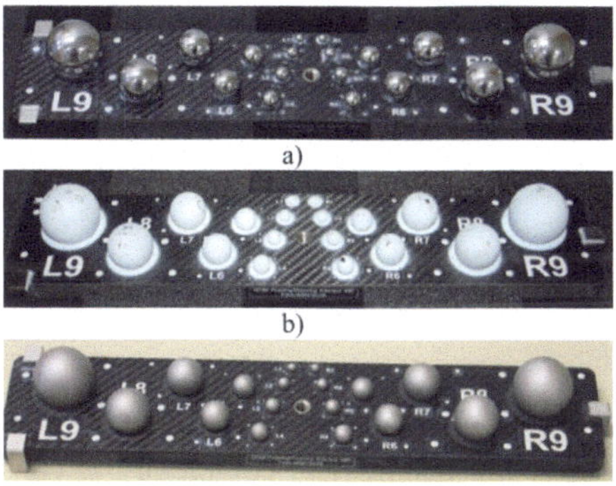

Fig. 3. Examples of spheres plates coated or treated in order to reach Lambertian conditions [3].

Although, basic geometries such as spheres and gauge block are not well representative of the geometrical complexity characterizing the real industrial cases. With the aim to cover most of criticalities related to 3D optical instruments when measuring complex artefacts, other reference objects have been designed in [17] to aid a better understanding of 3D optical scanner main factors: resolution, orientation, illumination effect, sensitivity to surface colour, material and finishing.

Other artefacts, closer to the real cases, were developed and are discussed in the next sections.

3 Artefacts for Performance Verification of Freeform Shaped Parts

The measurement of freeform shaped objects (Fig. 4) is a topic of great discussion in literature. It represents, indeed, a complex measuring task for several reasons. Firstly, a unique and overall accepted definition does not exist: a freeform geometry is usually described as a geometry not referable to any known simple geometry.

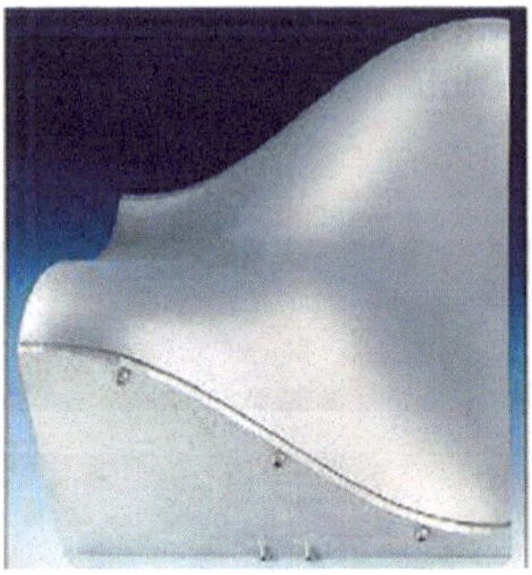

Fig. 4. General purpose freeform artefact [15].

Another fundamental issue is related to the instruments actually capable of measuring those objects. If just the geometrical complexity is considered, a non-contact measuring system should be selected as the best choice, thanks to its capability to acquire large amounts of points in short time, regardless the complexity of the object's shape. Although, there is a big issue related to the traceability of those systems, which strongly depends on the considered measuring task. For complex surfaces, comparative measurement between optical based scanners and CMMs is the most implemented way to ensure traceability, since, generally, from a metrological point of view, CMMs are still considered the best choice for dimensional verifications with low uncertainties. Nevertheless, their traceability is not completely assured when measuring freeform objects, due to the difficulties in determining the contact direction and point. For this reason, the use of artefacts showing a freeform shapes as a result of composition of basic solid geometries, (e.g. spheres, planes, cones, etc.…), enables a meaningful CMM measurement and, as a consequence, a meaningful optical scanner/CMM results comparison. One example is the NPL free form artefact (Fig. 5) [16] which has been already tested in [16, 18, 19]. This artefact was designed in order to highlight all the main weaknesses of optical based systems. It allows to conduct a double analysis: considering each single basic shape, such as sphere, cone, torus and their relations (distances), or considering the resulting composed shape as a freeform shape. Regarding the former, it is possible to compare results obtained from the optical scanners with highly reliable CMM results; for the latter, a 3D comparison (Fig. 6) can be carried out for comparing optical scanners and CMM values, but considering the previous highlighted limitation of the CMMs when used for measuring freeform parts.

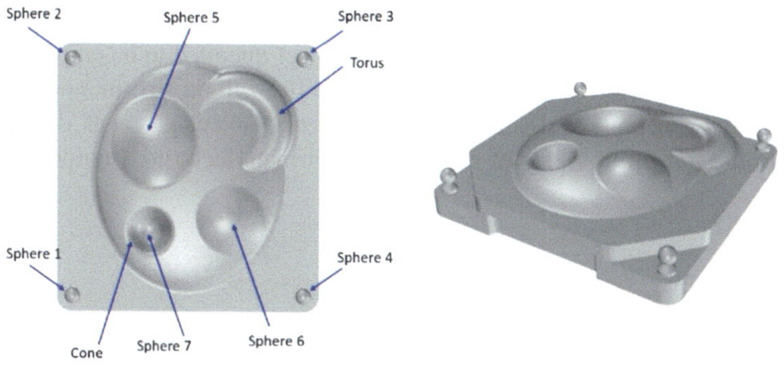

Fig. 5. Photograph and schematic representation of NPL freeform artefact [19].

In [19] the NPL freeform artefact has been successfully used for the performance verification of instruments suitable for in line measurements of additive repair processes.

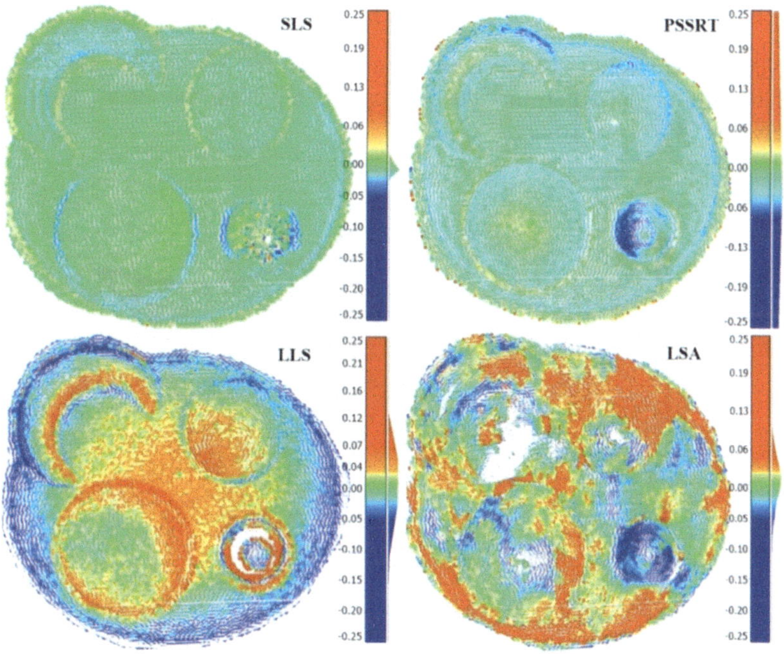

Fig. 6. Errors coloured maps of four different optical based scanners: Structured Light Scanner, SLS, Photogrammetric Scanning System with Rotary Table, PSSRT, Laser Line Scanner, LLS and Laser Scanning Arm, LSA [19].

In literature, there are also examples of reference objects featuring both prismatic shapes and freeform shapes. In [20] a reference object intended to be employed for the evaluation of the performance of several contactless digitizers in computer-aided inspection has been presented (Fig. 7). The reference object part was designed with attention to the industrial sector of production tool manufacturing (moulds and dies).

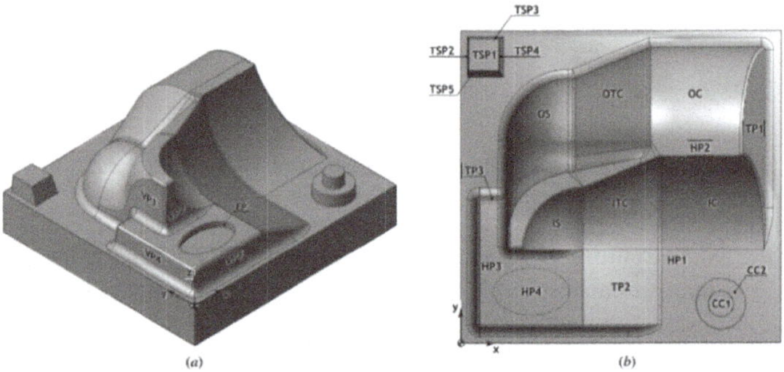

Fig. 7. Artefact proposed in [20]

Several classic features and a couple of sculptured free-form geometries (NURBS) were rationally organized so that the part can be representative of a wide range of products (Fig. 8).

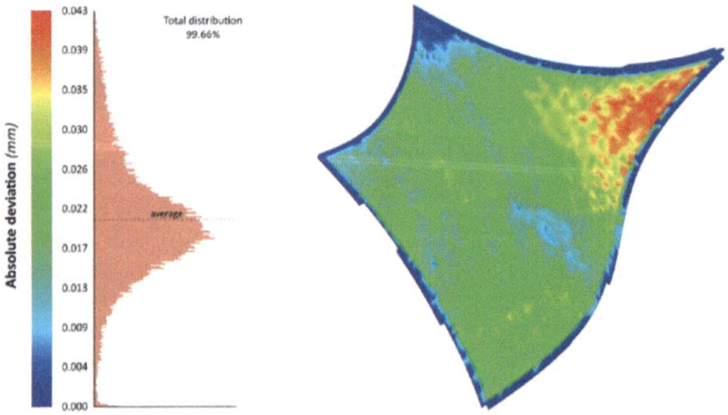

Fig. 8. Example of a coloured map of deviations evaluated on a NURBS surface [20].

4 Prismatic Artefacts

In literature, there are many examples of prismatic artefacts. In this section some of the most significant have been described.

4.1 Step Gauges

Step gauges are typically obtained by assembling typical gauge blocks made of steel or ceramic, since they have been developed for the performance verification of Coordinate Measuring Machines. Although, these kinds of materials are very difficult to acquire with an optical-based system due to their optical properties, or, as well, with a CT scanner due to their high density. Polymers represent a good alternative even if they present well known drawbacks, in terms of stability over time and machinability with sufficient accuracy and surface quality [21]. Step gauge artefacts were originally designed in [22] and subsequently adopted to characterize and correct systematic errors in a CT scanning system [23–25]. The step gauge geometry is well suited for detecting and correcting systematic deviations since it features unidirectional as well as bidirectional lengths. The former are suitable for scale correction and they can be used for assessing the accuracy of a measuring system. Bidirectional lengths take into account the "probing" effect and they can be used for detecting effects due to the interaction between the measuring instrument and the optical properties of the workpiece (Fig. 9).

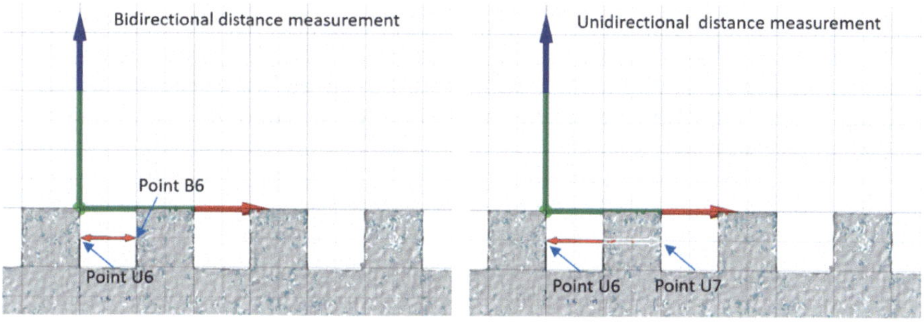

Fig. 9. Examples of unidirectional and bidirectional length definition on a scanned surface.

Step gauges have been recently implemented for several purposes.

A miniature step gauge characterized by high surface cooperativeness and suitable for the verification of optical based scanners was introduced for the first time by replication using a bisacryl material for dental applications (Luxabite) [26]. Afterwards, miniature step gauges manufactured using different materials, such as aluminium, steel and polymers, were successfully used for correcting systematic errors in CT scanning [25]. In connection with studies related to Computed Tomography, Polyphenylene sulfide (PPS) with 40% of glass was found to be a good material, featuring low form errors, similar to those obtained using aluminium and steel, good thermal stability, low density, and good surface cooperativeness [25].

In [27], a miniature step gauge made of black polyphenylene sulfide (PPS) was used for the performance verification of three different optical scanners: a structured light scanner (SLS), a laser line scanner (LLS), and a photogrammetry-based scanner (PSSRT), having comparable resolutions and working volumes. The same sample was used by the authors for the evaluation of the reduced depth of field due to the increase of the magnification level, typical of measuring systems suitable for small objects. Its geometry structured in steps resulted to be well suited for this kind of tests and allowed to conduct a quantitative analysis of the consequences of the reduced depth of field on the reconstruction quality and accuracy.

One of the biggest limits of the optical-based techniques is their dependency on optical surface characteristics, because of the possible different interaction with the measuring system [17]. This topic has become of fundamental importance, since the usage of polymeric materials has strongly increased in many manufacturing fields.

In [2, 28] the interaction among 3D optical scanners and objects realized with different materials and colours was investigated. The interaction is mainly related to the translucency of some materials, such as polymers, which is a category of materials ever more demanded in industry. Translucency causes a subsurface scattering effect that has been modelled for certain cases and affect the dimensional accuracy of measurements, as well as, the quality of reconstruction. In this work [2], the investigation involved 3D optical scanners exploiting different measuring principles: laser, structured light based on phase shifting and Gray code, and photogrammetry. This was done in order to study how the subsurface scattering effect can vary with the materials, colours, and, also, the measuring principle used. With this purpose, five miniature step gauges made of different polymers and different colours (Fig. 10) were used for evaluating the effects due to their different optical properties and, according to the results obtained, translucency of materials has a dual and significant effect: a measurement bias and an increase of the uncertainty due to the different quality of reconstruction retrieved. Thus, it has to be considered properly, for each case.

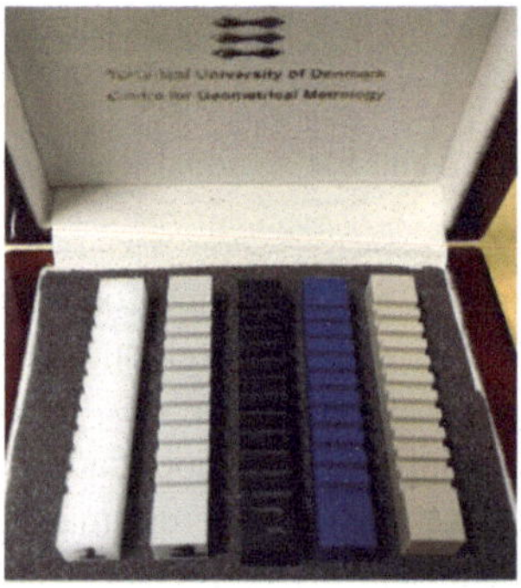

Fig. 10. Polymer step gauge used in [2].

4.2 Other Artefacts

Other artefacts have been designed for the performance verification of specific optical based systems (Fig. 11). In [29] performance verification test objects have been designed and measured with a laser line scanner mounted on a CMM head.

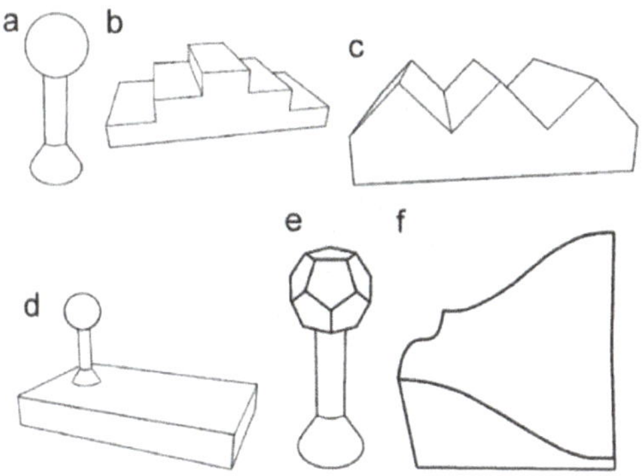

Fig. 11. Possible artefacts for verification of laser line scanners: (a) steps; (b) edges, (c) sphere-plane combination, (d) faceted sphere, (e) double-curved surface [29].

Other reference objects have been specifically designed for photogrammetric scanners, such as the one reported in [14], which is a low-cost mechanical artifact for the metrological verification of photogrammetric measurement systems. It is mainly composed of spheres and cubes manufactured in different sizes. A set of circular targets are fixed on these elements to aid the photogrammetric reconstruction.

In [30] four pyramidal artefacts with complex surface and sub millimeter features were used for the validation of a Photogrammetric Scanning System with Rotary Table (Fig. 12). The pyramidal geometry, and the presence of various features such as holes of different sizes, chamfers and grooves, make the samples particularly useful to test the suitability of the photogrammetric scanner or the reconstruction of complex objects with submillimeter features. Since, the proposed artefacts were made in aluminium characterized by high reflectivity, chemical etching was performed in order to make the object surface more cooperative.

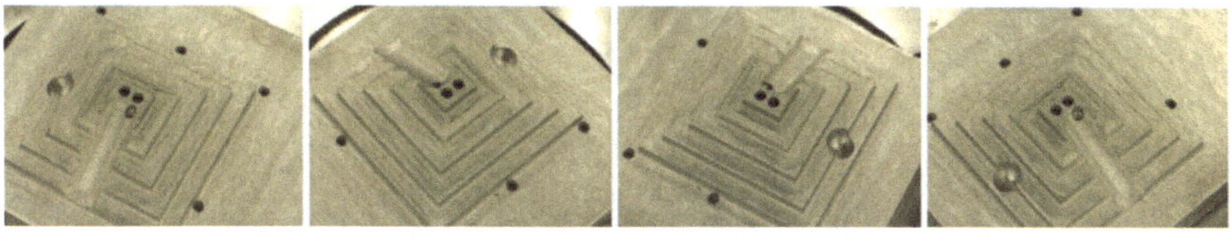

Fig. 12. Example of pyramidal artefacts used for validating a photogrammetric scanner with rotary table when measuring complex shapes with sub-millimeter features [30].

More recently, in [31] the implementation of a 3D reference object, in the form of a staircase artefact, for the performance verification of the same Photogrammetric Scanning System with Rotary Table has been evaluated (Fig. 13). In this work, the staircase geometry was used for a double aim: the estimation of the external orientation, *scale adjustment*, which is a peculiar characteristic of the photogrammetric scanner and for the uncertainty assessment of the same implemented scanner.

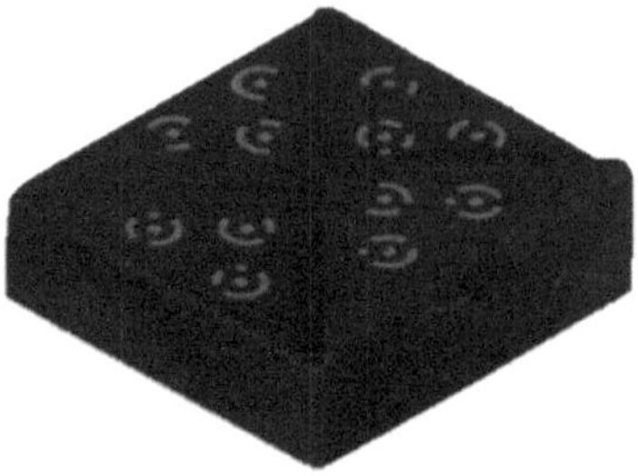

Fig. 13. Staircase artefact [31]

The effectiveness of this reference artefact was proved through the reconstruction a pyramidal artefact. The reference object used for the external orientation computation was designed considering characteristics which aid the photogrammetric reconstruction, such as the presence of clearly visible edges and a good surface texture, where with the term "texture" the photogrammetric definition is considered.

The reference object was then designed being inspired by the one developed in [32], a staircase-like artifact, with converging steps used to calibrate a 3D SEM instrument.

Another category of artefacts are the ones used for machine tool verification. An example is reported in [18], where, in order to assess the capability of a photogrammetric measurement system in a metrology assisted robotic machining application, two artefacts were chosen: the NPL freeform artefact, and the prismatic machine tool test artefact NAS 979, better known as *circle-diamond-square* (Fig. 14). In the next future, measuring instruments are called to be used directly in the production cells. This leads to a double purpose of the testing artefacts: the machine tool verification and the evaluation of capabilities of the measuring instruments called to assess the conformity of the manufactured components in the same production cell.

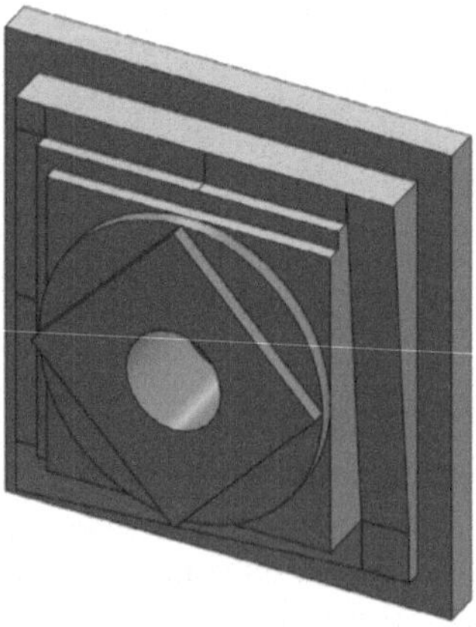

Fig. 14. Circle diamond square NAS 979 [18].

5 Conclusion

Some of the most interesting artefacts used for the performance verification of optical based scanners have been presented in this paper, considering different kind of geometries, freeform and prismatic and highlighting the main purposes and results.

This is a topic of great interest due to the potentiality of these 3D scanners with respect to the more traditional CMMs. In particular, the ones considered as testing instruments in this paper, are 3D scanners adopted for the dimensional verification of components in the close range, with measuring range below one meter and structural resolution on the order of cents of millimeters or below. The accuracy and uncertainty of those systems, considering the measuring tasks analysed in literature and reported in this work, is comprised within 0.020 mm.

Using optical scanners, the measuring task has to be accurately specified every time since the measurement result is strictly related to it, and it is really difficult to characterize the scanners with a single performance value. With the use of the reported artefacts, freeform and prismatic, optical based scanners have been tested considering different kinds of geometries and materials and they registered very good results with low uncertainty in most cases. Moreover, the use of proper geometries, such as the step gauge geometry featuring unidirectional and bidirectional lengths, allows to better understand the optical interactions with the materials in order to correct for them. Further works have to be conducted in order to improve the performance of 3D optical scanners and the knowledge about their main sources of error.

References

1. Galantucci, L.M., Guerra, M.G., Lavecchia, F., Dassisti, M.: Additive manufacturing: new trends in the 4th industrial revolution. In: Lecture Notes in Mechanical Engineering, pp. 153–159 (2019). https://doi.org/10.1007/978-3-030-18180-2_12
2. Guerra, M.G., Gregersen, S.S., Frisvad, J.R., De Chiffre, L., Lavecchia, F., Galantucci, L.M.: Measurement of polymers with 3D optical scanners: evaluation of the subsurface scattering effect through five miniature step gauges. Meas. Sci. Technol. **31**, 015010 (2020). https://doi.org/10.1088/1361-6501/ab3edb
3. Beraldin, J.A., Mackinnon, D., Cournoyer, L.: Metrological characterization of 3D imaging systems: progress report on standards developments. In: International Congress of Metrology, vol. 3, pp. 1–21 (2015). https://doi.org/10.1051/metrology/20150013003
4. ISO 10360-7:2011: Geometrical product specifications (GPS) — Acceptance and reverification tests for coordinate measuring machines (CMM) — Part 7: CMMs equipped with imaging probing systems (2011)
5. ISO 10360-8:2013: Geometrical product specifications (GPS) – Acceptance and reverification tests for coordinate measuring systems (CMS) – Part 8: CMMs with optical distance sensors (2013)
6. VDI-Standard: VDI/VDE 2634 Blatt 2: Optical 3-D measuring systems - Optical systems based on area scanning (2012)
7. VDI-Standard: VDI/VDE 2634 Blatt 3: Optical 3D-measuring systems - Multiple view systems based on area scanning (2008)

8. ISO/IEC Guide 98-3: Uncertainty of measurement - Part 3: Guide to the expression of uncertainty in measurement (GUM:1995) (2008)
9. ISO 14253-2:2011: Geometrical product specifications (GPS) – Inspection by measurement of workpieces and measuring equipment – Part 2: Guidance for the estimation of uncertainty in GPS measurement, in calibration of measuring equipment and in product verification (2011)
10. ISO 15530-3:2011 Geometrical product specifications (GPS) – Coordinate measuring machines (CMM): Technique for determining the uncertainty of measurement – Part 3: Use of calibrated workpieces or measurement standards (2011)
11. Müller, P., Hiller, J., Dai, Y., Andreasen, J.L., Hansen, H.N., De Chiffre, L.: Estimation of measurement uncertainties in X-ray computed tomography metrology using the substitution method. CIRP J. Manuf. Sci. Technol. **7**, 222–232 (2014). https://doi.org/10.1016/j.cirpj.2014.04.002
12. Sims-Waterhouse, D., Piano, S., Leach, R.: Verification of micro-scale photogrammetry for smooth three-dimensional object measurement. Meas. Sci. Technol. **28**, 055010 (2017). https://doi.org/10.1088/1361-6501/aa6364
13. De Chiffre, L., Hansen, H.N., Andreasen, J.L., Savio, E., Carmignato, S.: Geometrical Metrology and Machine Testing. Textbook. Polyteknisk Forlag (2015)
14. González-Jorge, H.: Verification artifact for photogrammetric measurement systems. Opt. Eng. **50**, 073603 (2011). https://doi.org/10.1117/1.3598868
15. Robson, S., Beraldin, J.-A., Brownhill, A., MacDonald, L.: Artefacts for optical surface measurement. In: Videometrics, Range Imaging, and Applications XI, vol. 8085, p. 80850C (2011). https://doi.org/10.1117/12.882702
16. Acko, B., McCarthy, M., Haertig, F., Buchmeister, B.: Standards for testing freeform measurement capability of optical and tactile coordinate measuring machines. Meas. Sci. Technol. **23** (2012). https://doi.org/10.1088/0957-0233/23/9/094013
17. Dury, M.R., Woodward, S.D., Brown, S.B., McCarthy, M.B.: Surface finish and 3D optical scanner measurement performance for precision engineering. In: Proceedings of ASPE 2015 Annual Meeting, pp. 419–423 (2015)
18. Barnfather, J.D., Goodfellow, M.J., Abram, T.: Photogrammetric measurement process capability for metrology assisted robotic machining. Meas. J. Int. Meas. Confed. **78**, 29–41 (2016). https://doi.org/10.1016/j.measurement.2015.09.045
19. Guerra, M.G., Lavecchia, F., Maggipinto, G., Galantucci, L.M., Longo, G.A.: Measuring techniques suitable for verification and repairing of industrial components: a comparison among optical systems. CIRP J. Manuf. Sci. Technol. **27**, 114–123 (2019). https://doi.org/10.1016/j.cirpj.2019.09.003
20. Iuliano, L., Minetola, P., Salmi, A.: Proposal of an innovative benchmark for comparison of the performance of contactless digitizers. Meas. Sci. Technol. **21** (2010). https://doi.org/10.1088/0957-0233/21/10/105102
21. Cantatore, A., Angel, J., De Chiffre, L.: Material investigation for manufacturing of reference step gauges for CT scanning verification. In: 12th International Conference of the European Society for Precision Engineering and Nanotechnology, EUSPEN, pp. 129–132 (2012)
22. Cantatore, A., De Chiffre, L., Carmignato, S.: Investigation on a replica step gauge for optical 3D scanning of micro parts. In: 10th International Conference of the European Society for Precision Engineering and Nanotechnology (2010)
23. Stolfi, A., De Chiffre, L.: CT crown for on-machine scale calibration in Computed Tomography. In: Proceedings of the 16th International Conference of the European Society for Precision Engineering and Nanotechnology, EUSPEN (2016)

24. Stolfi, A., De Chiffre, L.: 3D artefact for concurrent scale calibration in Computed Tomography. CIRP Ann. Manuf. Technol. **65**, 499–502 (2016). https://doi.org/10.1016/j.cirp.2016.04.069
25. Angel, J., De Chiffre, L., Kruth, J.P., Tan, Y., Dewulf, W.: Performance evaluation of CT measurements made on step gauges using statistical methodologies. CIRP J. Manuf. Sci. Technol. **11**, 68–72 (2015). https://doi.org/10.1016/j.cirpj.2015.08.002
26. De Chiffre, L., Carmignato, S., Cantatore, A., Jensen, J.D.: Replica calibration artefacts for optical 3D scanning of micro parts. In: Proceedings of the 9th International Conference of the European Society for Precision Engineering and Nanotechnology, EUSPEN, pp. 352–355 (2009)
27. Guerra, M.G., De Chiffre, L., Lavecchia, F., Galantucci, L.M.: Use of miniature step gauges to assess the performance of 3D optical scanners and to evaluate the accuracy of a novel additive manufacture process. Sensors **20** (2020). https://doi.org/10.3390/s20030738
28. Wilm, J., Madruga, D.G., Jensen, J.N., Gregersen, S.S., Brix, M.E., Guerra, M.G., Aanæs, H., De Chiffre, L.: Effects of subsurface scattering on the accuracy of optical 3D measurements using miniature polymer step gauges. In: Proceedings euspen's 18th International Conference & Exhibition, Venice, IT, pp. 449–450 (2018)
29. Van Gestel, N., Cuypers, S., Bleys, P., Kruth, J.P.: A performance evaluation test for laser line scanners on CMMs. Opt. Lasers Eng. **47**, 336–342 (2009). https://doi.org/10.1016/j.optlaseng.2008.06.001
30. Galantucci, L.M., Pesce, M., Lavecchia, F.: A powerful scanning methodology for 3D measurements of small parts with complex surfaces and sub millimeter-sized features, based on close range photogrammetry. Precis. Eng. (2015). https://doi.org/10.1016/j.precisioneng.2015.07.010
31. Lavecchia, F., Guerra, M.G., Galantucci, L.M.: Performance verification of a photogrammetric scanning system for micro-parts using a three-dimensional artefact: adjustment and calibration. Int. J. Adv. Manuf. Technol. **96**, 4267–4279 (2018)
32. De Chiffre, L., Carli, L., Eriksen, R.S.: Multiple height calibration artefact for 3D microscopy. CIRP Ann. Manuf. Technol. **60**, 535–538 (2011). https://doi.org/10.1016/j.cirp.2011.03.054

An Approach of Development Smart Manufacturing Metrology Model as Support Industry 4.0

Slavenko M. Stojadinović[1(✉)], Vidosav D. Majstorović[1], Dragan Djurdjanović[2], and Srdjan Živković[3]

[1] Faculty of Mechanical Engineering, Production Engineering Department, University of Belgrade, Kraljice Marije 16, 11120 Belgrade, Serbia
sstojadinovic@mas.bg.ac.rs

[2] Department of Mechanical Engineering, University of Texas, Austin, TX, USA

[3] Coordinate Metrology Lab, Military Technical Institute, Belgrade, Serbia

Abstract. The framework for smart manufacturing metrology model (S3M), are based on integration of digital product metrology information through metrological identification, application artificial intelligence techniques and generation of global/local inspection plan for coordinate measuring machine (CMM). S3M has an extremely expressed requirement for better control, monitoring and data mining. Limitations still exist in data storages, networks and computers, as well as in the tools for complex data analysis, detection of its structure and retrieval of useful information. This paper will present recent results of our research on building of S3M as support Industry 4.0. Presented approach to S3M development includes four levels: (i) mathematical model of the measuring sensor path, which establishes a connection between the coordinate systems; (ii) generating the needed set of information to integrate the given tolerances and geometry of the parts by applying an ontological knowledge base; (iii) the application of AI techniques such as ACO and GA to optimize the measurement path, numbers of measuring part setup and configuration of the measuring probes; (iv) simulation of measurement path for a collision check. After simulation of the measurement path and visual checks of collisions, the path sequences are generated in the control data list for appropriate CMM. The experiment was successfully carried out on the examples of prismatic part and two turbine blades or its free-form measuring surfaces.

Keywords: Smart metrology · Industry 4.0 · Inspection planning · Free-form surfaces · Prismatic parts · CMM

1 Introduction

In today's rapidly changing world, globalization, products customized to customer requirements and automation play a decisive role in the development of the industry, especially mechanical engineering. Introducing advanced technologies and techniques that will change products, processes and supply chains this industry is at the top of Industry 4.0. This industry also enables even greater connectivity through IC

L. Wang et al. (Eds.): AMP 2020, LNME, pp. 190–204, 2020.
https://doi.org/10.1007/978-3-030-46212-3_13

technologies, enabling producers to maintain their competitive benefit and respond flexibly and quickly to customer requirements [1].

Industry 4.0 in manufacturing plays a key role in three areas [2, 3]: (a) smart supply chains - greater coordination and flow of information in real time, enabling better tracking of goods and raw materials in an integrated business planning model and production. This provides new models for coordination and collaboration between supply chains; (b) smart manufacturing - the use of data analytics and new manufacturing techniques and technologies (such as autonomous robots, multi-purpose production lines and augmented reality) helps to improve quality and accelerate production. This enables new business models such as mass customization, and (c) smart products - rapid innovation and faster delivery times to the market are made possible by data acquisition about the product, along with user feedback, collected through social networks on the Internet. This data also enables remote diagnostics and predictive maintenance.

Industry 4.0 is an intensive digital transformation of producing and other industries in a connected environment of data, people, processes, services, systems and the Internet of Things (IoT) - industrial resources assets with the generation, use and reuse of information that can be applied as a way and means to accomplish smart industries and ecosystems, based on industrial innovation and collaboration.

Industry 4.0, as a German strategic initiative, aims to create intelligent (smart) factories where manufacturing technologies are upgraded and transformed into SFS, IoT and cloud computing [4–6]. Such factories, in addition to the key roles mentioned above consist of components such as smart production, smart metrology and smart machine tools.

Generating of detail model from the quality measurements process in manufacturing requires the development of dedicated framework, like to the example CP3M. Based on testing model of CP3M, we have encountered the generation of a number of huge data sets that required processing, finding of correlations between data and extraction of useful information to be shared between CP3M modules [7].

The contribution of this paper refers to development smart manufacturing metrology model as support I4.0 concept based on the existing CP3M [7] and the application of AI techniques such as EO, ACO, and GA to optimize the measurement path, the number of measuring part setups, as well the configuration of the measuring probes.

Testing of the developed model in practice was performed on a prismatic parts manufactured on machine tool HMC 500 and two turbine blades manufactured using AM technology on the SLS machine. The geometry of the blades was first checked on a CMM NIKON Altera 10.7.6 using non-contact method and in the second part on a CMM DEA Epsilon 2304.

2 Outline of the Concept

In the presented 5C CPS architecture [8] in-process quality control represents key asset. Namely, modern understanding of measurements of quality in manufacturing is that their purpose is to enable adequate monitoring and tracking relevant process parameters, based on which any deviations away from their nominal behavior can be corrected via manual intervention, or automatically. Development of CPS offers new opportunities to accomplish this function via detailed models (ideally, "twin models") of the underlying process and product. Timely generation of these models requires extraction of useful information from bulk data using big data analytics and different data mining methods. The increasing role of big data leads to continuous growth of research in these fields including [9, 10]: (a) machine learning methods such as different kinds of regression (support vector machines, neural networks, ANOVA) to gain insight into trends within the data, (b) pattern recognition methods including supervised classification and clustering, which structure big data sets, and (c) expert knowledge, for example using a lookup table.

Our model S3M primarily refers to the cloud manufacturing (CM), the part that refers to the cloud of manufacturing metrology [7]. It is an approach to develop (smart) manufacturing metrology as an integral part of CM (smart factory) with the using of CMM, as support of Industry 4.0. The structure of our research model based on the existing CP3M is presented on Fig. 1, and includes two basic components: physical and cyber. Physical level consists from machined parts (prismatic part and turbine blades) and real CMM.

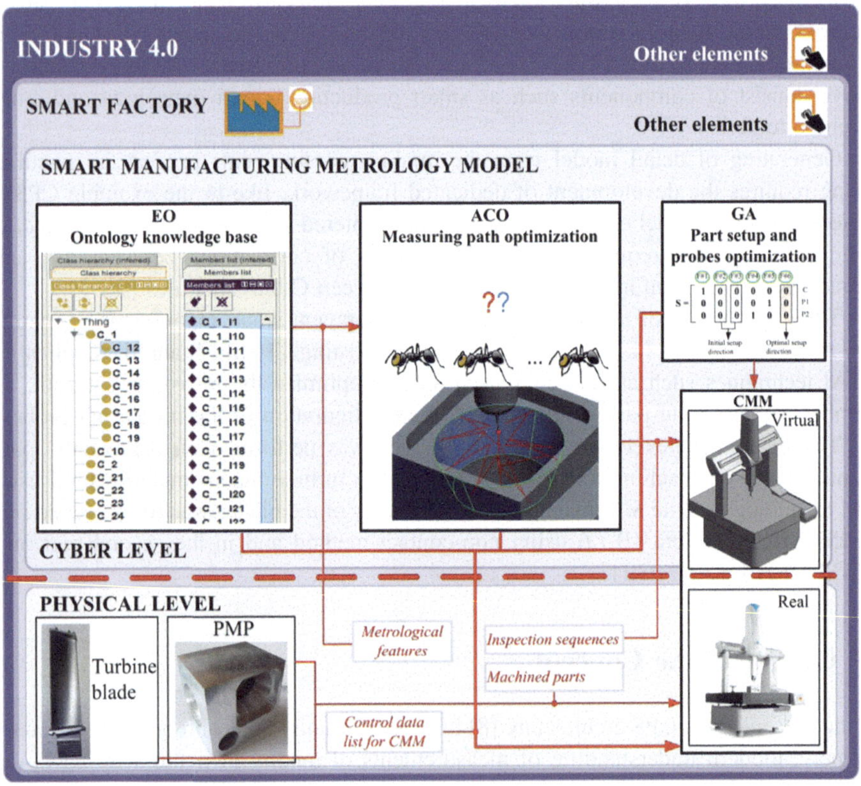

Fig. 1. S3M framework

The S3M covers all elements of planning, preparation and realization of parts inspection on CMM, using the developed elements and structure links from the Fig. 1:

- EO module for definition, recognition of geometrical features from CAD model of the measurement part and integration by ontology knowledge base, where we used it for definition GD&T of metrological features;
- ACO module for building of measuring path optimization as first part of intelligent inspection process planning, that contains methods for prismatic part and freeform surfaces application;
- GA module for optimal part setup and probes optimization, as second part of intelligent inspection process planning.
- Link between EO, ACO, GA and real CMM is control data list for CMM that is transferred to CMM using cloud technology;
- Link between ACO and virtual CMM is inspection sequences for optimal part setup and optimal configuration of probes.
- Link between EO and ACO is metrological features which involve in certain standard type of tolerance.

Overall output S3M is module for analysis of results and generation of the final measuring reports on real CMM stored on cloud. Cloud services within the organization provide the necessary information for integration of knowledge and data from various phases in product design and manufacturing/metrology into inspection planning, and make available information about inspection results to all interested parties in product lifecycle.

3 A Proposed Approach of Development S3M

As mentioned, the presented approach to S3M development includes four levels: (i) mathematical model of the measuring sensor path, which establishes a connection between the coordinate systems; (ii) generating the needed set of information to integrate the given tolerances and geometry of the parts by applying an ontological knowledge base; (iii) the application of AI techniques such as ACO and GA to optimize the measurement path, numbers of measuring part setup and configuration of the measuring probes; (iv) simulation of measurement path for a collision check.

3.1 Mathematical Model

The basic element from which the development start is the mathematical model presented in [11, 12]. Its primary role is to establish links between coordinate systems and generate an initial (point-to-point) measurement path that will be later optimized in purpose to shorten length of path and traveling time of the measurement probe. The basic equation of the model is:

$${}^{M}\mathbf{r}_{P_i} = {}^{M}\mathbf{r}_{W} + {}^{W}\mathbf{r}_{F} + {}^{F}\mathbf{r}_{P_i} = {}^{M}\mathbf{r}_{F} + {}^{F}\mathbf{r}_{P_i}$$

According to [11, 12] generating point-to-point measurement path defines distribution of two sets of points: (i) set of measuring points, and (ii) set of nodes points. Distribution of measuring points for different geometric features such as plane, circle, hemisphere, cylinder, etc. is obtained by modifying Hamersley [13] sequences. An example of formulas for calculation of measuring points coordinates $P_i(s_i, t_i, w_i)$ in Cartesian coordinate system for a plane according [11, 12] is given as follows:

$$s_i = \frac{i}{N} \cdot a$$

$$t_i = \left(\sum_{j=0}^{k-1} \left(\left[\frac{i}{2^j} \right] \mathrm{Mod}2 \right) \cdot 2^{-(j+1)} \right) \cdot b$$

$$w_i = 0$$

where is: a[mm] - x-axis constraint value; b[mm] - y-axis constraint value.

According to [11, 12] set of node points implies two sets $P_{i1}(s_{i1}, t_{i1}, w_{i1})$ and $P_{i2}(s_{i2}, t_{i2}, w_{i2})$, where is $i = 0, 1, 2, \ldots, (N-1)$ and N – number of measuring points. Sub-set $P_{i1}(x_{i1}, y_{i1}, z_{i1})$ presents points for the transition from fast to slow feed. The distance between points $P_{i1}(x_{i1}, y_{i1}, z_{i1})$ and $P_i(x_i, y_i, z_i)$ is presented (Fig. 2) by d_1 - slow feed probe path, and the distance between points $P_{i2}(x_{i2}, y_{i2}, z_{i2})$ and $P_{i1}(x_{i1}, y_{i1}, z_{i1})$ is d_2 - rapid feed probe path.

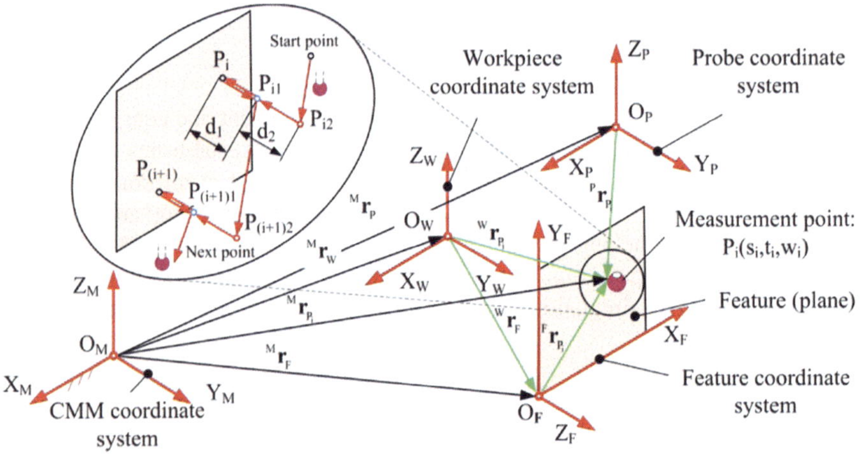

Fig. 2. Mathematical model of S3M (adapted from [11, 12])

3.2 Implementing the Geometrical Features in the Software Protégé

Software Protégé is a free, open sours ontology editor and knowledge – base framework, based on Java. Protégé implements a set of knowledge – modeling structures and actions that support the creation, visualization, and manipulation of ontologies in various representations data formats [14]. In this paper is used Protégé – OWL editor that supports the web ontology language, as most recent development in standard ontology language in purpose to development engineering ontology for domain coordinate metrology. OWL ontology includes description of classes, properties and individuals.

According to [15–17] the implementation of metrological features in Protégé includes: (i) modeling classes, (ii) modeling of class hierarchy, (iii) modeling of the individuals, (iv) modeling the classes and individuals properties (object and data properties). In Fig. 3 is shown part of the OntoGraf, which the ontological describes measuring part in relation on coordinate system of the part.

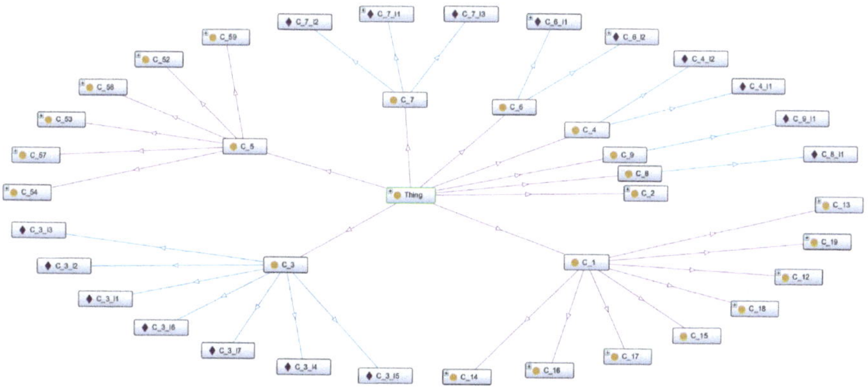

Fig. 3. A part of the OntoGraf - ontological description of the geometrical primitives in relation on coordinate system of the measuring part [16].

3.3 Optimization Models

As mentioned, S3M incorporates a model for optimizing the measurement path based on ant colony and models for optimizing the number of measuring part setup and probe configurations based on GA.

3.3.1 ACO for Measuring Path

Application of ants colony optimization (ACO) in a coordinate metrology is based on the solution of travelling salesmen problem (TSP), where the set of cities that the salesman should pass through with the shortest possible path corresponds to the set of points of a minimal measuring path length [18]. According to [19] ACO is based on Eq. (1) for calculation of the measuring probe path during the measurement on N measuring points:

$$\min\{D_{tot}\} = K + \left\{ \sum_{i=0}^{N-1} \begin{array}{l} (\min\left\{\left|\overrightarrow{P_{i1}P_{(i+1)2}}\right|\right\} \vee \min\left\{\left|\overrightarrow{P_{i1}P_{(i+1)1}}\right|\right\} \vee \\ \vee \min\left\{\left|\overrightarrow{P_{i(i+1)}P_{(i+1)2}}\right|\right\} \vee \min\left\{\left|\overrightarrow{P_{i2}P_{(i+1)1}}\right|\right\}) \end{array} \right\} \quad (1)$$

where $K = N \cdot (2 \cdot d_1 + d_2)$ presents constant part of path with constant lengths, $d_1 = \left|\overrightarrow{P_{i1}P_i}\right|$ and $d_2 = \left|\overrightarrow{P_{i2}P_{i1}}\right|$ as presented in mathematical model chapter. In order to solve Eq. (1) and thus obtain the optimal path, it is necessary to solve TSP.

According to [20, 21] TSP can be represented by a complete weighted graph $G = (N, A)$ (Fig. 4) with N being the set of nodes representing the cities, and A being the set of arcs. Each arc $(i, j) \in A$ is assigned a value (length) d_{ij}, which is the distance between cities i and j, with $i, j \in N$. The goal in TSP is to find a minimum length Hamiltonian circuit of the graph where a Hamiltonian circuit is a closed path visiting each of the $n = |N|$ nodes of G exactly once [20], so that an optimal solution to the TSP is a permutation π of the node indices $\{1, 2, \ldots, n\}$.

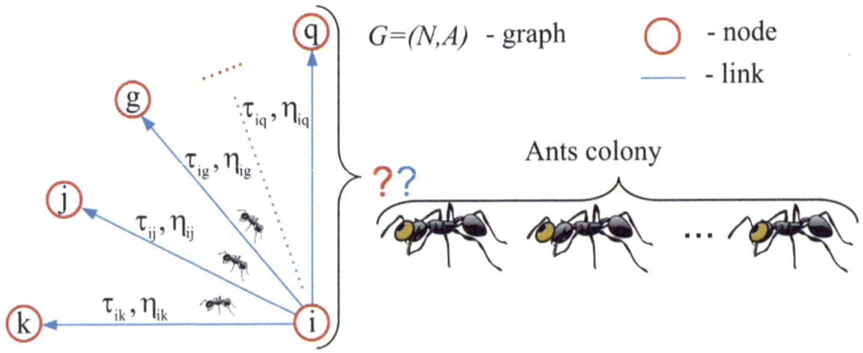

Fig. 4. Principe of optimization path model [18]

After the conducted circulations and permutations, the decision on which path is the shortest is obtained based on the criteria of the maximum number evaporated of ants colony pheromones, i.e. the values of the weight coefficients τ_{ij} and η_{ij}. Coefficient η_{ij} represents the heuristic information or the influence of the distance between two nodes of a graph, while τ_{ij} probability that an ant leaving to node i visit as next node j.

3.3.2 GA for PMP Setup and Configuring of Probes

As it is known, inspection on three-axis CMM can be performed from three orthogonal directions corresponding to the axes X, Y and Z. From these directions, can be derived six directions corresponding to the axes of the machine +X, −X, +Y, −Y, +Z and −Z. From the standpoint of access to the feature it is introduced the term feature approach direction (FAD), while from the standpoint probes it is introduced probe approach direction (PAD). The FADs are shown in Fig. 5b). They define possible directions of access to the features and are used to analyze the setup of the measuring part. PADs are shown in Fig. 5a). They define the possible directions of access the probe and are oriented opposite to the FAD. Due to the setup of the measuring part at the working table of CMM, one of the PADs is lost, so that of PADs can be a maximum of 5.

In order to apply GA, it is necessary to define the Boolean matrices of setup S and configuration C. The elements of the matrix S links to the FADs and can be 0 or 1. For example, according to Fig. 5c) element of the matrix S (C,F#1) takes value 1 because to the cylinder C can be accessed from FAD#1. Similarly, the element of the matrix S (F, F#4) takes the value 0 because to the cylinder C cannot be accessed from the FAD # 4.

Analogous to the filling of elements of the matrix S, the configuration matrix C is also filled to use PADs. It should be noted that the number of rows of both matrices is equal to the number of features which creating tolerances. In this case, it is two types of tolerance and three features.

Optimal solutions for the case of measuring parts setup are obtained by GA model [12] and are represented by zero-columns. Optimal solutions for the case of probe configuring (measuring heads) are also obtained by mentioned GA model, but represented by the unit-column.

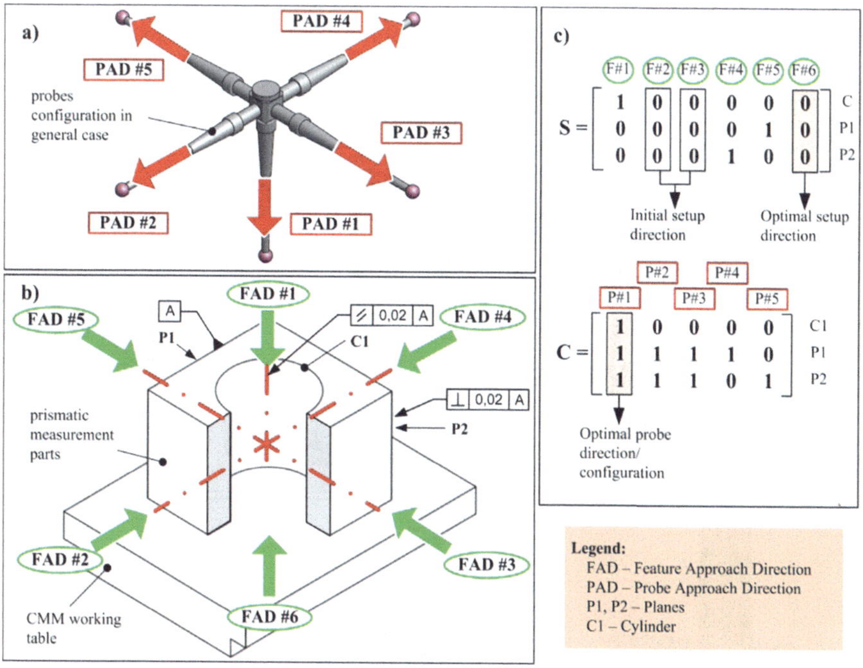

Fig. 5. PMP setup and configuring of probes model of S3M

3.4 Simulation Model

The simulation of the measurement path is based on [22] and is carried out in order to configure the off-line CMM environment for programming and program verification prior to the measurement process, as well as to avoid collision of the measuring sensor with fixture and part.

The PTC Creo software and the CMM sub-module within it used to generate the measurement path. According to [22] for generating a measurement path, are need to be realized activities: (i) loading of reference CAD part for inspection; (ii) modeling of components and assembly of CMM; (iii) defining of measurement operation and selecting of task' coordinate frames; (iv) configuring of measuring probes; (v) selection of metrological (inspection) features according to specified tolerance; (vi) setting of measuring parameters; (vii) import of measurement points, given as a result of the output of the MatLab visualization code for the metrological features; (viii) generation of measuring path; (ix) simulation of measuring path on configured Virtual CMM including probe.

The basic purpose of configured virtual CMM is to collision check the programmed measurement path in a CAD/CAM environment and generates an output file suitable for further processing or post-processing [22].

A detail virtual CMM model in PTC Creo software [22], with basic components, assembly and kinematic links between moving and non-moving components is shown in Fig. 6. Moving kinematic links (sliders) are used for all translational movements (X, Y, Z), Fig. 6a). They allow the movement of CMM components within the permissible limits for each axis and realize the programmed path of the measuring probe as end effector.

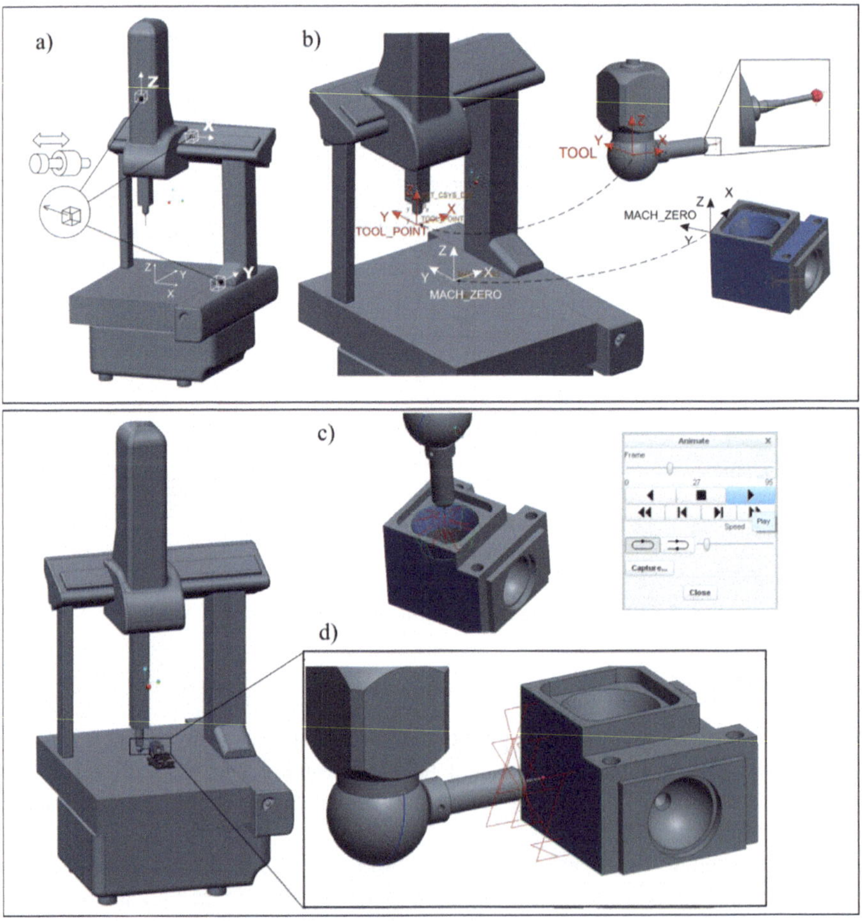

Fig. 6. A CAD model of CMM and simulation of measurement path: a) CMM assembly with kinematic types of link; b) coordinate frames, c) simulation of measurement path for a plane surface, and d) simulation of measurement path for a truncated cone (adapted from [22]).

According to [22] and marks on Fig. 6b) it is necessary to define coordinate frames of:

- CMM working table (MACH_ZERO),
- workpiece (MACH_ZERO),
- sensor holder (*TOOL_POINT*), whose axes have the same direction and orientation as the axes of the coordinate frame of the workpiece, and
- measurement probe (TOOL), if it is necessary modeling a non-standard (does not exist in the software probes base) measuring probe

Matching the MACH_ZERO coordinate frame of workpiece with the MACH_ZERO working table coordinate frame, enables the measurement part setup on the working table of configured virtual CMM during simulation. The same procedure is for matching of coordinate frames TOOL and TOOL_POINT for setting of probe in probe holder.

Summarized, of this simulation verified the measurement path based on a modified Hamersley algorithm for distributing measurement points for basic geometric features. The simulation of measurement path is implemented on a configured virtual CMM in CAD/CAM environment. After the simulation, the measurement path was verified and saved in a CL file (DMIS program) for further distribution in the purpose of executing the measurement program on real CMM.

4 Experiment and Results

Testing of the developed model in practice was performed on two turbine blades manufactured using AM (Additive Manufacturing) technology: on the SLS (Selective Laser Sintering) machine of a well-known German manufacturer. Turbine blades are mainly manufactured using forging, casting, and recently using some AM technologies. In all these cases, fir-tree root, the part that serves for precise positioning on the rotor, is processed by cutting (milling, grinding). Suction side and pressure side are not machined, they remain the same as they were obtained by casting, forging or AM.

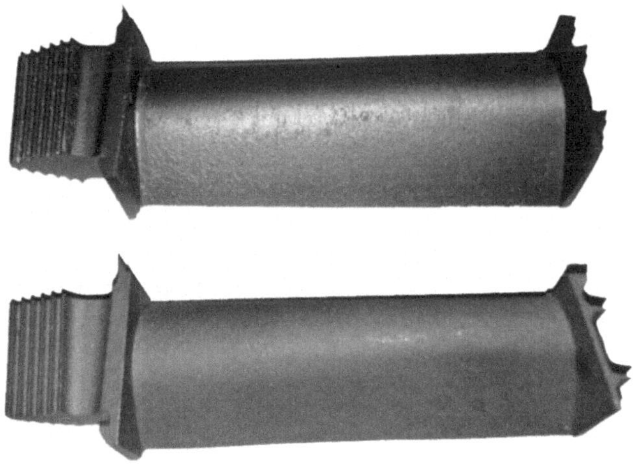

Fig. 7. Photograph of the turbine blades

Both blades shown in Fig. 7 they are identical to the nominal geometry. These blades have different surface quality, as well as the deviations of fir-tree root. Deviations are the result of changing the parameters of the SLS process in order to obtain the best possible parameter values that lead to the specified geometric requirements.

The geometry of the blades was first checked on a Coordinate Measuring Machine using non-contact method (CMM: NIKON Altera 10.7.6, motorized probe head Renishaw PH10M, laser scanner LC15Dx, software CAMIO8) and in the second part using contact method (CMM: DEA Epsilon 2304, motorized probe head Renishaw PH10M, touch trigger probe TP20, software Wilcox PC-DMIS 2019). Measurement of the blade by contact and non-contact methods was used for comparison and analysis of two different methods. Each method has its own advantages and disadvantages depending on the set metrological task.

Figure 8 (left) shown blade design; nominal geometry at CAD/CAM system (Siemens NX12) with all datum feature. Figure 8 (left) shows the geometric accuracy requirements defined by PMI (Product Manufacturing Information) symbols. In Fig. 8 (on the right), the blade geometry is blanked to better notice the control sections, the position of the coordinated system, and the reference planes.

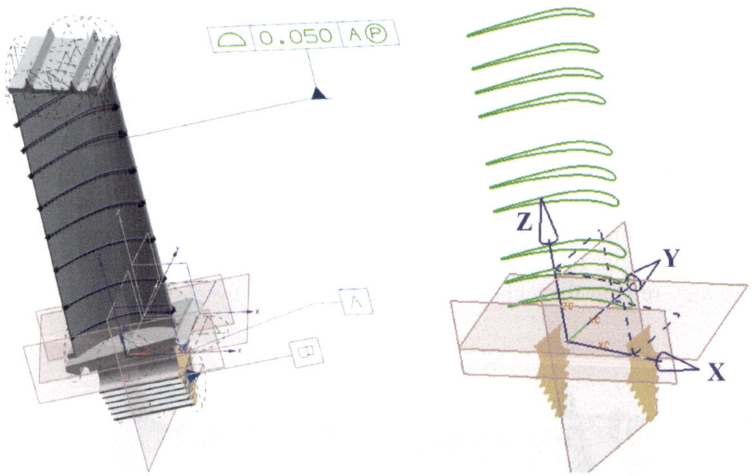

Fig. 8. Turbine blade nominal geometry and control sections

The blade was first measured by laser scanning, and the coordinate system was set as defined in the CAD model. The results of optical/laser scanning are shown in Fig. 9. The turquoise color is represented by a measured geometry, while the gray color represents the nominal geometry.

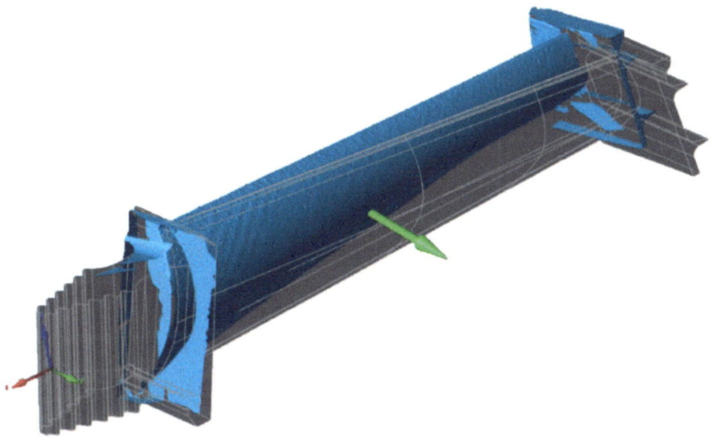

Fig. 9. Laser scanned geometry of the turbine blade

Figure 9 clearly shows that the geometry of the working area of the blade (pressure and suction sides) was moved away from the nominal geometry. Green arrow in Fig. 9 it shows the direction in which the measured geometry should be translated to match the nominal geometry. The main reason for these deviations is the fir-tree root, which must be processed by the cutting so that the position and orientation of the blade are consistent with the project requirements. It should be understood for turbine blades, the most important parameter is the gap between two adjacent blades, i.e. the pressure side of one blade and the suction side of the neighboring blade (Fig. 10).

Fig. 10. Turbine blade positioned in the auxiliary tool

Figure 11 shows the final measurements of the control sections on the CMM with contact probes; a left-isometric view with a selected 3 control sections total of 10 control sections (3 of 10 for image clarity). Figure 11 on the right shows the results of measuring the reference cross section Z = 10 mm.

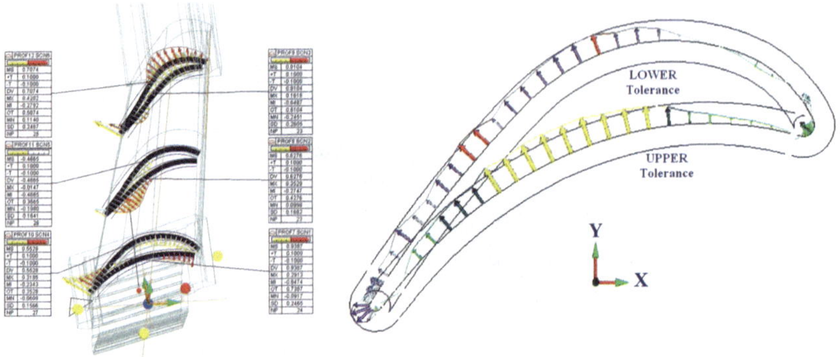

Fig. 11. Turbine blade measured result; control section Z = 10 mm

The layout of the nominal cross section is shown realistically, the measured contour, as well as the upper and lower tolerance field are displayed enlarged to more clearly the direction of deviation. The image shows that the entire measured cross-section geometry is located in a defined toleration field. Finishing the upper part of the turbine blade is performed so that the corrected deviations are shown in Fig. 9 which existed before the cutting of the fir-tree root.

The described procedure clearly indicates the complexity of manufacturing and coordinate Metrology of turbine blades. Setting the coordinate system on aerodynamic surfaces is crucial for successful completion of the work, as described in the monograph [23]. Optimization of the number of control sections, as well as the number of control points of the airfoil turbine blades for each of the sections is explained in [24]. For turbine blades of large dimensions and longer chords, the control sections must be located in cylindrical sections [25].

5 Conclusion

The output from the simulation on virtual CMM is a CL file (DMIS program). Generating this file and developing the appropriate postprocessor leaves the possibility to create a control data list for programming NUMMs different producers. As it known, CMMs are programmed in the language of their producer, therefore the proposed concept of this simulation and its output (file) could be useful in terms of the unification format of CMM programming languages and software.

Testing of the developed model in practice was performed on two turbine blades manufactured using AM technology on the SLS machine. The geometry of the blades was first checked on a CMM NIKON Altera 10.7.6 using non-contact method and in the second part on a CMM DEA Epsilon 2304. Measurement of the blade by contact and non-contact methods was used for comparison and analysis of two different methods. Each method has its own advantages and disadvantages depending on the set metrological task.

Acknowledgement. The presented research was supported by projects (TR 35022, TR 35007) of the Ministry of Education, Science and Technological Development of the Republic of Serbia.

References

1. The 4th Industrial Revolution: A Primer for Manufacturers, Oracle (2018). https://cloud.oracle.com/opc/saas/indmftg/reports/the-fourth-industrial-revolution-report.pdf. Accessed Nov 2019
2. Täuscher, K.: Business Models in the Digital Economy: An Empirical Study of Digital Marketplaces, Fraunhofer MOEZ, Fraunhofer Center for International Management and Knowledge Economy, Städtisches Kaufhaus Leipzig, Neumarkt, 9–19, 04109 Leipzig (2018). https://www.imw.fraunhofer.de/content/dam/moez/de/documents/Working_Paper/Working_Paper_Digital_Marketplaces_final.pdf. Accessed Nov 2019
3. Zúñiga, R., et al.: The internet of things, factory of things and industry 4.0 in manufacturing: current and future implementations. In: Gao, J., et al. (ed.) Advances in Manufacturing Technology XXXI: Proceedings of the 15th International Conference on Manufacturing Research, Incorporating the 32nd National Conference on Manufacturing Research, 5–7 September 2017, University of Greenwich, UK, pp. 221–226. IOS Press Advances in Transdisciplinary Engineering. https://doi.org/10.3233/978-1-61499-792-4-221
4. Zhong, R., et al.: Intelligent manufacturing in the context of industry 4.0: a review. Engineering **3**, 616–630 (2017). https://doi.org/10.1016/J.ENG.2017.05.015
5. Lee, J., et al.: A cyber-physical systems architecture for industry 4.0-based manufacturing systems. Manuf. Lett. **3**, 18–23 (2015). https://doi.org/10.1016/j.mfglet.2014.12.001
6. Zezulka, F., et al.: Industry 4.0 – an Introduction in the phenomenon. IFAC-Pap. Line **49**(25), 008–012 (2016). https://doi.org/10.1016/j.ifacol.2016.12.002
7. Majstorović, V., Stojadinović, S., Živković, S., Djurdjanović, D., Jakovljević,, Ž., Gligorijević, N.: Cyber-physical manufacturing metrology model (CPM3) for sculptured surfaces – turbine blade application. In: The 50th CIRP Conference on Manufacturing Systems (2017). https://doi.org/10.1016/j.procir.2017.03.093. Procedia CIRP, 63:658–663
8. Lee, J., Bagheri, B., Kao, H.A.: A cyber-physical systems architecture for industry 4.0-based manufacturing systems. Manuf. Lett. **3**, 18–25 (2015)
9. Schmitt, R., Dietrich, F., Dröder, K.: Big data methods for precision assembly. In: 6th CIRP Conference on Assembly Technologies and Systems (CATS) (2016). Procedia CIRP, 44: 91–96
10. Babiceanu, R.F., Seker, R.: Big Data and virtualization for manufacturing cyber-physical systems: a survey of the current status and future outlook. Comput. Ind. **81**, 128–137 (2016)
11. Stojadinovic, S., Majstorovic, V., Durakbasa, N., Šibalija, T.: Towards an intelligent approach for CMM inspection planning of prismatic parts. Measurement **92**, 326–339 (2016)
12. Stojadinovic, S., Majstorovic, V.: An Intelligent Inspection Planning System for Prismatic Parts on CMMs. Springer, Cham (2019)
13. Lee, G., Mou, J., Shen, Y.: Sampling strategy design for dimensional measurement of geometric features using coordinate measuring machine. Int. J. Mach. Tools Manuf **37**(7), 917–934 (1997)
14. http://protege.stanford.edu/. Accessed Nov 2019
15. Stojadinovic, S., Majstorovic, V.: Developing engineering ontology for domain coordinate metrology. FME Trans. **42**(3), 249–255 (2014)
16. Stojadinovic, S., Majstorovic, V.: Towards the development of feature – based ontology for inspection planning system on CMM. J. Mach. Eng. **12**(1), 89–98 (2012)

17. Majstorovic, V., Stojadinovic, S., Sibalija, T.: Development of a knowledge base for the planning of prismatic parts inspection on CMM. Acta IMEKO **4**(2), 10–17 (2015)
18. Majstorovic, V.D., Stojadinovic, S.M., Durakbasa, N.M.: An in - process measurement inspection planning model for prismatic parts. In: Proceedings of the 13th International scientific conference MMA 2018 - Flexible Technologies, Novi Sad, Serbia, pp. 103–106 (2018)
19. Stojadinovic, S., Majstorovic, V., Durakbasa, N., Sibalija, T.: Ants colony optimization of the measuring path of prismatic parts on a CMM. Metrol. Meas. Syst. **23**(1), 119–132 (2016)
20. Dorigo, M., Stützle, T.: Ant Colony Optimization. The MIT Press, Cambridge (2004)
21. Dorigo, M., Blum, C.: Ant colony optimization theory: a survey. Theor. Comput. Sci. **344**, 243–278 (2005)
22. Stojadinovic, S.M., Zivanovic, S., Slavkovic, N.: Verification of the CMM measurement path based on the modified Hammersly's algorithm. In: Proceedings of the 12th International Conference on Measurement and Quality Control – Cyber Physical Issue, Belgrade, Serbia, pp. 25–38 (2019)
23. Živković, S.: Coordinate Metrology in Manufacturing of the Complex Spatial Forms with Applications to the Aerodynamic Surfaces, Scientific-Technical Information (monograph series). Military Technical Institute Belgrade, vol. LI N°2 (2014). ISBN 978-86-81123-68-3, ISSN 1820-3418
24. Šibalija, T., Živković, S., Fountas, N., Majstorović, V., Mačužić, J., Vaxevanidis, N.: Virtual optimization of CAI process parameters for the sculptured surface inspection. Procedia CIRP **57**, 574–579 (2016). https://doi.org/10.1016/j.procir.2016.11.099
25. Živković, S., Čerče, L., Kostić, J., Majstorović, V., Kramar, D.: Reverse engineering of turbine blades kaplan's type for small hydroelectric power station. Procedia CIRP **75**, 379–384 (2018). https://doi.org/10.1016/j.procir.2018.04.037

Intelligent Process Planning for Smart Factory and Smart Manufacturing

Mijodrag Milošević[1(✉)], Mića Đurđev[2], Dejan Lukić[1], Aco Antić[1], and Nicolae Ungureanu[3]

[1] Faculty of Technical Sciences, University of Novi Sad, Novi Sad, Serbia
mido@uns.ac.rs
[2] Technical Faculty “Mihajlo Pupin”, University of Novi Sad, Zrenjanin, Serbia
[3] IMTech, Technical University of Cluj-Napoca, Cluj-Napoca, Romania

Abstract. The goal of the Industry 4.0 is the Smart factory which provides flexible and adaptive production processes in complex production conditions. Smart factory is a solution for manufacturing conditions that have hyper-dynamic character and are rapidly changing. The automation and constant optimization of production are inevitable and enable maximal utilization of workforce and production resources. The main task of technologies and services within the Smart factory is the implementation of artificial intelligence in all aspects of production. In this way, the smart manufacturing is achieved where the tasks are focused on finding optimal solutions in the preparation of production as well as the prediction of errors before they occur in production stages. Smart manufacturing relies on the concept of Cloud manufacturing in which different services are based on artificial intelligence. Smart services utilize various intelligent tools such as nature-inspired metaheuristics, search algorithms whose implementation in manufacturing has grown in the recent period. In this paper, three modern nature-inspired metaheuristic algorithms will be briefly introduced as an efficient tool in intelligent process planning optimization and their performance will be presented on three experimental studies.

Keywords: Industry 4.0 · Smart factory · Smart manufacturing · Intelligent process planning · Nature-inspired metaheuristics

1 Introduction

The final goal of the Industry 4.0 is the smart factory, i.e. intelligent manufacturing environment in which all manufacturing resources and logistic systems are organized without human intervention. Smart factory utilizes industrial internet of things and cloud technology to connect to real and virtual world, Fig. 1. In this way, cyber-physical system is integrated at all levels of manufacturing therefore enabling monitoring and reconfiguration of a production process [1]. Connecting embedded production systems and dynamical business and engineering processes enables cost-effective manufacturing according to personalized customer demands. In that case, one of the most important components is smart manufacturing which presents the

L. Wang et al. (Eds.): AMP 2020, LNME, pp. 205–214, 2020.
https://doi.org/10.1007/978-3-030-46212-3_14

utilization of intelligent technologies and solutions for optimization of production processes. Optimization relies on knowledge, services and applications that are integrated within the cloud manufacturing environment.

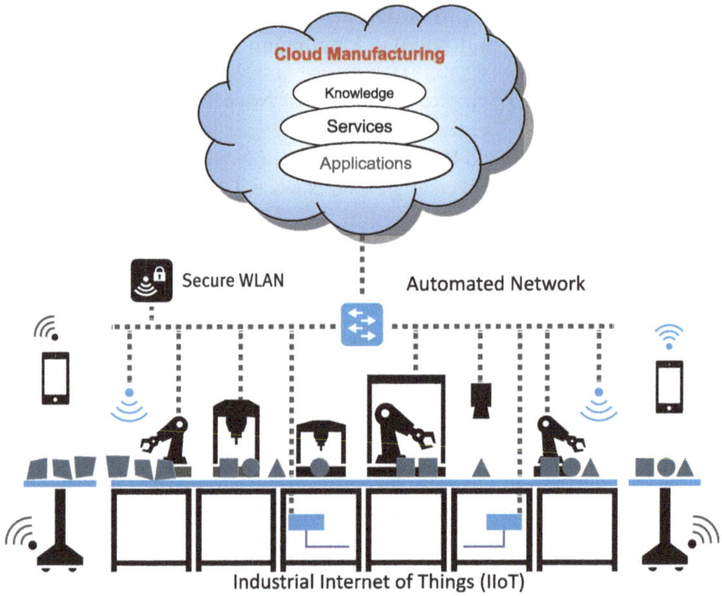

Fig. 1. Smart factory–link between cloud manufacturing and smart manufacturing technologies.

Cloud services in smart manufacturing are autonomous and intelligent and often utilize multi-agent technologies, various methods of self-organization and coordination between agents as well as function block-based methods [2]. These services are used when solving numerous tasks concerned with the production process and are based on the utilization of various methods of artificial intelligence. One of the main tasks is the preparation of production and generation and optimization of process plans. Intelligent process plans are based on generative process planning which include the utilization of AI methods, primarily nature-inspired metaheuristic algorithms.

As more advanced and complex approach, generative process planning minimizes the employment of planners and enables rapid generation of process plans by utilizing the advantages of decision-making logics, algorithms and wide use of artificial intelligence methods. Four main groups of decision-making activities within the generative process planning can be emphasized such as machining features recognition, operations selection, machines and cutting tools selection and setup plan generation. In order to obtain feasible and optimal process plans for an observed part or product these activities have to be considered simultaneously. Therefore, intelligent process planning problem had to be modelled in an optimization manner so that appropriate methods could be used for their solving.

2 Intelligent Process Planning Optimization

Process planning as a crucial function of manufacturing systems is primarily reflected by the need for detailed prediction, planning, preparation and organization of manufacturing activities with the purpose of meeting high market and customer demands for efficient production, development and business. The scope of process planning covers the development and design of products, planning and development of documentation and information for manufacturing, as well as all the measures and functions that ensure and control the realization of manufacturing. The systematic approach to process planning is the basic prerequisite for economical and productive manufacturing.

One of the main properties of process plans is their multiple alternativeness that can be found in almost all planning stages. Alternative process plans are mostly affected by the type of raw material, type and sequence of operations, manufacturing resources and techno-economical effects [3]. In this paper, we considered the intelligent process planning problem which is generally decomposed to the problem of selecting optimal machines, tools and tool approach directions required for each machining feature of a considered part or a product, as well as the problem of finding the optimal sequence of machining operations [4]. The alternatives in this sense belong to the macro process planning as one of the stages of production preparation. With the focus on prismatic parts, apart from the mentioned two tasks it is inevitable to mention the presence of precedence constraints which alleviate generation of feasible process plans [5]. These constraints are most often formulated as matrices, graphs or networks [6].

The problem of optimizing intelligent process planning belongs to the group of combinatorial problems considering the fact that the number of alternatives can be very large and therefore requires high computational power. The process of solving these problems boils down to the search for solutions within the discrete or finite set of various alternatives. The intelligent process planning problem is an NP-hard polynomial problem meaning that the problem dimensions, e.g. number of machines or tools, affect the demands for computational time and memory (time and space complexity) [7].

3 Nature-Inspired Metaheuristic Algorithms

Metaheuristics are search algorithms used in optimization for finding the most suitable alternatives in a search space of possible solutions and in a reasonable time manner [8]. With the predetermined optimization criteria these methods are very efficient in solving hard optimization tasks. Metaheuristics are primarily characterized with simplicity, universality and stochasticity while the most important condition each metaheuristic need to fulfil is to achieve an appropriate balance between local and global search thereby defining two of their most important properties, intensification and diversification [9].

Another significant feature of metaheuristic algorithms is the diverse source of inspiration they get from social and natural behaviours that can be found in real life. Since the extensive research has been carried out in the field of metaheuristics and their applications so far, in this paper we will emphasize only particular nature-inspired algorithms. Figure 2 shows the illustrative representation of natural organisms that inspired development of metaheuristics.

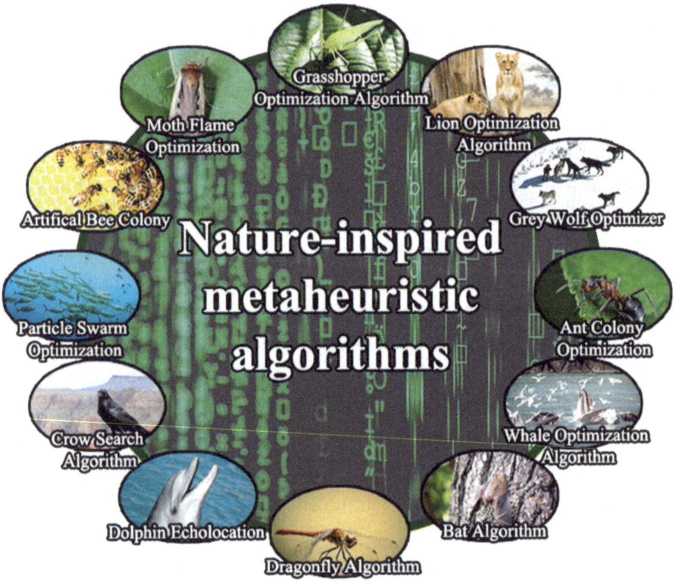

Fig. 2. Natural inspiration behind metaheuristic algorithms. (Images taken from pixabay.com and unisplash.com; shared with the courtesy of their authors)

Nature is enriched with various individuals who express certain intelligent behaviour mechanisms as well as individuals who express the same when being part of a larger group. Therefore, mutual interaction among living organisms such as fish, fireflies, birds, wolves, bees, ants who live in community are directed towards acquiring local information which is valuable for the community itself. Their social intelligence most often reflects on foraging or hunting behaviour. Also, intelligent behaviour of individual organisms can be noticed in creatures such as dolphins, crows or frogs. Special group of biologically-based solutions is inspired by evolution, genes, biogeography, flower pollination etc. Whether based on collective or individual intelligent mechanisms many authors from scientific community have developed various metaheuristic algorithms. Here, we will point out the contribution made by authors in [8, 10, 11].

The search procedure of nature-inspired metaheuristic algorithms is based on randomly generated population of individuals whether they represent fish, ants, bees or any of the abovementioned organisms. It is worth saying that population-based metaheuristics are generally more popular than single-based and therefore nature-inspired metaheuristic algorithms as population-based search techniques are characterized by social as well as individual intelligence. Algorithm's dynamics is determined by equations that characterize or emulate the observed natural behaviour on the basis of which the population evolves. In other words, these individuals are being evaluated using a predefined optimization criterion or criteria and the process is repeated for a predefined number of iterations. Most popular and also most relevant optimization criteria in intelligent process planning are manufacturing time and manufacturing cost. By adjusting input parameters, the search process can be directed towards promising regions of the search space.

So far, various metaheuristic implementations can be found in the scientific literature. Genetic algorithms, particle swarm optimization and ant colony optimization are the metaheuristics that have been most often adapted to the intelligent process planning problem. Here, we will focus on modern approaches that have been developed in recent years.

3.1 Grey Wolf Optimizer

The grey wolf optimizer (GWO) belongs to the swarm intelligence algorithms of new generation [10]. The inspiration of the GWO comes from the social hierarchy of grey wolves and their hunting mechanisms, such as tracking, chasing, pursuing, encircling, harassing and finally attacking the prey animal (Fig. 3a). Strict dominance hierarchy makes the pack of wolves a very dangerous and adaptable group. The alpha is the leader of the pack which makes most important decisions. The beta wolves are at the second level in the hierarchy while the delta wolves are at the third level. The lowest position is for omega wolves who are subordinate to all other wolves in the pack and are usually as scapegoats and given the last priority when feasting on prey.

3.2 Whale Optimization Algorithm

The first case of implementation to process planning optimization will be the metaheuristic proposed by the same author as the GWO algorithm. Whale optimization algorithm (WOA in the following) was originally developed for continuous optimization so additional adaptation to combinatorial optimization problem had to be considered [11].

WOA algorithm is inspired by the hunting style of humpback whales (Fig. 3b). Whales are known to live in social groups called pods and whales such as humpback whales hunt using a unique principle of encircling prey fish and then attack using the so called bubble-net mechanism. This mechanism consists of shrinking to an appropriate depth, then moving upwards following the spiral path and creating distinctive bubbles along the spiral path. These bubble prevent prey fish from escaping which ensures humpback whales can swallow them at the surface.

3.3 Crow Search Algorithm

Another nature-inspired metaheuristic is the crow search algorithm (CSA) proposed by [12]. This method is inspired by clever behaviour of crows or ravens which by following the movement of other birds can find their food hiding sources and therefore commit thievery (Fig. 3c). One specific and interesting idea that presents the crucial parameter in the CSA at the same time is crow's awareness probability. If crow has a slightly high awareness that makes her intelligent enough to avoid being pilfered by other crows. In other case, with low awareness they are victims of thievery. This makes the quick-witted game of strategy and deception in which a crow ends up being a food saver or a food loser. Although a population-based metaheuristic, the CSA is based on individual intelligence since crows do not pilfer other birds in flocks but individually. This is not a swarm intelligence algorithm such as the WOA.

(a) (b) (c)

Fig. 3. Inspiration for GWO, WOA and CSA algorithms.

4 Nature-Inspired Metaheuristics in Intelligent Process Planning

This section will give the short insight into the implementations of the previously described nature-inspired metaheuristics for optimization of intelligence process planning problem. All three algorithms have been coded in Matlab programming environment on PC with modern configurations. Each algorithm's search process ends after the predetermined number of iterations. The algorithms were run ten times and they obtained very good results but with further modifications their performance could be largely increased.

To test these methods, classical problem from the literature have been involved. Figure 4 shows solid models of prismatic parts that are frequently considered in case studies conducted for testing many different metaheuristic approaches to intelligent process planning. Since different approaches have been made, its complexity largely differs. The represented models require thorough consideration of precedence relationships among operations and therefore the involvement of precedence graphs and precedence matrices. In that sense, modified and adapted metaheuristic approaches have to be taken into account. So far, different approaches to intelligent process planning on the basis of precedence constraints have been proposed, such as those in [13–16].

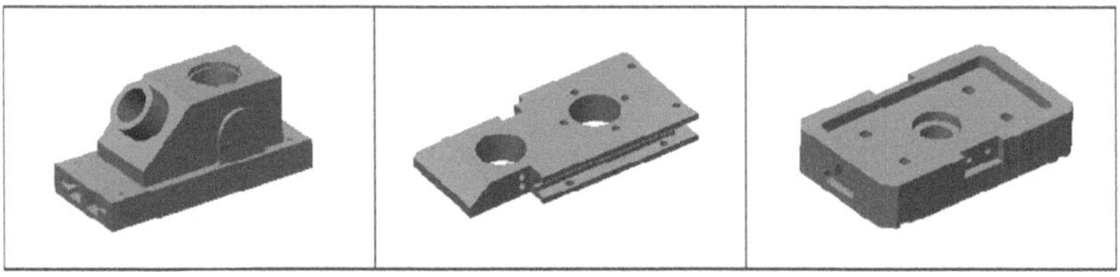

Fig. 4. Popular prismatic parts in intelligent process planning literature.

On the other side, AND/OR networks or graphs have also been introduced when dealing with intelligent process planning problem. The quality of this approach reflects in the fact that all flexibilities are included on one graphical representation making it the subject of many studies [6, 17]. The benefit of these networks is the fact that feasible process plans are generated in each iteration so precedence constraints do not have to be considered individually. Thus, in order to present the idea of how nature-inspired metaheuristic algorithms can be used for solving intelligent process planning problem within the smart factory concept, we have conducted three small case studies based on the AND/OR graphs which will be briefly described.

The first experiment was taken from [18]. The part requires 15 machining operations and only machine alternatives with machining times are included. The GWO, WOA and CSA algorithm were employed to solve this problem and the results proved that they are efficient methods for this case study. We adopted 100 search iterations and 30 search agents in this experiment. To improve efficiency, we included mutation strategy that influenced diversification of the search [19]. Table 1 gives insight into one of the optimal process plans that were generated using one of the nature-inspired metaheuristics. Minimal manufacturing time obtained by all three metaheuristics is 334 cost units.

Table 1. Optimal process plan for the first experiment.

Operation sequence	1	2	5	9	10	11	13	12	14	15
Machines	3	3	3	3	4	5	5	8	9	12
Processing time	46	36	6	12	50	25	35	36	40	30

The second and third experiment consider two cylindrical parts shown in Fig. 5 which were proposed in [6]. Apart from machine flexibility, these problems also cover cutting tool flexibility and tool approach direction flexibility making it more complex than the first one since the problem dimensions increase the number of alternative solution to the problem. The second experiment (upper example in Fig. 5) contains six features with 18 machining operations, 8 available machines and 12 tools while the third experiment (lower example in Fig. 5) for six features requires 14 machining operations with 6 available machines and 12 available tools. For these cases, we set the number of iterations to be 200 and the populations sizes of all three metaheuristics to be 60 search agents. The algorithms were repeatedly executed 10 times in order to provide credible results.

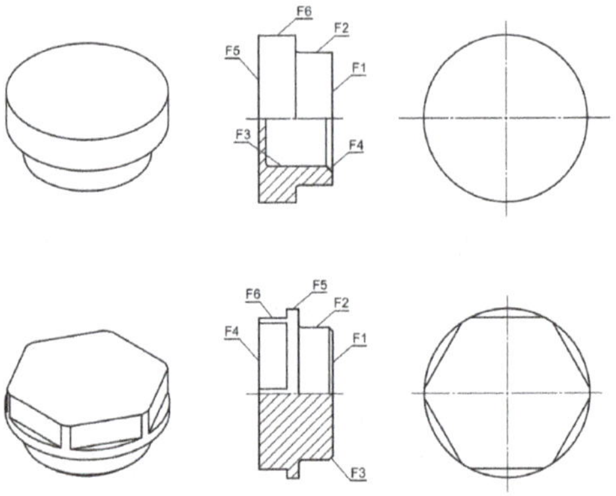

Fig. 5. Cylindrical part models. [6]

Tables 2 shows one of the optimal solutions to these problem instances. The minimal manufacturing time obtained by the GWO, WOA and CSA for the second is 74,8 cost units. To compare the performances of the GWO, WOA and CSA nature-inspired metaheuristics, we present the convergence curves of these algorithms which are given in Fig. 6. As it can be noticed, the CSA and WOA showed much better convergence compared to GWO which did not obtain optimal results in ten runs. The results of the CSA were most consistent compared to the GWO and WOA.

Table 2. Optimal process plan for the second experiment.

Operation sequence	1	2	5	6	7	8	10
Machines	1	6	6	6	1	1	1
Tools	1	10	6	8	1	1	1
Tool approach directions	+z	+z	+z	+z	+z	−z	−z
Processing time	2,3	3,2	3,4	13,4	0,7	2,3	0,5

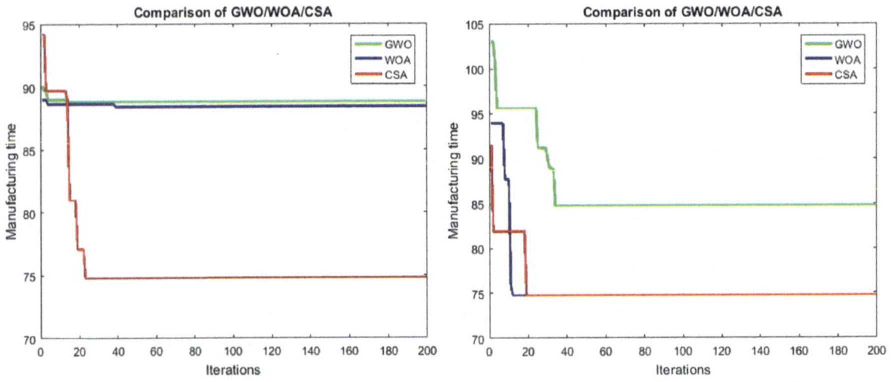

Fig. 6. Convergence curves of GWO, WOA and CSA for the second experiment.

As far as the third experiment is concerned, Table 3 shows one of the optimal results and Fig. 7 gives the comparison of convergence curves of the three proposed metaheuristics. Here, the GWO, WOA and CSA showed similar convergence towards optimum which in this case varies between 83,1 and 83,4 cost units.

Table 3. Optimal process plan for the third experiment.

Operation sequence	8	9	10	11	13	14
Machines	6	1	6	6	1	6
Tools	9	4	9	9	4	12
Tool approach directions	−z	−z	−z	−x	+z	+z
Processing time	2.8	0,4	6,8	2,9	1	0,5

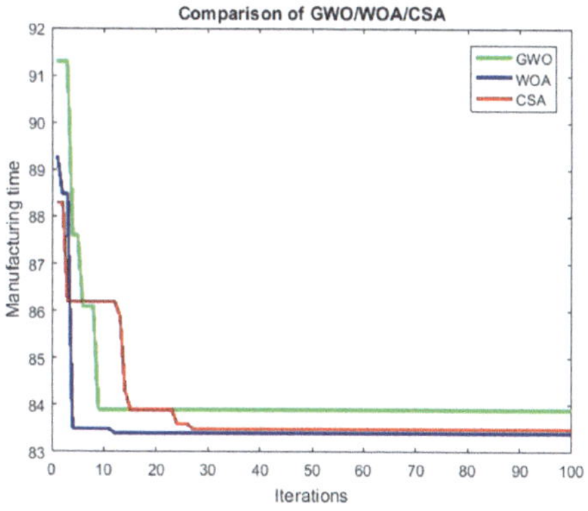

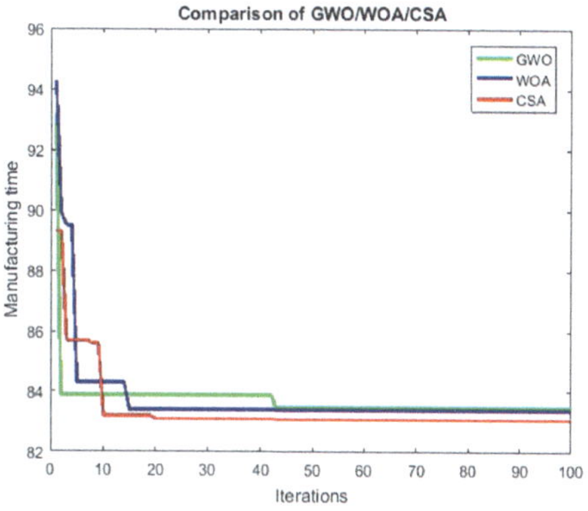

Fig. 7. Convergence curves of GWO, WOA and CSA for the third experiment.

5 Conclusion

Nature-inspired metaheuristic algorithms as smart services of artificial intelligence for intelligent process planning optimization within smart manufacturing and smart factory concepts were proposed in this paper. Intelligent process planning is therefore achieved by focusing on finding optimal process plans, i.e. optimal operation sequences and optimal selection of machines, tools and tool approach directions for each machining operation. The optimization of intelligent process planning mostly considers optimization criteria such as manufacturing time or manufacturing cost. In this paper, we gave brief biological background of three modern nature-inspired metaheuristic algorithms and employed them on intelligent process planning problem. Grey wolf optimizer, whale optimization algorithm and crow search algorithm proved to be efficient when solving simpler case studies. In order to boost their performances, we adopted one mutation strategy which improved their convergence rate. However, more complex approach towards intelligent process planning based on precedence constraints require additional modifications and improvements of the classical GWO, WOA and CSA.

References

1. Majstorović, V.D., et al.: Cyber-physical manufacturing in context of Industry 4.0 model. In: Lecture Notes in Mechanical Engineering, pp. 227–238 (2018)
2. Wang, X.V., Givehchi, M., Wang, L.: Manufacturing system on the cloud: a case study on cloud-based process planning. Procedia CIRP **63**, 39–45 (2017)
3. Denkena, B., Shpitalni, M., Kowalski, P., Molcho, G., Zipori, Y.: Knowledge management in process planning. Ann. CIRP **56**(1), 175–180 (2007)
4. Li, W.D., Ong, S.K., Nee, A.Y.C.: Integrated and Collaborative Product Development Environment – Technologies and Implementations. Series on Manufacturing Systems and Technology, vol. 2. World Scientific Publishing, Singapore (2006)
5. Lukić, D., Milošević, M., Erić, M., Đurđev, M., Vukman, J., Antić, A.: Improving manufacturing process planning through the optimization of operation sequencing. Mach. Des. **9**(4), 123–132 (2017)
6. Petrović, M.: Artificial intelligence in intelligent process planning. Ph.D. thesis, University of Belgrade, Mechanical Faculty (2016)
7. Rothlauf, F.: Optimization problems. In: Design of Modern Heuristics. Springer, Heidelberg (2011)
8. Yang, X.-S.: Engineering Optimization: An Introduction with Metaheuristics Applications. Wiley, Hoboken (2010)
9. Talbi, E.G.: Metaheuristics: From Design to Implementation. Wiley, Hoboken (2009)
10. Mirjalili, S.: Grey wolf optimizer. Adv. Eng. Softw. **69**, 46–61 (2014)
11. Mirjalili, S.: The whale optimization algorithm. Adv. Eng. Sofw. **95**, 51–67 (2016)
12. Askarzadeh, A.: A novel metaheuristic method for solving constrained engineering optimization problems: crow search algorithm. Comput. Struct. **169**, 1–12 (2016)
13. Dou, J., Li, J., Su, C.: A discrete particle swarm optimisation for operation sequencing in CAPP. Int. J. Prod. Res. **56**(11), 3795–3814 (2018)
14. Hu, Q., Qiao, L., Peng, G.: An ant colony approach to operation sequencing optimization in process planning. J. Eng. Manuf. **231**(3), 470–489 (2015)
15. Su, Y., Chu, X., Chen, D., Sun, X.: A genetic algorithm for operation sequencing in CAPP using edge selection based encoding strategy. J. Intell. Manuf. **29**, 313–332 (2015)
16. Milošević, M., Lukić, D., Đurđev, M., Vukman, J., Antić, A.: Genetic algorithms in integrated process planning and scheduling – a state of the art review. Proc. Manuf. Syst. **11**(2), 83–88 (2016)
17. Lian, K., Zhang, C., Shao, X.: Optimization of process planning with various flexibilities using an imperialist competitive algorithm. Int. J. Adv. Manuf. Technol. **59**, 815–828 (2011)
18. Lv, S., Qiao, L.: A cross-entropy-based approach for the optimization of flexible process planning. Int. J. Adv. Manuf. Technol. **68**, 2099–2110 (2013)
19. Huang, W., Hu, Y., Cai, L.: An effective hybrid graph and genetic algorithm approach to process planning optimization for prismatic parts. Int. J. Adv. Manuf. Technol. **62**(9), 1219–1232 (2011)

Model-Based Manufacturing System Supported by Virtual Technologies in an Industry 4.0 Context

Vesna Mandic(✉)

Faculty of Engineering Sciences, University of Kragujevac, Kragujevac, Serbia
mandic@kg.ac.rs

Abstract. Industry 4.0 concept of the new industrial revolution is based on the application of front-end and base technologies for producing digital solutions. Converging Smart Manufacturing and Smart Products with Big Data and Analytics plays a central role in implementing the I4.0 concept in today's industry. This paper presents the virtual components of the proposed Model-based Manufacturing System and their role in the I4.0 context. Two different industrial cases demonstrate the application and benefits of the MBM approach, which integrates virtual and rapid technologies for the design, analysis and validation of a product and its fabrication processes of sheet metal forming and forging.

Keywords: Model-based manufacturing · Industry 4.0 · Virtual manufacturing · Additive manufacturing · Virtual reality · Metal forming

1 Introduction

It is well known how challenging in the industry is to set up a new factory or product, in terms of time, material and human resources. If the whole process is carried out in traditional way, with numerous trial and error attempts in real production and by using production resources, then such approach is not competitive in today's increasingly demanding market. Namely, the I4.0 concept enables companies to adopt and put into practice the implementation of new approaches and techniques for digitalization, cloud computing, the Internet of Things and big data, in order to gain a competitive advantage [1]. New techniques are used to generate virtual and digitized environments to simulate real processes and systems, at all stages of development factory, product, production technology, even virtual quality control and behavioral prediction in real conditions, as well as in other stages of product life cycle. Only after validation of the virtual model it is possible to pilot verified solutions in physical production systems, which are also supported by software for fine-tuning and monitoring. Some examples available in the literature show that it is possible to set up an automotive part production unit in three days, instead of three months as it used to be, through 3D visualization of virtual plant, processes and interaction of workers and machines [2].

L. Wang et al. (Eds.): AMP 2020, LNME, pp. 215–226, 2020.
https://doi.org/10.1007/978-3-030-46212-3_15

In the future, the fourth industrial revolution, as a result of the introduction of the Internet of Things and Services into manufacturing, will establish global networks with shared production facilities and resources (Cyber-Physical Systems), which will consequently have significant improvements in industrial production, material use, supply chains and improved lives cycle management [3].

Industry 4.0 is conditioned by innovative technological advancements in the areas as listed below. For all of them digitization is common across the entire industrial environment and business activities, starting with business models, products and services, production systems, machines and workers [4, 5]:

1. ICT technology - for digitalization of information at all stages of the product life cycle, both within the company and beyond its facilities.
2. Cyber-physical systems - for monitoring and controlling physical processes and systems, including embedded sensors, robots and additive manufacturing devices [6, 7]
3. Network communications - for internal and external connection of machines, products, systems and people, via internet technology
4. Simulation, modeling and virtualization - for virtual and rapid development of product and manufacturing process, including virtual and augmented reality support for designers and workers
5. Collection of data, analysis and exploitation - for collecting production data and applying techniques for big data analysis.

Despite the fact that Industry 4.0 is considered to be new industry revolution, initiated in the year of 2011, where above mentioned emerging technologies are converging to produce digital solutions, there is still a lack of understanding of how companies should approach their comprehensive adaptation and application in the industry. In order to better understand the adaptation of I4.0 in companies, Frank et al. [8] proposed a conceptual framework for I4.0 within which all technologies are divided into front-end and base technologies. The first group includes technologies that are classified into four pillars: Smart Manufacturing, Smart Products, Smart Supply Chain and Smart Working. The second group is comprised of base technologies containing four elements: Internet of Things, Cloud Services, Big Data and Analytics. A study conducted in a sample of 92 companies found that the implementation of base technologies, especially Big Data and Analytics, is at a very low level, which is a future challenge for companies. It has also shown that Smart Manufacturing plays a central role in the implementation of the I4.0 concept in the industry and is highly integrated with Smart Products technologies.

The term digital twin is often used in the literature for an integrative approach that encompasses the physical product or process, the virtual product or process, and the associated data that link the physical and virtual worlds in the industry [9]. It is applied in all stages of the product life cycle, conceptual and detailed product design, design of production technologies, but also in the control itself. Future research should define the holistic approach of using digital twins in the entire process of product development and production, including the problem of standardization and efficient information flow.

A significant limiting factor in the implementation of the I4.0 concept for companies is the volume of information and data obtained in a cyber-physical system, especially through the application of simulations, modeling and data acquisition from manufacturing. Without a sophisticated approach to data analysis and software-assisted decision-making in the system, one can hardly feel the direct benefits of the I4.0 concept for the company. Therefore, research is focused on the development of Decision Support System based on CPS simulation and optimization [10].

The paper presents the results of applied research of the I4.0 concept carried out in specific industrial cases, which relate to simulations and modelling of production processes, so-called virtual manufacturing, additive manufacturing of products and tools, as well as advanced 3D visualization techniques through the use of systems and software for virtual reality. Their integrative application in rapid product development, validation of the recommended production processes for its manufacturing, optimization of influential process parameters on production outputs, visualization of the design solution is also shown through certain case studies.

2 Virtual Technologies in Model-Based Manufacturing System

Virtual technologies represent a whole set of interconnected engineering technologies for product and process design and are based on the digitization of real objects and the simulation of industrial processes with the acquisition of production data for the setting of input parameters for system modeling [11]. Figure 1 shows the components of a digital model-based manufacturing system for integrated product and process development. Digital model is centrally positioned with all associated information on product and tool design, technological processes, quality control requirements, production, assembly, maintenance conditions, which are generated in the product lifecycle [12].

Reverse engineering (RE), CAD/CAM/CAE (Computer Aided Design/Manufacturing/Engineering), Rapid Prototyping and Tooling (RP/RT), Virtual Manufacturing (VM), Virtual Reality (VR) have been identified as enabling technologies, whose integration within I4.0 context will create the environment where companies will be able to become more innovative and competitive.

Model-based manufacturing (MBM) implies technological merging of CAD/CAM/CAE, as VP&M (Virtual Prototyping and Manufacturing) methods, with RP/RT/RM as PP&M (Physical Prototyping and Manufacturing) methods. Virtual and rapid prototypes obtained in this way can be used for testing the functionality of product or assembly and different concepts in the early stage of design without expensive and long-term trial-and-error attempts in traditional design and production. It creates not only the model of a product/tool, but also the virtual simulation model of production processes in computer environment, known as virtual manufacturing (VM) approach. Based on nonlinear finite element analysis, it enables optimization of key factors of production which directly influence profitability, as an efficient way to test "what if" scenarios, to validate different concepts and optimize related parameters for shop floor. In brief, a capability to "manufacture in the computer" is so powerful tool which reduces the errors to the minimum, cuts the costs and shortens the time of product/tool design, as all modifications are made before the actual manufacturing process. Based on the digital model of a product or tool, it is possible, using specialized CAM software, to obtain an optimal fabrication strategy for CNC machines and to automatically generate NC code for tool movement.

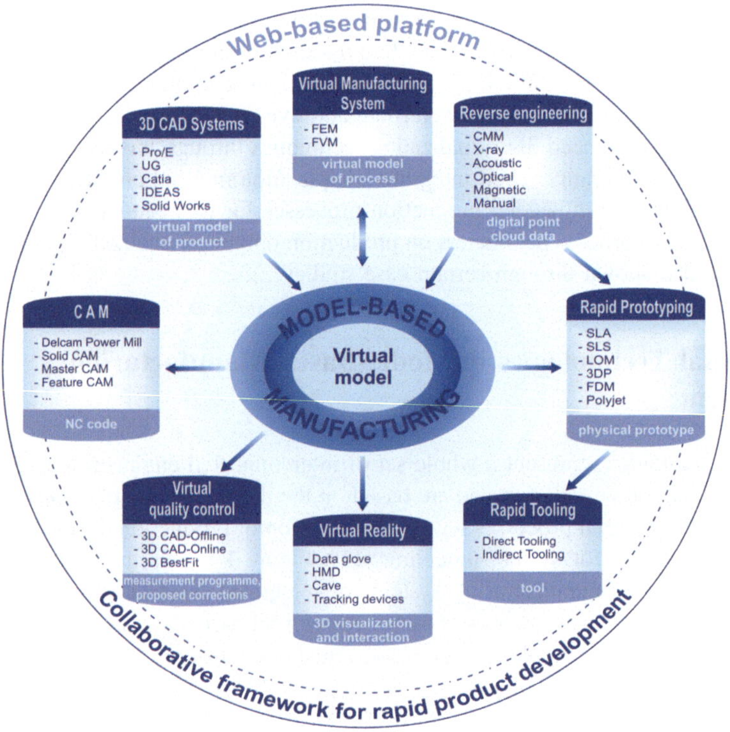

Fig. 1. Virtual technologies and their interconnections [12]

CAD and FEM/FVM models, which are basically digital models, can be obtained in physical form using rapid prototyping technologies and equipment. Modern additive manufacturing systems allow for rapid tooling approaches for production tools and dies. In addition to creating CAD models, 3D model of a product/tool can be also rapidly generated in digital form using reverse engineering, after that remodeled and exported to one of the systems for rapid prototyping/tooling, lately more and more known as Additive Manufacturing.

Model-based manufacturing approach also uses virtual reality (VR) as advanced technology for 3D presentation of model structure, composition and behavior as if it was physically manufactured. VR holds great potential in different engineering applications, such as product design, modelling, shop floor controls, process simulation, manufacturing planning, training, testing and verification thereby preventing mistakes of physical try-outs.

Thanks to the Internet of Things, nowadays it is possible to connect remote teams and companies in a collaborative environment in order to share models and analyze design solutions. All data regarding products, tools and associated manufacturing processes are thus made available in digital form, which can be used for further quality control in the production process, as well as for product life cycle modifications and improvements. If digital models are presented in a 3D environment through equipment and software for Virtual and Augmented Reality, such resources will provide a better understanding of the design solution and of the entire system, for all development professionals, engineers, designers, marketers, managers. It also provides easy maintenance and necessary training when digital resources are embedded in an augmented reality application.

3 The Role of Virtual Technologies in an I4.0 Context

3.1 Additive Manufacturing

Although the I4.0 concept is based on ICT technologies and digital models, Additive Manufacturing (AM), widely known as 3D printing or Rapid Prototyping, is considered to be a vital component that connects virtual environments with physical one. It combines quality and efficient production of customized products with sophisticated shapes and new materials, which are difficult to produce with traditional production. It enables an analysis of the product functionality within the assembly, checking of design solution, ergonomic and other functional testing. Its application exhibited reduction of the lead time for about 60% with respect to the traditional way. Currently, more industrial sectors (automotive, aerospace, biomedical, manufacturing, agriculture, healthcare, etc.) are adopting AM, incorporating greater flexibility and individualization of manufacturing processes, and connecting all processes by IoT [13–15].

An interesting multidisciplinary use case is a combination of electronics and mechanics in designing and rapid manufacturing of sensors and embedded electronics for humanoid robots. Flexible sensors for robots fingers can be designed and fabricated using AM in electronics as well as efficient motors (smaller, simpler, etc.) for achieving more natural robots behavior, for example face mimics [16].

Trends in the development of AM, from the aspect of the I4.0, relate to the development and application of new smart materials, then devices that would produce functional parts, even assemblies/machines, in a single step of fabrication. The third direction of development is related to the design issues limiting the AM process [17].

The choice of AM strategy, which includes parameters such as beam diameter and current, preheat temperature, is critical for the evolution of the microstructure. Predicting deviations in the process of cooling or post-processing and residual stresses using FEM simulations is a significant support for designers and AM operators [18]. The modeling of the AM process is not only a challenge to evaluate the final material properties and quality of the model, but also provides a basis for improving the production process. FEM simulations are very complex with multi-scale and multi-physics endeavor and parameter interaction in complex algorithms [19, 20].

3.2 Virtual Manufacturing

Virtual Manufacturing (VM) is a software-supported system for modelling, simulation and optimization of production processes used in product manufacturing. It generates the same information on the production environment and conditions that can be observed in a real manufacturing system. It enables the reduction of costs and time of product development, early evaluation of product alternatives and its fabrication operations, as well as producibility and affordability, in integrative simultaneous modelling and design of products, processes and resources. Virtual Manufacturing is, in a word, "production in a computer" [21–23].

Process modelling is based on nonlinear FEM (Finite Element Method) or FVM (Final Volume Method) analysis and simulation of all the processes in manufacturing technology of a certain product. Technology simulation makes possible for companies to optimize key factors which directly affect the profitability, like formability, final form and accuracy, level of residual stresses, reliability in exploitation, etc. [24].

Since virtual models of production processes, obtained through virtual manufacturing concept, are very flexible, they allow us to examine the impact of design changes, both product and tools geometry, as well as process parameters, on product quality and production costs. This way, it is possible to perform sensitivity analysis relatively quickly under the conditions of parallel numerical processing, and to identify the areas of the optimized design solution. Moreover, it is possible to predict failure and occurrence of defects in the product, optimal use of production equipment and tools, assessment of tool wear and its life, as well as fracture prevention. The optimal choice of relevant production parameters has positive consequences for reducing time-to-market, production costs, materials and tools, as well as increasing the final quality of the product [21].

Virtual Manufacturing is not only a tool for numerical simulation and optimization of production processes, but also a tool to support the PLM (Product Lifecycle Management) system, for making the right decisions by company management in the early stages of product development [25, 26]. The high complexity of models and analysis, which basically has many different types of data, can lead to problems in the flow of digital information and data. The PDM (Product Data Management) system offers a solution for reliable storage and monitoring of data, so that the right information and data are available in a timely manner and in the right location. Electronic data must be available in different formats, in order to be easily transmitted between subsystems of MBM systems, through appropriate interfaces (Fig. 1).

All this leads to the accumulation of a large amount of data which is not suitable for analysis and decision making. Therefore, recent researches in I4.0 are concerned with a structured and systematic reduction of data gathered during production processes. Systems for the automatic control and integrated consideration of communication in the MBM system with a networked components are being developed [27, 28].

3.3 Virtual and Augmented Reality

Virtual Reality (VR) can be defined as a simulation within which computer graphics are applied to create a realistic-looking world, where the synthetic world is not static but responds in a certain way to user response and modifies in the real-time environment. There is a great need for reality in presentation within a wide range of fields, from education, art, medicine, to virtual production [11, 29].

The implementation of the immersive environment in I4.0 is realized in three ways: Virtual Reality (VR), Augmented Reality (AR) and Mixed Reality (MR) [30]. AR can be defined as a computer graphics technique where virtual symbols are superimposed on a real image of the external world [31]. The application of these systems in the industry is especially useful for assembly and maintenance processes, when training and instructions for workers are prepared mainly in the AR technique. For this purpose, expensive VR systems do not need to be used, but for example see-through glasses equipped with camera and small projectors on lenses, or mobile devices like tablets or smartphones. Immersive interfaces and workers' experience are combined to achieve better and more efficient procedures than in the conventional approach.

4 Industrial Application Cases

4.1 MBM Application in Sheet Metal Forming Process

The main objective of the presented case study is to apply an integrated approach to the arbitrarily selected product or product component in the application of multiple virtual technologies in the re-engineering of technological processes in sheet metal forming processes and verification of the proposed tool design. The sheet metal handle, used in the manufacture of different types of cookware, is obtained by processes of blanking, punching, deep drawing and bending of sheet metal. The last bending and closing operation of the handle may be unstable, depending on the workpiece shape in the previous deep drawing/bending operation, and further conditioned by sheet anisotropy.

If the technology development and the tool design are based only on the designer's experience, having numerous physical prototypes of tools and try-outs is inevitable. Virtual product development and optimization of technological processes by virtual manufacturing significantly reduce development time and costs. In established MBM environment based on digital models of the handle and tools, designers can propose and validate several alternatives for product and tools design.

As it is presented in Fig. 2, the applied integrated MBM approach comprised the following technologies [11, 12]:

- Reverse Engineering, for scanning of blank shape and free surfaces of handle, using the multi-sensor coordinate measurement machine WERTH VideoCheck IP 250, which is equipped with three sensors: optical, laser and fiber contact sensor,
- CAD modelling for generating 2D model of sheet blank and 3D models of the handle and tools, based on RE data and created tools for recommended production technology
- Virtual Manufacturing for FEM simulation and verification of proposed technology and tool design by using Simufact.forming software
- Rapid Prototyping for physical verification of obtained digital FE model of handle as VM result, by application of the PolyJet technology and the 3D printer ALARIS 30

- Quality control for comparison between real metallic part and RP model of the handle obtained by FEM simulation, using CMM WERTH and its optical and laser sensors, and
- Virtual Reality for 3D visualization of virtual models in MBM system, as well as for interaction with virtual models; for this purpose VR application was developed by use of the following software and hardware components: Wizard VR Toolkit program, 5DT Data Glove, Wintracker, magnetic 6DOF tracking device.

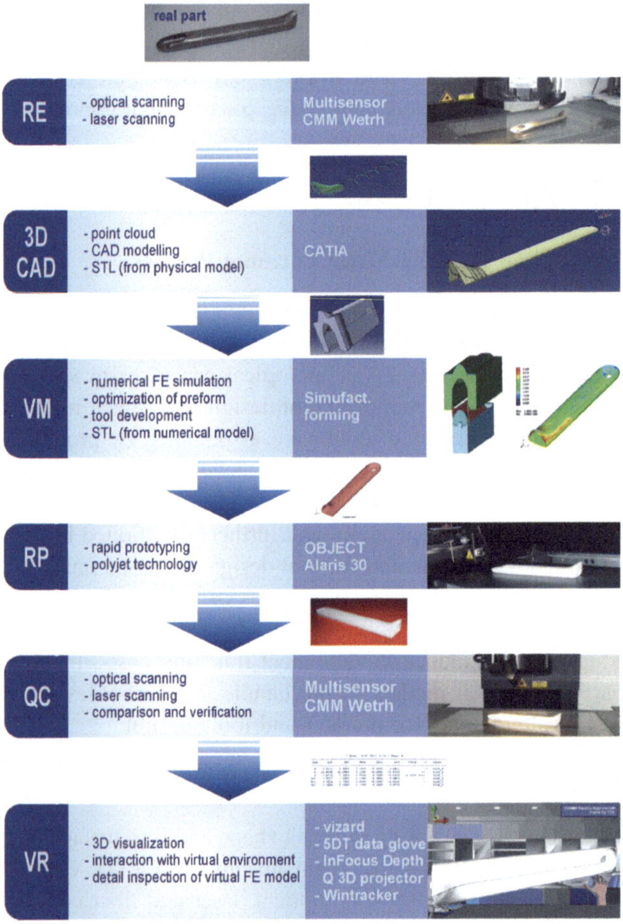

Fig. 2. Integrated MBM approach in sheet metal forming industrial case [12]

Advantages and possibilities of the MBM approach were demonstrated through the presented industrial case, by applying the CAD/CAM/CAE, VM, RP/RM and VR techniques. It was shown that engineering design and development can be very successfully realized, with respect to quality, costs and time savings, by application of the virtual/rapid prototyping/manufacturing technologies.

4.2 MBM Application in Forging Process of the Artificial Hip Stem

Similar industrial case is the development of customized implant of artificial hip, with wide application areas in medicine [32]. The reverse engineering technologies as computed tomography (CT) and magnetic resonance imaging (MRI) have evolved over the years as a beneficial tool in medical diagnostics. Both mentioned scanners generate a certain number of 2D cross-sections (slices) of tissues. There were several proposed techniques to reconstruct 3D objects from 2D DICOM sliced images. By this approach it can be obtained customized shape and dimensions of artificial hip, that is its stem, and it could be generated customized CAD model of hip and its stem.

As it is shown in Fig. 3, CAD model of stem is base for forging technology optimization through numerical simulations (VM) and testing of tool design. After verification of proposed technology, CAM modelling is applied for NC code development for CNC machining of tool components. RP model is used to test prosthesis assembly, and finally coordinate metrology on multi-sensor CMM for certification requirements. The whole cycle in product development, from idea to certified product, that is implant, is covered as total solution for industry application.

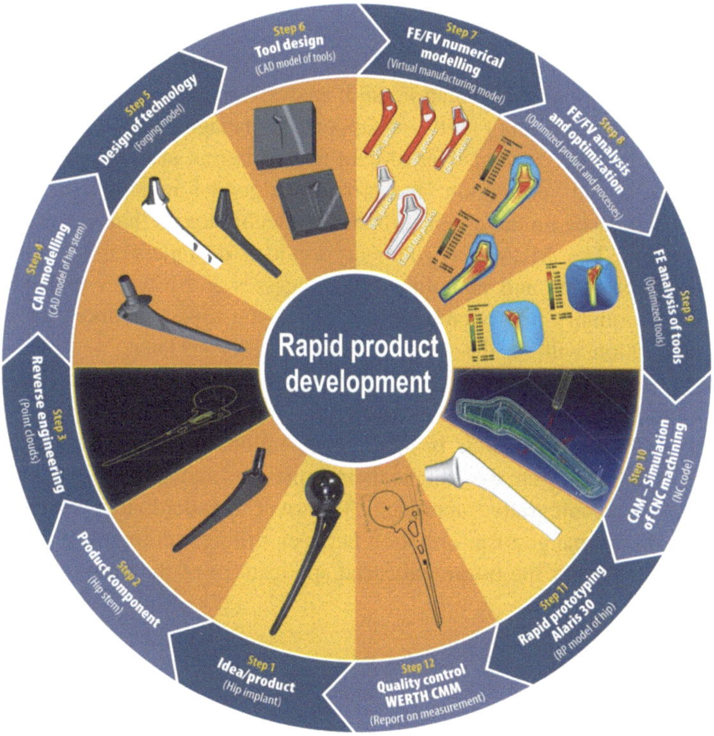

Fig. 3. Integrated MBM approach in design of customized implant and its fabrication processes

By identification of geometrical properties of the hip stem applying reverse engineering technology on CT or MGI devices (step 3), it was possible to develop its CAD model using software CATIA and its modules Generative Shape Design and Part Design (step 4). In step 5, the technology of forging a hip stem, of 2CrNiMo18143 alloy, was designed in two operations of preform and final forging. This was preceded by the design of a CAD model of stem forging.

The FVM simulation of both forging operations was implemented in Simufact.-forming software, to validate the tool design and proposed technology, especially the preform shape. After validation of the design solution, a CAM strategy for CNC machining of tools was prepared.

In addition to the VM validation results of the technology, the virtual forged model obtained by FVM simulation was exported as an STL model and additionally printed using PolyJet technology on a 3D ALARIS printer. For the final measurement and verification of dimensions, a multi-sensor CMM WERTH machine was used, especially its optical and laser sensors.

The presented case confirms the benefits of implementing an integrative BMB approach for the development of medical products tailored to the target patient.

5 Conclusions

The paper deals with a set of virtual technologies, whose integration provides the MBM environment for modeling, simulation and analysis of product and its fabrication processes, with the provision of software support for managing product data throughout the life cycle. It enables easy transfer of data from different system components, from design to analysis and verification, and provides a good foundation for virtual engineering approach based on the application of base technologies that are integral part of the I4.0 concept. The MBM environment provides designers and engineers with visualization of products and manufacturing processes and a better understanding of them, leading to quality improvements, shortening the time to market, providing the right design solution without the need for a costly redesign later.

Through the presented case studies, MBM approaches for two industrial examples of different products and technologies have been demonstrated.

The trend of future research is in the application of other physical and cyber components of MBM system to evaluate their limitations and advantages, which should be overcome by an integrative approach. Moreover, standardizing the interface and data form for the exchange of information between different components of the MBM system is a challenge for the future industrial applications of the proposed approach in the I4.0 context.

References

1. Castelo-Branco, I., Cruz-Jesus, F., Oliveira, T.: Assessing Industry 4.0 readiness in manufacturing: evidence for the European Union. Comput. Ind. **107**, 22–32 (2019)
2. Roland Berger Consultants, INDUSTRY 4.0, The new industrial revolution, How Europe will succeed (2014). http://www.iberglobal.com/files/Roland_Berger_Industry.pdf. Accessed 01 Mar 2020
3. Recommendations for implementing the strategic initiative INDUSTRIE 4.0, Final report of the Industrie 4.0 Working Group (2013). http://alvarestech.com/temp/RoboAseaIRB6S2-Fiat/CyberPhysicalSystems-Industrial4-0.pdf. Accessed 15 Feb 2020
4. Industry 4.0 - Digitalisation for productivity and growth, Briefing, Members' Research Service, European Parliament September Parliament (2015). https://www.europarl.europa.eu/RegData/etudes/BRIE/2015/568337/EPRS_BRI(2015)568337_EN.pdf. Accessed 25 Jan 2020
5. Alcácer, V., Cruz-Machado, V.: Scanning the Industry 4.0: a literature review on technologies for manufacturing systems. Eng. Sci. Technol. Int. J. **22**, 889–919 (2019)
6. Konstantinov, S., Ahmad, M., Ananthanarayan, K., Harrison, R.: The cyber-physical e-machine manufacturing system: virtual engineering for complete lifecycle support. In: The 50th CIRP Conference on Manufacturing Systems (2017). Procedia CIRP 63, pp. 119–124
7. Majstorovic, V., Stojadinovic, S., Zivkovic, S., Djurdjanovic, D., Jakovljevic, Z., Gligorijevic, N.: Cyber-physical manufacturing metrology model (CPM3) for sculptured surfaces – turbine blade application. In: The 50th CIRP Conference on Manufacturing Systems (2017). Procedia CIRP 63, pp. 658–663
8. Frank, A.G., Dalenogare, L.S., Ayala, N.F.: Industry 4.0 technologies: implementation patterns in manufacturing companies. Int. J. Prod. Econ. **210**, 15–26 (2019)
9. Wagner, R., Schleich, B., Haefner, B., Kuhnle, A., Wartzack, S., Lanz, G.: Challenges and potentials of digital twins and Industry 4.0 in product design and production for high performance products. Procedia CIRP **84**, 88–93 (2019)
10. Salama, S., Eltawil, A.B.: A decision support system architecture based on simulation optimization for cyber-physical systems. Procedia Manuf. **26**, 1147–1158 (2018)
11. Mandic, V.: Virtual engineering. Mechanical Engineering Faculty of the University of Kragujevac, Kragujevac, Serbia (2007). (in Serbian)
12. Mandic, V., Cosic, P.: Integrated product and process development in collaborative virtual engineering environment. Tech. Gaz. **18**(3), 369–378 (2011)
13. Bikas, H., Stavropoulos, P., Chryssolouris, G.: Additive manufacturing methods and modelling approaches: a critical review. Int. J. Adv. Manuf. Technol. **83**, 389–405 (2016)
14. Shahrubudina, N., Leea, T.C., Ramlana, R.: An overview on 3D printing technology: technological, materials, and applications. In: 2nd International Conference on Sustainable Materials Processing and Manufacturing (SMPM 2019) (2019). Procedia Manufacturing 35, pp. 1286–1296
15. Horst, D.J., Duvoisin, C.A., de Almeida Vieira, R.: Additive manufacturing at Industry 4.0: a review. Int. J. Eng. Tech. Res. (IJETR) **8**(8), 3–8 (2018)
16. Stojanovic, G., Mandic, V., Curcic, M., Vasiljevic, D., Kisic, M., Radosavljevic, N.: Combining rapid prototyping techniques in mechanical engineering and electronics for realization of a variable capacitor. Rapid Prototyp. J. **20**(2), 115–120 (2014)
17. Dilberoglua, U.M., Gharehpapagha, B., Yamana, U., Dolena, M.: The role of additive manufacturing in the era of Industry 4.0. Procedia Manuf. **11**, 545–554 (2017)
18. Lunback, A., Lindgren, L.-E.: Finite element simulation to support sustainable production by additive manufacturing. Procedia Manuf. **7**, 127–130 (2016)

19. Metallic Additive Manufacturing Process Simulations, ESI Group. https://www.esi-group.com/software-solutions/virtual-manufacturing/additive-manufacturing
20. Rosenthala, S., Hahna, M., Tekkayaa, A.E.: Simulation approach for three-point plastic bending of additively manufactured Hastelloy X sheets. In: 47th SME North American Manufacturing Research Conference, Pennsylvania, USA (2019). Procedia Manufacturing 34, pp. 475–481
21. Mandic, V., Eric, D., Adamovic, D., Janjic, M., Jurkovic, Z., Babic, Z., Cosic, P.: Concurrent engineering based on virtual manufacturing. Tech. Gaz. **19**(4), 885–892 (2012)
22. Sharma, P.: Concept of virtual manufacturing. GRD J. Global Res. Dev. J. Eng. **2**(6), 2455–5703 (2017)
23. Dobrescu, R., Merezeanu, D., Mocanu, S.: Process simulation platform for virtual manufacturing systems evaluation. Comput. Ind. **104**, 131–140 (2019)
24. Virual Manufacturing, The Next Revolution in Global Manufacturing, MSC.Software Corporation (2001). http://www.mscsoftware.com/assets/1776_vm_2001.pdf
25. Souza, M.C.F., Sacco, M., Porto, A.J.V.: Virtual manufacturing as a way for the factory of the future. J. Intell. Manuf. **17**, 725–735 (2006)
26. Ahmeda, M.B., Saninb, C., Szczerbickic, E.: Smart virtual product development (SVPD) to enhance product manufacturing in Industry 4.0. In: 23rd International Conference on Knowledge-Based and Intelligent Information & Engineering Systems (2019). Procedia Computer Science 159, pp. 2232–2239
27. Hagenah, H., Schulte, R., Vogel, M., Hermann, J., Scharrer, H., Lechner, M., Merklein, M.: 4.0 in metal forming – questions and challenges. In: 12th CIRP Conference on Intelligent Computation in Manufacturing Engineering, CIRP ICME 2018 (2019). Procedia CIRP 79, pp. 649–654
28. Bauza, M.B., Tenboer, J., Li, M., Lisovich, A., Zhou, J., Pratt, D., Edwards, J., Zhang, H., Turch, C., Knebel, R.: Realization of Industry 4.0 with high speed CT in high volume production. CIRP J. Manufact. Sci. Technol. **22**, 121–125 (2018)
29. Djukic, T., Mandic, V., Filipovic, N.: Virtual reality aided visualization of fluid flow simulations with application in medical education and diagnostics. Comput. Biol. Med. **43**(12), 2046–2052 (2013)
30. Roldán, J.J., Crespo, E., Martín-Barrio, A., Peña-Tapia, E., Barrientos, A.: A training system for Industry 4.0 operators in complex assemblies based on virtual reality and process mining. Robot. Comput. Integr. Manuf. **59**, 305–316 (2019)
31. Ceruti, A., Marzocca, P., Liverani, A., Bil, C.: Maintenance in aeronautics in an Industry 4.0 context: the role of augmented reality and additive manufacturing. J. Comput. Des. Eng. **6**, 516–526 (2019)
32. Mandic, V., Stefanovic, M., Gavrilovic, Z.: Development of the forging technology for producing the artificial hip stem through application of virtual manufacturing. In: Proceedings of XII International Conference KODIP 2014, Budva, Montenegro, pp. 96–105 (2014)

New Trends in Machine Design Within Industry 4.0 Framework

Radivoje M. Mitrović[1], Ivana Atanasovska[2](✉), Natasa Soldat[1], and Zarko Miskovic[1]

[1] Faculty of Mechanical Engineering, University of Belgrade, Belgrade, Serbia
[2] Mathematical Institute of the Serbian Academy of Sciences and Arts, Belgrade, Serbia
iatanasovska@mi.sanu.ac.rs

Abstract. The basic framework of the Industry 4.0 as an approach and philosophy implies the introduction of new procedures and principles, as well as the development and improvement of machine systems towards the introduction of a high level of manufacturing digitalization. In order to meet the achievements of Industry 4.0, very significant changes are also required in the design, monitoring and maintenance phases of machinery. Therefore, all the research and advancements in the field of calculation and design of machine elements and assemblies have to be observed within the paradigm of Industry 4.0. This paper outlines the main research tasks and aims within the determination of basic principles and methods for suiting and improving machine elements and systems in the context of the requirements of the Industry 4.0. Also, the part of the paper gives the description of the new improved methodology and systems for monitoring the parameters of rolling bearings during their operation, which would significantly contribute to the prediction of the possible failure of the rolling bearings and lead to important savings. The developed methodology is a sample for new trends in the context of the vision of Industry 4.0 machinery and respects the requirements for safety, energy efficiency and reliability.

Keywords: Industry 4.0 · Machine Design · Safety · Efficiency · Ball bearings

1 Introduction

The interest of the experts and scientists about the developments and implementations in the framework of the new industrialization, called Industry 4.0, has increased in recent years. The Industry 4.0, as a term, philosophy, and in a general sense, as a set of new and completely different approaches, has been gaining more and more support from year to year and is becoming a dominant trend. The most adequate terms associates with Industry 4.0 indisputably are: Digitalization, Factories of the Future, Computer Systems and Smart Machinery. But the scope of the new paradigm, referred to as Industry 4.0 or the fourth industrial revolution, is certainly much broader and enters all the pores of modern engineering, even everyday life, [1, 2]. In this context, we must not dismiss the security and economic aspects, for which in the coming period are expected to bear the greatest benefits of the new industrialization achievements.

L. Wang et al. (Eds.): AMP 2020, LNME, pp. 227–238, 2020.
https://doi.org/10.1007/978-3-030-46212-3_16

IT engineers, electronics and manufacturing engineers were already recognized their roles and involvements in Industry 4.0 trends and in the development of the Factories of the Future (FoF). The development of computer systems, as well as of the mathematical methods which follow the development of high-performance systems for data transfer, data processing, decision support and optimization, has contributed to the creation of new opportunities in all areas of manufacturing and industry and has in some ways led to the appearance of a new approach called Industry 4.0 [3]. The development of electronic devices and new materials are also research areas whose major advances in recent decades have laid the basis for new industrialization. All these aspects have already been discussed and published, and in this paper we can appoint them as a starting point for considering the development and research in Machine Design within Industry 4.0.

2 Machine Design Within Industry 4.0 Framework

The introduction of Industry 4.0 in Machine Design and vice versa can and must be discussed in interaction, comprehensively, in detail, but in same time critically. All research in this area must be considered within the framework of methods and objectives that will contribute to the improvement of safety, reliability, energy efficiency and environmental protection. A representation of the implementation and the impact of Industry 4.0 on Machinery and Machine Design are shown on Fig. 1.

Also, for defining the research goals of the Industry 4.0 framework in Machine Design, the impact of the expected improvements on generally defined economic goals, energy savings, sustainable development and the environment, as well as on education and social peace must not be excluded. The advancement of industry development and the introduction of a high level of digitalization and robotization in manufacturing facilities and other manufacturing activities can have a very critical effect on reducing the general level of employment and the appearance of a lack of adequately educated staff. Therefore, in line with the development of industry within the paradigm of Industry 4.0, intensive attention must be paid to adequate education, changes in educational programs, as well as additional education of working-age human resources, and human-machine-human interaction in general [4]. Another aspect that should not be forgotten is the ethical, social and national codes, which must protect the privacy and emotional status of people, [5]. Therefore, as an inevitable part of development, great attention is paid to all forms of electronic data protection, and major research in cybersecurity.

The issue of applying Industry 4.0 and all its undeniable benefits to raising standards in human lives and significantly increasing profits, leads to the conclusion that every step in the implementation of Industry 4.0 must be detail viewed, and that different experts besides engineers must be involved, such as economists, psychologists, teachers and others.

In this paper, the main focus is on defining possible directions for the application of Industry 4.0 in the field of Machine Design. However, the authors have tried to define their research in the context of all above discussed observations and conclusions.

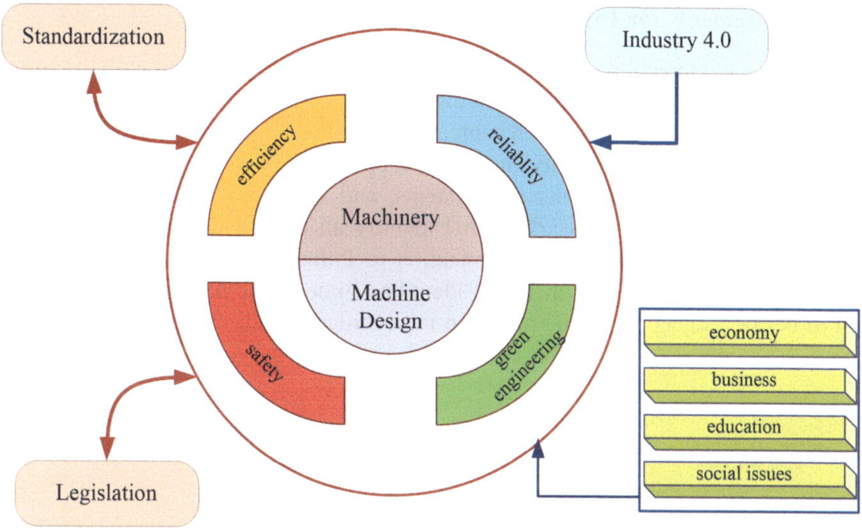

Fig. 1. Machine Design within Industry 4.0 framework.

Particular attention in this context deserves standardization: developing, harmonizing and implementing new technical standards and national legislation that will accompany the development of Smart Devices, Smart Machinery and Smart Factories, [3].

The machinery safety requirements, which are already subject of special attention for more than a decade, should also be mentioned. The holders of the mandatory safety rules, which have been incorporated into a large number of national regulations, have noticed in recent years the need to amend the current Directive on machinery (Directive 2006/42/EC of the European Parliament and of the Council of 17 May 2006) in the context of an Industry 4.0 perspective. The amending the Directive on machinery by the TC199 Machine Safety Technical Committee within the ISO (the International Organization for Standardization) is underway. Also, Technical Comity TC199 is working to adopt new standards of critical importance for safety in Industry 4.0 [6], such as ISO/DIS 21260 "Safety of machinery - Mechanical safety data for physical contacts between moving machinery or moving parts of machinery and persons" [7], which is adopted in collaboration with ISO/TC299 (Robotics) and which specifies limits for physical contacts between the machine or parts of the machine and humans that are caused by the movement of the machine as part of its intended use or foreseeable misuse. Technical Comity ISO/TC199 has also issued Technical report IS/TR 22100-4 "Safety of machinery - Relationship with ISO 12100. Part 4: Guidance to machinery manufacturers for consideration of related IT security (cyber security) aspects" [8], which consists of two main parts: one that analyzes the roll of IT Security for Machinery Safety, while the other is the practical guidance for considering IT security aspects by manufacturers.

The basic requirements and standards related with the European Framework Directive on Safety and Health at Work (Directive 89/391 EEC) must also be considered within the possible implications of the application of Industry 4.0 to occupational safety and health, [9].

2.1 Requirements for Digitalization in Machine Design

Considering the interaction of R&D in Machinery and Machine Design with the current tendencies of Industry 4.0, it is clear that adapting all disciplines and research fields to the new digitalization goals is inevitable. As a high degree of such development is already achieved practically only in the field of CAD/FEA systems for design and calculation of machine elements and systems, and CAM (Computer Added Manufacturing) systems, it is necessary to first identify all the requirements that Machine Design must fulfill in terms of customization to Industry 4.0 principles in the next period. The algorithm shown on Fig. 2 identifies some of the basic requirements that need to be the subject of special attention in accordance with the specified goal.

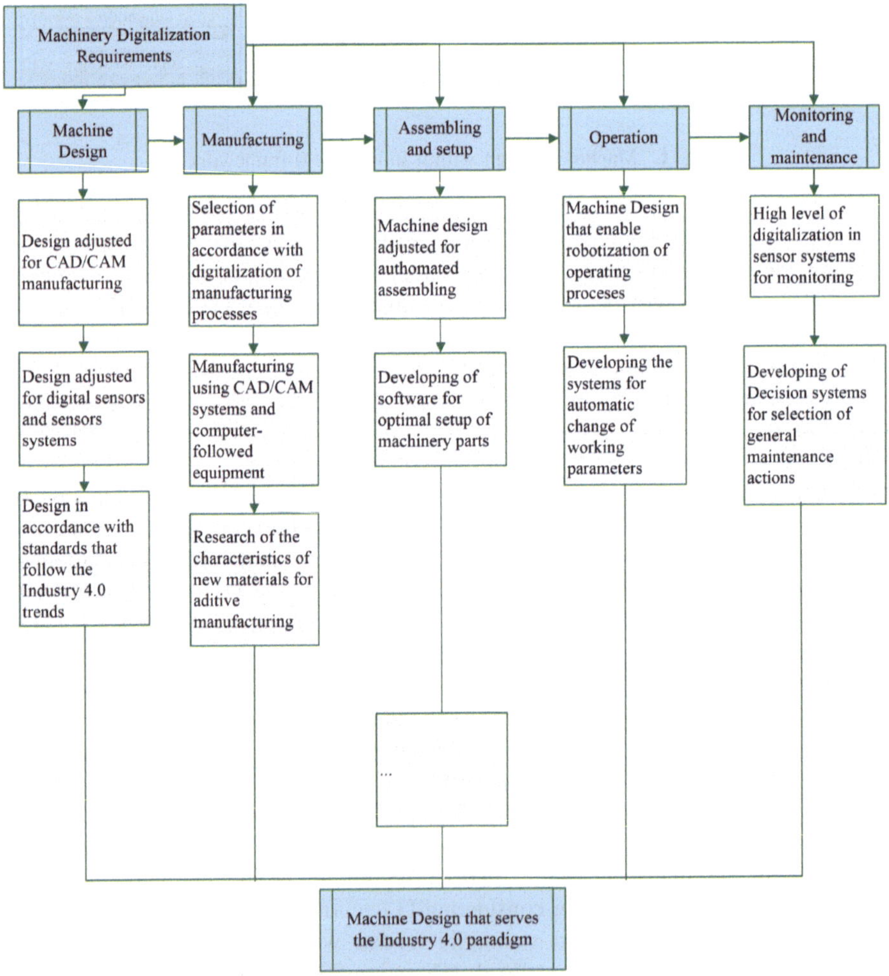

Fig. 2. Requirements for digitalization in Machine Design.

2.2 Benefits of Digitalization in Machine Design

Comparable to defining the requirements in front of the Machine Design experts, the benefits that can be count on in this area by introducing the benefits of digitalization, smart sensors, data transfer, data processing, and other Industry 4.0 capabilities can be identified. An overview of the basic benefits in this area is shown by the algorithm given on Fig. 3.

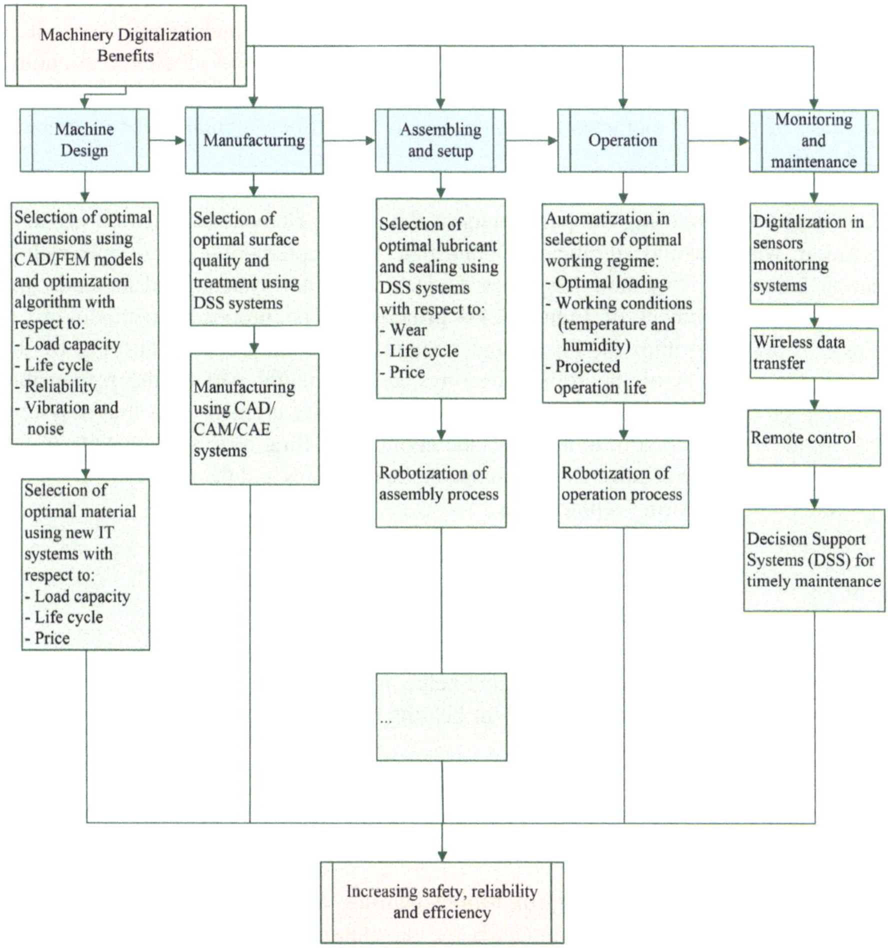

Fig. 3. Digitalization benefits in Machine Design.

The algorithms shown on Fig. 2 and Fig. 3 can serve as guidelines for the development of sub-disciplines within Machine Design and must be complemented and discussed in further interactions. The existing of the undeniable expected benefits of implementing the capabilities of Industry 4.0, but also the need for great commitment and new multidisciplinary research by scientists and experts in the discussed field stem from the analysis of the presented algorithms.

3 Rolling Bearings Monitoring and Damage Prediction Within Industry 4.0 Framework – The New Methodology

Following the above postulates, results and observations, it is clear that significant improvements and savings can be achieved in many areas and stages of the operation life of machine elements: design, manufacturing, operation and maintenance. It is indisputable that the specific benefits at the same time in terms of efficiency, economy and safety can be expected through the development of new systems for monitoring the condition of machine elements and systems and their timely maintenance or replacement, [10].

This part of the paper defines and describes the new developed methodology and the system for monitoring the parameters of the rolling-element bearings during their operation, which would significantly contribute to the prediction of possible rolling bearing failure. This methodology uses devices, computer systems and methods that follow the development of Industry 4.0 principles. The proposed methodology is shown by the algorithm on Fig. 4 and contains the continuous monitoring of the operating parameters of the rolling bearings as an integral part of the responsible assembly or structure, and continuous processing of measurement data into a system that would be developed in order to decide about the rolling bearing's capacity in the condition of possible occurrence of damage at the raceways and the requirement for the load reduction or bearing replacement.

The basic steps of this methodology are briefly described below:

- Basic design of rolling-element bearing
- Finite Element Analysis of rolling-element bearing
- Manufacturing and assembly of rolling-element bearing
- Mathematical prediction for Dynamic behavior of rolling-element bearing
- Sensor monitoring of rolling-element bearing
- Decision Support System
- Maintenance action

3.1 Basic Design of Rolling-Element Bearing

Figure 5 shows the step (phase) of rolling bearing design with all the influencing factors. In addition to existing standards for calculations and applicable national legislative, a design in according with the Industry 4.0 must focus on the application of new materials, more precise systems for parts manufacturing and finishing, and therefore more accurate considering of measures and tolerances. The trends in the bearings design are thus directed towards the savings and optimization of all parameters of the rolling bearings, from the choice of type, dimensions, operating characteristics, considering the necessary safety conditions and adapting to new computer-monitored manufacturing systems and proactive maintenance systems. Therefore, the necessary changes in the shape of the bearing elements for the purpose of simple and precise automatic manipulation during assembling and replacement, as well as for the installation of sensors and other measuring devices for monitoring the parameters must be foreseen in the earliest design stages of rolling bearings.

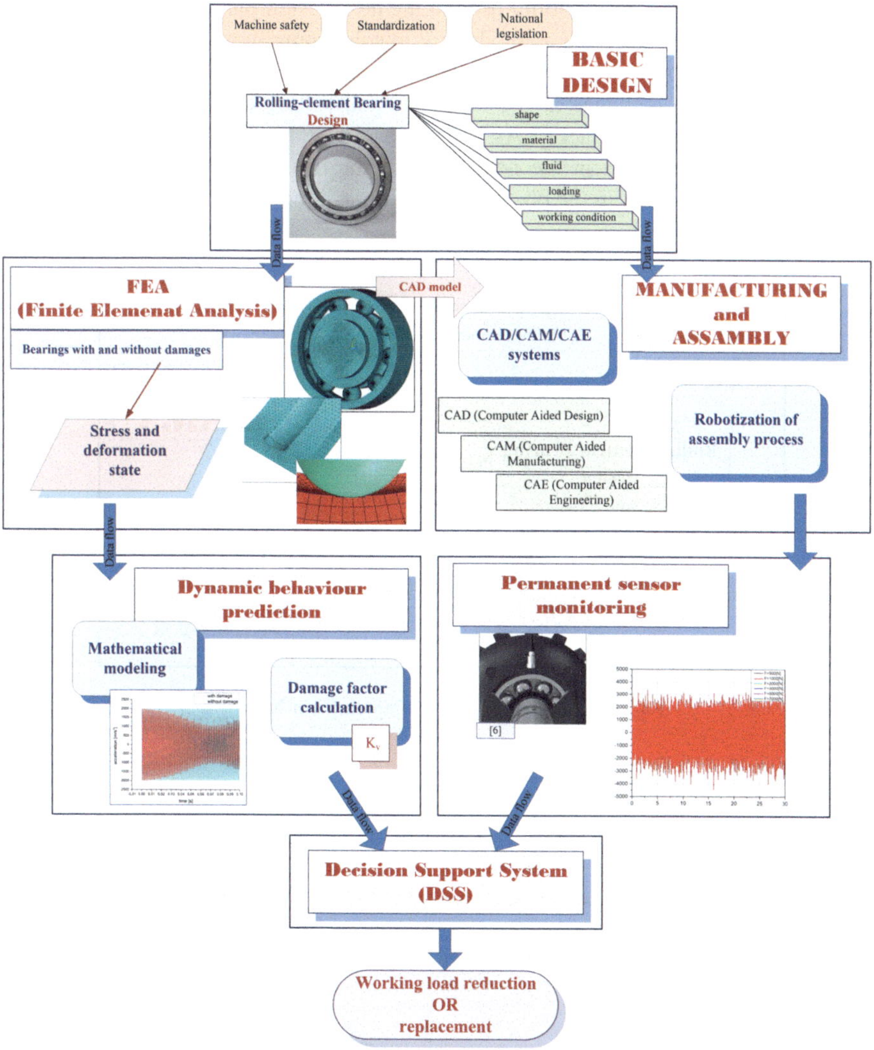

Fig. 4. Methodology for damage prediction on rolling-element bearing

3.2 Finite Element Analysis of Rolling-Element Bearing

An indispensable part of all modern calculations and systems for ensuring the reliability, efficiency and safety of machine elements and systems in recent decades has been the calculation and analysis by the finite element method. This phase shown on

Fig. 6 is also an important part of the methodology developed and described. In this case, the FEA (Finite Element Analysis) plays a multiple role. First, as a recognized method, it is the verified approach to control the load capacity of the selected rolling bearing type and dimensions for specific operating purposes. Also, as can be seen from the diagram shown on Fig. 4, the CAD model of the rolling bearing from this phase of the methodology can be successfully used in the next design phase. However, the FEA has also a important role in this methodology in calculating the stress and deformation state of the rolling bearings in case of varying the load conditions, and taking into account the possible damages at the raceways and rolling elements. High-precision geometry modeling and contact FEAs at this stage guarantee the very accurate results obtained [11, 12], required in the next steps of the methodology. A series of these results is the starting point in the phase of mathematical modeling of the rolling bearing dynamic behavior.

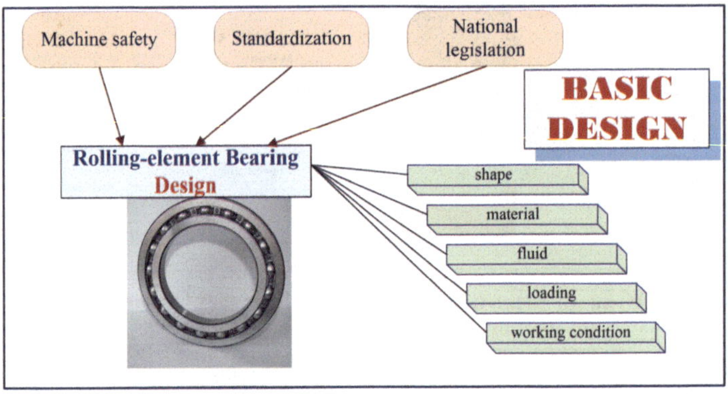

Fig. 5. Basic design of rolling-element bearing

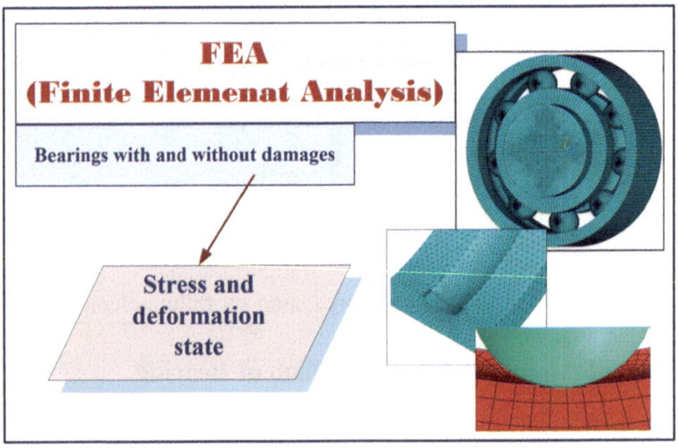

Fig. 6. Finite Element Analysis of rolling-element bearing

3.3 Manufacturing and Assembly of Rolling-Element Bearing

The previously described rolling bearing design and modeling phases follow the future of manufacturing brought by Industry 4.0 and have to be complementary with CAD/CAM/CAE systems and the complete digitalization of rolling bearing manufacturing and assembling. CAD (Computer Added Design)/CAM (Computer Added Manufacturing)/CAE (Computer Added Engineering) systems are high-level integrated computer and digital systems, which have been developed and applied decades ago. The development of these systems can be considered as one of the basic preconditions and precursors to the complete digitalization that Industry 4.0 has been aspiring for in recent years. Without these systems and adequate engineering education for developing and using the advantages of the Industry 4.0 would be unthinkable. Therefore, it is not unusual that the engineers and experts for these systems were among the first to be involved in the realization of the goals of the Industry 4.0 and in the development of Smart Manufacturing and Smart Factories. Accompanied by decades of development of robots and robotic systems in the industry, these systems are the basis for the manufacturing and assembling of all machine elements, as well as of the rolling bearings, which is shown in the developed methodology by a special phase (step), given on Fig. 7.

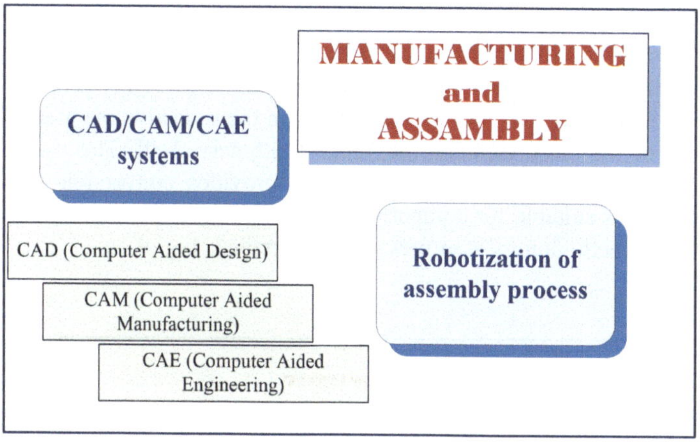

Fig. 7. Manufacturing and assembly of rolling-element bearing

3.4 Mathematical Prediction for Dynamic Behavior of Rolling-Element Bearing

A special contribution of the developed methodology is given by the phase of mathematical modeling of the dynamic behavior of the rolling bearings, Fig. 8, which is detail described in the previous works of the authors [11–15]. This phase involves the development of a program for mathematical calculation (prediction) of vibrations under given operating conditions and for the cases when the damage exists at raceways of inner and/or outer bearing ring and/or rolling elements. Also, the introduction of a new impact factor, its definition and calculation are provided. The data obtained in this step represents the input to the final phase (step) - decision making about the required action in the framework of the predictive maintenance.

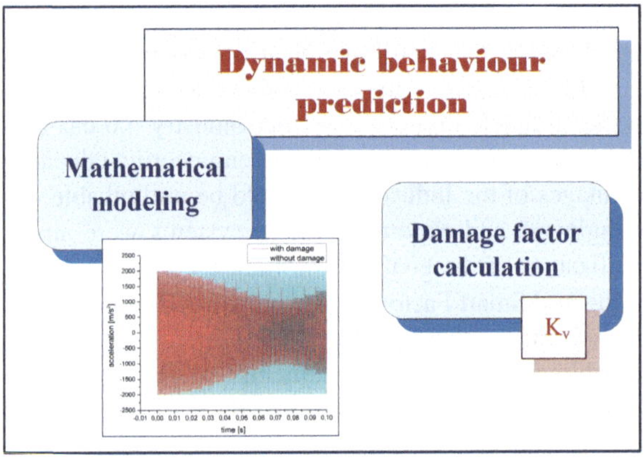

Fig. 8. Mathematical prediction for dynamic behavior of rolling-element bearing

3.5 Sensor Monitoring of Rolling-Element Bearing

The step of permanent sensor monitoring implies the use of highly sensitive vibration sensors fully digital and remotely monitored within the proactive maintenance system, which is one of the basic elements of Smart Machinery, [10]. The use of specially developed and customized computer programs provides conversion of the output signals to variables suitable for comparison with the values obtained by mathematical modeling in the methodology step previously described, Fig. 9.

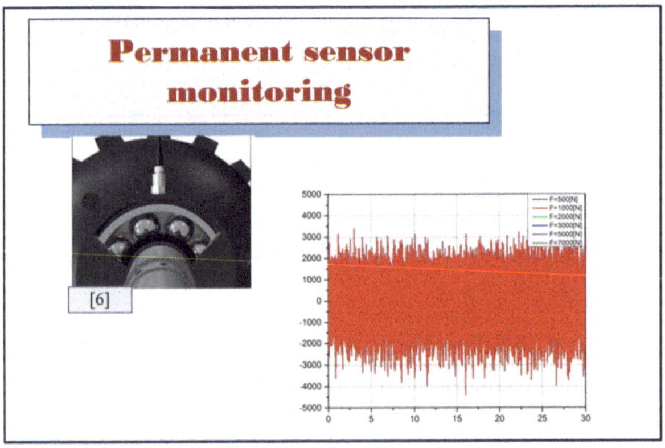

Fig. 9. Sensor monitoring of rolling-element bearing

3.6 Decision Phase

The last but not least important or simpler step in the developed methodology is to decide on the maintenance action to be taken in order to achieve a high level of safety and savings. This phase involves developing a Decision Support System (DSS) that will contain all the necessary expertise and adapt to the new developments of Machine learning, as one of the most important disciplines of contemporary Artificial Intelligence.

4 Conclusions

The research results presented in this paper clearly indicate that the intensive research and improvements in the field of calculation and design of machine elements and systems are necessary. Thus, this area of mechanical engineering, considered closed for significant shifts and achievements for decades, is opening up as a new research field with great potential, above all in terms of economic savings, reliability and safety. In this paper the basics of the necessary directions for the development of machine elements and systems in order to their adaptation to the main principles of the Industry 4.0 are defined. An overview of the potential benefits of applying Industry 4.0 in this area has also been research and given. Definitely, the identification of more benefits can be expected in the future.

Developing a system of proactive and predictive maintenance has been recognized as one of the major contributing disciplines in next years. A proposal for a methodology and system for monitoring the parameters of the rolling bearings during their operation, which would significantly contribute to the prediction of possible rolling bearing failure, was developed and presented in the paper. It should be emphasized that the proposed methodology defined continuous monitoring of the condition of the rolling bearings, primarily as an integral part of the responsible assembly or structure, and continuous processing of measurement data into a system that would be developed in order to decide on the condition of the rolling bearings in the event of possible occurrence of damage at raceways or rolling elements, and the requirements for reduction the load or bearing replacement. Benefits of such a developed system can be expected in terms of safety, cost reductions due to the failure of large and significant machines and plants, or a longer out of work of the operation due to replacement of failed parts. The proposed methodology is based on the application of state-of-the-art computer systems as well as digital manufacturing and maintenance processes. In the view of the expected increasing development of these areas, its implementation and improvement can be expected, too.

Acknowledgment. Parts of the presented research are funded by the Ministry for education, science and technological development of the Republic of Serbia, through the Faculty of Mechanical Engineering of University of Belgrade and the Mathematical Institute of the Serbian Academy of Sciences and Arts, Belgrade, Serbia.

References

1. Majstorovic, V., Mitrovic, R.: Industry 4.0 programs worldwide. In: Proceedings of the 4th International Conference on the Industry 4.0 Model for Advanced Manufacturing (AMP 2019). Lecture Notes in Mechanical Engineering, Belgrade, pp. 78–99. Springer (2019)
2. Rojko, A.: Industry 4.0 concept: background and overview. iJIM **11**(5), 77–90 (2017)
3. Waschbusch, L.: IoT Basics: What Does Industry 4.0 Mean? (2019). https://www.spotlightmetal.com/iot-basics-what-does-industry-40-mean-a-842216/?cmp-=go-aw-art-trf-SLM_DSA-20180820&gclid=Cj0KCQiAvc_x%E2%80%A6
4. Gorecky, D., et al.: Human-machine-interaction in the Industry 4.0 era. In: 12th IEEE International Conference on Industrial Informatics (INDIN), pp. 289–294 (2014)
5. Badri, A., et al.: Occupational health and safety in the industry 4.0 era: a cause for major concern? Saf. Sci. **109**, 403–411 (2018)
6. Steiger, G.: New digital production technologies - challenges for machinery safety standardisation. https://www.euroshnet.eu/fileadmin/Redaktion/Presentations/1_Steiger_Digi-talisation_in_the_machinery_sector-implications_on_safety.pdf. Accessed 31 Jan 2020
7. ISO/DIS 21260. Safety of machinery — Mechanical safety data for physical contacts between moving machinery or moving parts of machinery and persons (2018)
8. ISO/TR 22100-4. Safety of machinery — Relationship with ISO 12100. Part 4: Guidance to machinery manufacturers for consideration of related IT-security (cyber security) aspects (2018)
9. Negre, T.: Safety 4.0 - Smart Machines & Factories (2017). https://smartmachinesandfactories.com/news/fullstory.php/aid/173/Safety_4.0.html. Accessed 31 Jan 2020
10. Canito, A., et al.: An architecture for proactive maintenance in the machinery industry. In: Ambient Intelligence—Software and Applications—8th International Symposium on Ambient Intelligence (ISAmI 2017). Advances in Intelligent Systems and Computing, vol. 615, pp. 254–262. Springer (2017)
11. Mitrović, R., et al.: Effects of operation temperature on thermal expansion and main parameters of radial ball bearings. Therm. Sci. **19**(5), 1835–1844 (2015)
12. Soldat, N., et al.: The methodology for analyzing the radial ball bearing vibrations. Trans. FAMENA **44**(1), 13–28 (2020)
13. Atanasovska, I., et al.: Nonlinear dynamics as a tool in selection of working conditions for radial ball bearing. In: Nase sa izracunavanjem vibracija, IUTAM Bookseries 37 - IUTAM Symposium on Exploiting Nonlinear Dynamics for Engineering Systems, pp. 49–58. Springer (2020)
14. Mitrović, R., et al.: Dynamic behaviour of radial ball bearing due to the periodic variable stiffness. J. Mach. Des. **7**(1), 1–4 (2015)
15. Atanasovska, I.: The mathematical phenomenological mapping in non-linear dynamics of spur gear pair and radial ball bearing due to the variable stiffness. Int. J. Non-Linear Mech. **73**, 114–120 (2015)

A Cloud Computing Model for Achieving Competitiveness of Domestic Enterprises

Dejan Đorđević[1], Dragan Ćoćkalo[1], Mihalj Bakator[1](✉), Srđan Bogetić[2], Miloš Vorkapić[3], and Cariša Bešić[4]

[1] Technical Faculty "Mihajlo Pupin", University of Novi Sad, Đure Đakovića bb, Zrenjanin 23000, Republic of Serbia
mihalj.bakator@uns.ac.rs

[2] Belgrade Business and Arts Academy of Applied Studies, Kraljice Marije 73, 11000 Belgrade, Republic of Serbia
srdjan.bogetic@bbs.edu.rs

[3] Institute of Chemistry, Technology and Metallurgy - Centre for Microelectronic Technology and Mono-Crystals, University of Belgrade, Njegoševa 12, 11000 Belgrade, Republic of Serbia
worcky@nanosys.ihtm.bg.ac.rs

[4] Faculty of Technical Science in Ĉaĉak, University of Kragujevac, Svetog Save 65, 32000 Ĉaĉak, Republic of Serbia
car.besic@gmail.com

Abstract. Amidst the globalization of markets and within the framework of the fourth industrial revolution - Industry 4.0, domestic enterprises face challenges when it comes to business and market performance. Domestic enterprises lack necessary productivity and product quality in order to compete on the global market. In this paper cloud computing solutions in various industries are analyzed. The goal is to develop a cloud computing model for improving the competitive ability of domestic enterprises. The model integrates several cloud-based solutions and takes into consideration the business metrics of domestic enterprises. Furthermore, future trends in cloud computing advances and its application are discussed. The paper also presents suggestions and guidelines for improving competitiveness of domestic enterprises through the application of cloud computing solutions. The results of this study significantly contribute to the existing body of literature and provide a solid basis for future research in the domain of cloud computing application in industries in developing countries.

Keywords: Domestic enterprises · Cloud computing · Industry 4.0 · Competitiveness · Model

1 Introduction

The globalization of markets has put a tremendous pressure on enterprises when it comes to achieving and maintaining a competitive position on the international market. The negative effects of globalization on the business performance of enterprises are even more noticeable in Serbia, where the lack of modern equipment, low productivity, and low product quality additionally contribute to poor competitive ability of domestic

L. Wang et al. (Eds.): AMP 2020, LNME, pp. 239–249, 2020.
https://doi.org/10.1007/978-3-030-46212-3_17

enterprises. In order to increase their competitive ability, domestic enterprises, and enterprises overall, have to develop new, high quality products which will satisfy the needs and wishes of customers. In addition to the pressure put on by the globalized markets, enterprises have to transform, adapt and to conduct business within the frameworks of the fourth industrial revolution - Industry 4.0 (Popkova et al. 2019). Some of the main technologies which characterize Industry 4.0 are Radio Frequency Identification (RFID); social product development cyber-security; 3D printing; cloud computing and its industrial applications (cloud manufacturing, cloud logistics); Internet of Things (IoT); Internet of Value (IoV); advanced robotics; wireless sensors; and other advanced technologies (Lu and Xu 2019). This paper focuses on cloud computing applications for increasing competitive ability of enterprises. Cloud computing can be viewed as an integrated whole, on-demand network model, where computing resources are shared. These resources are configurable and can take the form of networks, mass storage, servers, applications and services. Cloud computing as a model had, and still has, a tremendous impact on information-communication technologies (ICT), and it "cemented" its position as highly scalable, applicable and adaptable technology in the domain of information technologies (IT). Cloud computing has found its place in various industries in the form of cloud manufacturing, cloud logistics, Internet of Things, cloud security and other business solutions (Zhou et al. 2018). Now, for example, the benefits of switching from traditional enterprise systems to cloud enterprise systems depends on several factors such as compatibility, financial support, industry pressure and support from vendors (Liu et al. 2017). It can be argued that cloud computing and its implementation can positively affect business performance, but also that the improvement depends on other business metrics as well.

In this paper the application of cloud computing technologies in achieving competitiveness of domestic enterprises in Serbia is analyzed. The main goal is to develop a theoretical model based on previous findings in the domain of cloud manufacturing, cloud infrastructure, cloud logistics and other integrated cloud-based solutions. Based on the developed model, suggestions and guidelines for improving the competitiveness of domestic enterprises within the framework of Industry 4.0, are discussed. The whole aim of the paper can be integrated into the following proposed hypothesis: *H: Implementing cloud computing-based solutions can increase the competitiveness of domestic enterprises*. Further, the testing process of the proposed hypothesis is guided by the following research questions: *What are the main issues of domestic enterprises when it comes to international competitiveness? How and what type of cloud computing solutions can be applied within various industries? How can cloud computing affect the business performance of domestic enterprises?*

The whole paper is consisted of four main sections (excluding the *Introduction* and *Conclusion* sections). First, the paper addresses domestic enterprises and their lack of competitive ability on the international market. Afterwards, a more detailed review of cloud computing technology solutions are analyzed. Further, a model is developed and potential trends in the domain of cloud computing advancement in Serbia are discussed. From here, suggestions and guidelines for improvement are presented. Finally, conclusions are drawn and future research is suggested.

2 Serbian Economy and the Competitiveness of Domestic Enterprises

Globalization and Industry 4.0 are dictating new market paradigms. In Serbia, the majority of SMEs lack adequate technologically advanced manufacturing equipment, productivity, product and service quality and ineffective use of knowledge and organizational infrastructure (Ćoćkalo et al. 2019). The lack of productivity and quality in domestic enterprises negatively affects their competitiveness on the international market. The low quality, not innovative and not technologically enriched products, as well as, their selling for a uncompetitive price can't increase the competitive ability of domestic SMEs. Therefore, domestic enterprises have to establish platforms (in this current paper the focus is on cloud platforms) for R&D projects which will answer the demand from the globalized market. Furthermore, current, foreign investments are labor-intensive, thus they don't majorly contribute to technological development (Ćoćkalo et al. 2019). Another problem, may come from a more macro, government-level. Namely, while West European countries invest around 2% of their GDP into R&D (Radulescu et al. 2018), Serbia invests only around 0.93% (World Bank Group 2019). Additionally, taken into consideration that Serbia has a significantly lower overall GDP, it is evident that there is not enough development and modern technology application. It is necessary to establish a structured, long-term project which will enable, and motivate SMEs to implement and apply modern technologies with the goal to increase their competitiveness, thus increasing the competitiveness on a national level as well. The lack of competitiveness on a national level is also evident in the Competitiveness reports published by the World Economic forum. The indicators used in the determination of the ranks were: Institutions, Infrastructure, ICT, Macroeconomic stability, Healthcare, Skills, Products market, Labor market, Financial system, Market size, Business dynamics, and Innovation capability. The data on competitiveness ranks are presented in Table 1.

Table 1. Competitiveness ranks of Serbia, its neighbors and developed EU countries.

	2015	2016	2017	2018	2019
Serbia	84	90	78	65	72
Croatia	77	74	74	68	63
Slovenia	59	56	48	35	35
North Macedonia	60	68	/	84	82
Bosnia and Hercegovina	111	107	103	91	92
Montenegro	70	82	77	71	73
Romania	53	62	68	52	51
Bulgaria	54	50	49	51	49
Hungary	63	69	60	48	47
Albania	93	80	75	76	81
Austria	23	19	18	22	21
Germany	4	5	5	3	7
United Kingdom	10	7	8	8	9
France	22	21	22	17	15
Total number of countries evaluated that year	*140*	*138*	*137*	*140*	*141*

(Source: WEF reports from 2014 to 2019)

Based on Table 1 it can be seen that in the last five years Serbia slowly achieved slightly higher competitiveness rankings. However, among the presented countries, only North Macedonia, Montenegro and Bosnia and Hercegovina have lower competitiveness ranks. Overall, this indicates that there is potential, but also that there is a lack of momentum, which further indicates that dramatic changes are necessary for a stronger momentum in the right direction which is broad implementation of modern ICT in SMEs, with the help of government incentives with the goal to achieve competitiveness on the global market.

3 Cloud Computing Applications in Various Industries

Cloud computing technologies present a significant way of increased data sharing across a company, within the framework of Industry 4.0. This data sharing can result in with improved system performance in the form of increased flexibility, adaptability and agility. In the same study it was noted that there are existing cloud-based manufacturing systems, however due to various parts of an enterprise (different modules, subsystems, machines, tools, devices etc.). Some of the main advantages of cloud computing include shared resources for applications, servers, and networks; higher performance through bigger networks which allow higher processing capabilities; reducing costs through pay-per-use business models where costs are linear to the resources used; no need for maintenance as maintenance is carried out by the cloud-service providers; massive scalability options and adaptable infrastructure; self-regulating and self-organizing of resources; fast implementation measured in days, sometimes just hours; large number of providers which leads to competition-driven higher quality service; sustainability as cloud infrastructures don't require massive data centers, thus limiting the negative effects of excessive power consumption; virtualization which enables separating different aspects of business from one another (logistics from manufacturing, packaging etc.); and the possibility to utilize Internet technologies and applications (Radwan et al. 2017).

Further, lets analyze the various applications of cloud-based technologies in various industries. One of the widely spread main cloud application is cloud manufacturing, which can be viewed as a service oriented model which distributes manufacturing resources with the goal to reduce costs and improve productivity. This approach can be implemented in a way where the cloud platform can manage multiple manufacturing task and procedures simultaneously (Liu et al. 2017). In the same research it was pointed out that this type of approach to manufacturing task management is usually supported by the logistics department. The cloud manufacturing architecture consists of four main layers. These are the manufacturing resource layer (production facility, design center, shipping facility, operational management center); virtual resource layer (contains the cloud platform); service layer; and application layer which is accessed by the user (Esposito et al. 2016).

Besides, managing manufacturing tasks, cloud manufacturing approaches may integrate robotic applications with the goal to reduce labor costs and to increase productivity. Within this, cloud robotic application approach, the communication routes integrate the cloud database, robots and sensors, and human resource knowledge

(Wang et al. 2017). Basically, this means that the human knowledge is communicated on the cloud platform and a remote control is established with a specific manufacturing robot (robotic hand, welding hand, stamping, laser engraving etc.). Now, the main advantage of this approach is that the whole process is monitored and evaluated in real time, thus timely optimization can occur. Information in the form of sound or image are processed along with other raw data. So far, it is evident that cloud manufacturing offers immensely more options when it comes to scalability, adaptability, and re-configuration of manufacturing processes. Cloud manufacturing can be viewed as advanced model of cloud-based solutions and it can integrate Internet of Things, service-oriented technologies, and virtualization (Majstorovic and Mitrovic 2019).

Further, manufacturing task schedule management, and real-time manufacturing tracking and optimization, cloud computing can also be applied in the process of predictive maintenance. This process includes the measurements and metric data from the production process. Collects them and processes them in order to timely detect potential machine or equipment failure (Wang et al. 2015). Cloud computing platforms are also suited for other types of maintenance, including corrective maintenance, and preventive maintenance (predetermined and condition based) (Schmidt and Wang 2016). Some of the applied ICT are Supervisory Control and Data Acquisition (SCADA), Condition monitoring (CM), Enterprise Resource Planning (ERP) and other technologies (Schmidt and Wang 2016). It can be argued that, cloud manufacturing, and cloud maintenance can integrate a large variety of tools and technologies, hence the exact combination of these tools and applications depends on the type, size, and goals of the enterprise. Also, cloud manufacturing platforms can be used as a collaboration tool between SMEs in regards of different products and services (raw material, transport, execution procedures, remote distance manufacturing etc.) (Zhou et al. 2018).

Another approach to improving manufacturing efficiency is through cloud logistics and the integration of cloud manufacturing with RFID (Radio Frequency Identification) and Internet of Things (IoT). This approach also includes Big Data analytics and is the basis of IoT-based cloud manufacturing (Zhou et al. 2018). Improved logistics capability can be achieved through a logistics resource sharing network, or a logistics product service system (LPSS). This type of system utilizes RFID in various stages of product manufacturing as well as with production vehicles and robots. This way, an "interaction" is maintained between the manufacturing equipment and semi-products or finished products. Similarly to the previously reviewed cloud-based solutions, cloud logistics is also "fueled" by information from various manufacturing stages and processes. Additionally, cloud computing can be used for efficient and effective knowledge management and intellectual capital allocation within an enterprise (Bakator et al. 2016). This way intellectual capital-based innovations could be made in a timely manner, and the overall intensity of new innovations could increase business performance.

It can be argued that cloud manufacturing is heavily based on information distribution, and this information often contains sensitive business metric data, therefore adequate cloud security has to be established. One of the main issues with cloud manufacturing is the potential risk of breaching, retrieving and/or modifying sensitive data. Therefore, there are several potential solutions for this issue of security such as encryption (but still vulnerable of malicious attack from insiders); cryptographic keys (must be effectively managed), fast and timely identification of breaches (Esposito et al. 2016). In the majority (68,8%) of SMEs, managers rely on expert evaluation of security threats to the cloud platform. However, this approach can significantly increase managing costs, and may not be profitable for enterprises with low revenue streams.

Overall, cloud computing-based solutions can be implemented in various industries (transport, healthcare, construction, education, telecommunication, computer, agriculture and others). Some of the main types of services within cloud computing are storage and backup solutions; enterprise resource planning systems; web-based e-mail services; computer networks; customer relationship management; software developing and testing tools; business intelligence and analytics; online service software; and similar technological, cloud-based solutions (Radwan et al. 2017).

4 A Cloud-Computing Based Model for Improving Competitiveness

The cloud computing-based model for improving the competitive ability of domestic enterprises in Serbia relies on the existing findings in literature and practice. The model is generic in nature as the goal is to be suitable for wide variety of enterprises in various industries. During the development of the model, literature addressing cloud manufacturing application systems (CMASs) (Guo 2015); on-demand cloud manufacturing service (Lu and Xu 2019); collaborative cloud manufacturing, resource selection in distributed manufacturing (Wang et al. 2017); smart manufacturing systems (Zheng et al. 2018), was taken into consideration, as well as the reviewed studies in the previous section. The developed model is shown on Fig. 1.

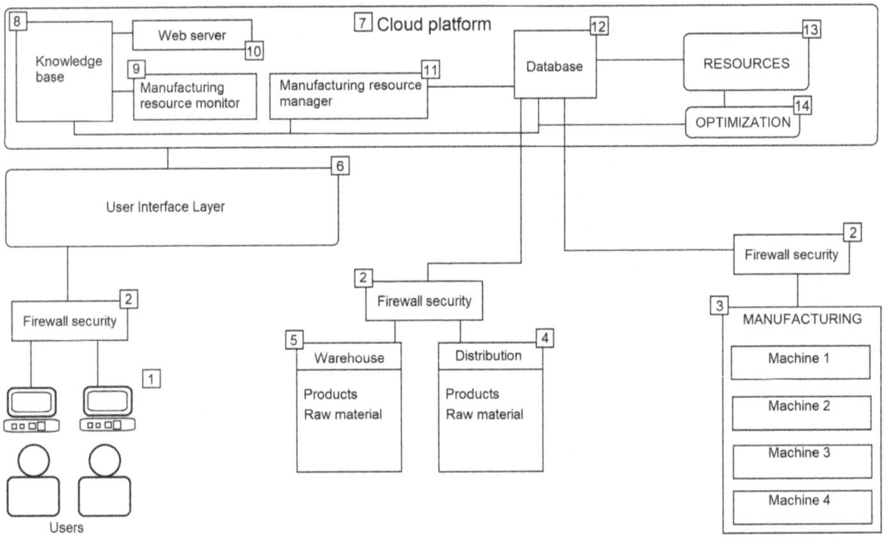

Fig. 1. Cloud computing-based model for improving competitiveness.

The model on Fig. 1 includes fourteen main elements and additional details regarding each element are: **Users** - administrators, managers, employees, operators, technicians. Through a device (desktop pc, smartphone, tablet device etc.) the end-user accesses the User Interface Layer which graphically presents the options, data (inputs, outputs), and configuration of the Cloud platform. From here, significant insight into the manufacturing process, and all other business activities can be monitored and managed. **Firewall security** - Authentication and access allowance to the UIL and Cloud platform. The firewall can be in the form of software and hardware. Every input and output is scanned through this "filter" in order to increase security and reduce the risk of data-breach. **Manufacturing** - the manufacturing process which includes employees and machines. Here, RFID could be used for some parts of a product, or finished product for every machine, effectively collecting data on how many products were produced on which machine, with quality control (QC) data and other important metrics. **Distribution** - product shipments, or resource shipment, truck route tracking are available within the cloud platform. This allows the managers to effectively plan out distribution schedules. **Warehouse** - similarly to the distribution and manufacturing processes, within the warehouse, RFID can be used to track orders, delivered products and pending for delivery products. **User interface layer** - the main layer where the end-users (technicians, employees, managers etc.) access data and information within the cloud platform **Cloud platform** - the core of the cloud computing model. It integrates several databases (knowledge data, data from manufacturing, warehouse and other), resources, and the optimization module. **Knowledge base** - includes data and information collected from employees. It includes organized, ready-to-serve information which can be used for decision-making, innovation development and other R&D activities **Manufacturing resource monitor** - includes modules which track RFID products in the manufacturing process. **Web server** - includes applications regarding Internet websites and other online applications. **Manufacturing resource manager** - an integral part of the manufacturing process. This module allows effective and efficient. **Database** - stores data from every module and allows access to this data for authorized personnel. **Resources** - can be in the form of networks, servers, storage, memory allocation units. **Optimization** - an important part of the cloud-based solution. Through the optimization module resources are allocated in accordance of prioritized modules which are the integral parts of the model as a whole. Further, the trends in cloud computing in Serbia and domestic enterprises are analyzed. The newest available data on the usage (in %) of cloud-based solutions in Serbian and EU SMEs, is presented in Table 2.

Based on Table 2 it can be seen that compared to the EU, Serbia has significantly lower percentage of enterprises which use a cloud-based solution. What is even more concerning, that the majority of the data from 2018, from Serbia, was unreliable, therefore data from 2017 is also presented. Further, data on the use of ICT in SMEs in Serbia is given in Table 3.

Based on Table 3 it can be seen that there is a positive trend in several segments of where the percentage of enterprises who have a website rose by a few percent. Based on the presented data, and the reviewed literature, a trend-line graph is presented on Fig. 2, depicting the competitiveness of Serbia, and future competitiveness taking into consideration the potential of cloud computing technology implementation within domestic enterprises.

Table 2. Cloud-based solutions application in Serbian and EU SMEs in 2018 (Eurostat).

2018	Serbia	EU-28
Cloud computing services used over the Internet	15%	26%
E-mail cloud computing	Data not reliable for 2018, 5% in 2017	18%
Cloud computing office software	Data not reliable for 2018, 4% in 2017	14%
Enterprise database hosting	Data not reliable for 2018, 5% in 2017	13%
Storage of files	Data not reliable for 2018, 5% in 2017	18%
Finance and accounting software	Data not reliable for 2018, 3% in 2017	10%
Customer relationship management services	Data not reliable for 2018, 1% in 2017	8%
High cloud computing services	Data not reliable for 2018, 1% in 2017	6%

(Source: Eurostat 2018)

Table 3. ICT usage in domestic enterprises from 2015 to 2019.

	Enterprise size	2015	2016	2017	2018	2019
Enterprises who use a computer for their business activities	10–49 employees	100	99.6	100	99.4	100
	50–249 employees	100	100	100	99	100
	250+ employees	100	100	100	100	100
Enterprises who use the Internet in their business activities	10–49 employees	98.9	99.8	99.7	99.7	99.7
	50–249 employees	100	99.9	99.6	100	100
	250+ employees	100	100	100	100	100
Enterprises which have a website	10–49 employees	71.2	77.9	76.9	80.1	80.5
	50–249 employees	88	89.8	92.1	90.4	93.7
	250+ employees	92.9	93.9	93.6	94.8	93.1

(Source: Statistical Office of the Republic of Serbia 2018)

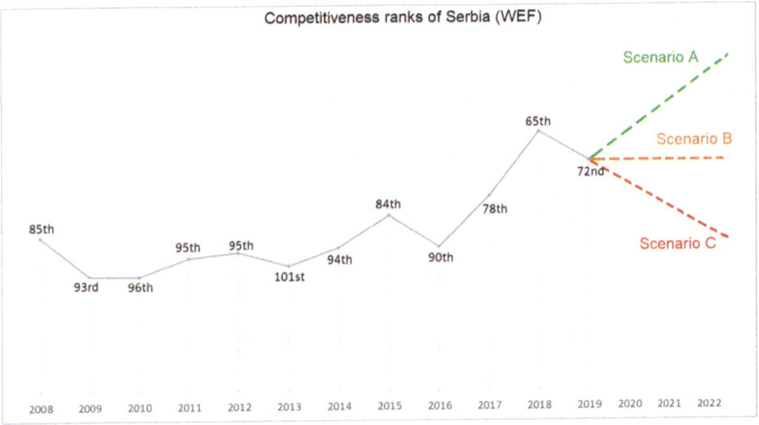

Fig. 2. Competitiveness of Serbia and the potential of cloud computing technology implementation

On Fig. 2, it can be seen that there are three potential scenarios proposed, when it comes to future competitiveness ranks of Serbia according to the World Economic Forum. In the context of cloud computing-based technologies, Scenario A is the most optimistic one where a large number of domestic enterprises would implement a cloud-based solution. In addition, clusters between enterprises could be formed where information and support is shared through the cloud platform. This approach would increase innovation frequency, and overall more effective R&D projects would arise from mutual collaboration between SMEs. This would further result in lower production costs, higher quality products, which are innovative compared to the competition, thus effectively increasing the competitive ability of not only one, but several enterprises which are part of that specific cluster. The second scenario, scenario B, is the case where enterprises don't fully accept the "cloud-way" of conducting business, and only implement simple, low-cost cloud-based solutions which are convenient, but not "powerful" enough in order to increase the value which is distributed to the customer. The third, the worst-case scenario, scenario C, is the situation where domestic enterprises give resistance to change and digitalization of business activities.

5 Suggestions and Guidelines

Based on the developed cloud computing model and literature review the following suggestions and some guidelines for improving the competitiveness of domestic enterprises through cloud computing-based solutions are proposed:

- domestic enterprises have to analyze the international market and to obtain information on trending products and technologies;
- based on the obtained information from the international market, innovations throughout the business must occur;
- establishing and implementing a cloud-based solution can improve business performance but also can eat into profits, thus enterprises should consider the best cost-to-benefits ratio when implementing a cloud computing technology;
- through cloud platforms multiple enterprises could collaborate by sharing information and by mutual support of specific business activities (ex. transport logistics, manufacturing equipment scheduling for mutual benefits etc.);
- clustering through cloud platforms could be simplified and made more effective, as it reduces bureaucratic procedures;
- the government should incentivize the digitalization of enterprises and provide financial and infrastructural support;
- cloud computing solutions should be viewed as a necessity for achieving competitiveness and managers and employees have to embrace it, rather than to resist its inevitable implementation;

Overall, domestic enterprises have to commit to advanced ICT. The managers of SMEs need to understand that the globalized markets ask for high-tech, high-innovation products, and the only way to be more competitive is to be more innovative, while product quality is an imperative, and there should be no compromise in the domain of quality. well-integrated cloud computing solution can improve not only product quality, but also enhance other business processes making them more effective and efficient.

6 Conclusion

Globalized markets have taken their toll on SMEs who were and are not prepared for the constant changes regarding market segmentation and fragmentation and competition intensity. In this paper the potential of cloud computing for improving competitiveness was analyzed. The three main research questions which guided this paper were:

What are the main issues of domestic enterprises when it comes to international competitiveness? The research results indicate that low productivity, low product quality, old manufacturing equipment, the lack of modern management tools and techniques, the lack of modern technology application and the lack of intellectual capital are the main issues of domestic enterprises which have to be addressed.

How and what type of cloud computing solutions can be applied within various industries? Cloud computing as a model is a generic term and it can take virtually a massive number of applications. The main applications of cloud-based solutions addressed in this paper were cloud manufacturing, cloud logistics, cloud security, and robotic applications.

How can cloud computing affect the business performance of domestic enterprises? Based on the literature review, it can be proposed that the cloud computing technologies if implemented in an adequate manner, can positively affect business processes and overall business performance.

Further, the results of the conducted research indicates that the proposed hypothesis: "Implementing cloud computing-based solutions can increase the competitiveness of domestic enterprises." is failed to be rejected. The research data and conceptual study support the hypothesized proposition. Now, it can be concluded that domestic enterprises and Serbia overall, have solid potential when it comes to implementing modern ICT. The paper provides a significant contribution to the existing body of literature in the domain of cloud computing solutions in developed countries. In addition, it presents a solid basis for future research. In this future research, a nation-wide survey could be conducted regarding the use of cloud computing technologies in domestic SMEs.

Acknowledgement. This work is a part of the current project TR-35017 funded by Ministry of Education, Science and Technological Development of the Republic of Serbia.

References

Bakator, M., Terek, E., Nikolić, M.: The use of cloud computing technology in knowledge management and intellectual capital allocation. In: Paper presented at the XV International Conference Change Management in Socio and Economic Systems, 2016, Voronezh State University, Economic Faculty Department of Economics and Organization Management, Voronez, Russia, pp. 7–17 (2016). ISBN 978-5-00044-436-8

Ćoćkalo, D., Đorđević, D., Bogetić, S., Bakator, M., Bešić, C.: Competitiveness of domestic enterprises in changing markets and Industry 4.0. In: International Conference on the Industry 4.0 model for Advanced Manufacturing, pp. 113–127. Springer, Cham, June 2019

Esposito, C., Castiglione, A., Martini, B., Choo, K.-K.R.: Cloud manufacturing: security, privacy, and forensic concerns. IEEE Cloud Comput. **3**(4), 16–22 (2016). https://doi.org/10.1109/mcc.2016.79

Eurostat. Cloud computing services use in SMEs (2018). https://appsso.eurostat.ec.europa.eu/nui/show.do?dataset=isoc_cicce_use&lang=en. Accessed 12 Dec 2019

Guo, L.: A system design method for cloud manufacturing application system. Int. J. Adv. Manuf. Technol. **84**(1–4), 275–289 (2015). https://doi.org/10.1007/s00170-015-8092-0

Liu, Y., Xu, X., Zhang, L., Wang, L., Zhong, R.Y.: Workload-based multi-task scheduling in cloud manufacturing. Robot. Comput.-Integr. Manuf. **45**, 3–20 (2017). https://doi.org/10.1016/j.rcim.2016.09.008

Lu, Y., Xu, X.: Cloud-based manufacturing equipment and big data analytics to enable on-demand manufacturing services. Robot. Comput.-Integr. Manuf. **57**, 92–102 (2019). https://doi.org/10.1016/j.rcim.2018.11.006

Majstorovic, V.D., Mitrovic, R.: Industry 4.0 programs worldwide. In International Conference on the Industry 4.0 model for Advanced Manufacturing, pp. 78–99. Springer, Cham, June 2019

Popkova, E.G., Ragulina, Y.V., Bogoviz, A.V.: Industry 4.0: Industrial Revolution of the 21st Century. Springer International Publishing, Berlin (2019)

Radulescu, M., Fedajev, A., Sinisi, C., Popescu, C., Iacob, S.: Europe 2020 implementation as driver of economic performance and competitiveness. panel analysis of CEE countries. Sustainability **10**(3), 566 (2018). https://doi.org/10.3390/su10020566

Radwan, T., Azer, M.A., Abdelbaki, N.: Cloud computing security: challenges and future trends. Int. J. Comput. Appl. Technol. **55**(2), 158 (2017). https://doi.org/10.1504/ijcat.2017.082865

Research and development expenditure. World Bank Group. https://data.worldbank.org/indicator/GB.XPD.RSDV.GD.ZS. Accessed 13 Dec 2019

Schmidt, B., Wang, L.: Cloud-enhanced predictive maintenance. Int. J. Adv. Manuf. Technol. (2016). https://doi.org/10.1007/s00170-016-8983-8

Statistical Office of the Republic of Serbia. The use of information-communication technologies (2018). https://data.stat.gov.rs/?caller=SDDB. Accessed 13 Dec 2019

Wang, X.V., Wang, L., Mohammed, A., Givehchi, M.: Ubiquitous manufacturing system based on cloud: a robotics application. Robot. Comput.-Integr. Manuf. **45**, 116–125 (2017). https://doi.org/10.1016/j.rcim.2016.01.007

World Economic Forum (WEF). The Global Competitiveness Reports from 2014 to 2019. http://reports.weforum.org. Accessed 13 Dec 2019

Zheng, P., Wang, H., Sang, Z., Zhong, R.Y., Liu, Y., Liu, C., Xu, X.: Smart manufacturing systems for Industry 4.0: conceptual framework, scenarios, and future perspectives. Front. Mech. Eng. **13**(2), 137–150 (2018). https://doi.org/10.1007/s11465-018-0499-5

Zhou, L., Zhang, L., Ren, L.: Modelling and simulation of logistics service selection in cloud manufacturing. Procedia CIRP **72**, 916–921 (2018). https://doi.org/10.1016/j.procir.2018.03.197

Strategic Approach for Robotics Development in SME by Value Linkage Concept

Hideaki Hohnoki(✉)

COHO Consulting, Chiba 261-0013, Japan
hideakihohnoki@s6.spaaqs.ne.jp, akihohnoki@gmail.com

Abstract. Small and Medium Enterprises (SME) in Japan has been facing the difficulties of global competition caused by manufacturing commoditization as well as by collapsing affiliation system known as "Keiretsu". Effective measure for this SME issue would be regarded as potential remedy not only for Japan but also for other countries including Germany and Serbia because of their high percentage of SME ratio in industries.

This study attempts to make positive effects on the said issue by providing the new perspectives for SME to obtain driving force to create a survival path with utilizing digital technologies. This approach also covers robotics industries which has strong possibilities for SME to strengthen the basis for versatile growth and to facilitate SME to implement Industry 4.0.

Keywords: Digital manufacturing · Mass customization · Robotics · PLM · SME · Additive manufacturing · 3D printer

1 Introduction

SME (Small and Medium Enterprises) ratio in industry in Japan is 99.7% and a considerable number of them has been suffering from current global competition. One of the effective ways to overcome this situation is to utilize recent digital technologies and the previous paper introduces a new perspective called Value Linkage Concept with the digital technologies. This concept provides the way to set the target for new path in the market for SME by adding value to existing items in order to create new service, to focus on customized area and to reflect and to incorporate the realities of the market.

This study attempted to apply Value Linkage Concept to the field of robotics and its applied result shows that the concept is fully applicable to the development aspect of robotics field. Furthermore, it is also discussed that the robotics field itself has some potential opportunities suited for SME. This applied concept in robotics field are expected also to cover SME in Serbia which has the same high percentage SME ratio such as 99.7% as in the case with Japan and Germany. This study provides new perspective to make positive effects for the purpose of contributing to the issues stated below:

a. To promote the development of lifestyle support robot by SME: The perspective will assist SME to find a path or clue of new development work which helps

L. Wang et al. (Eds.): AMP 2020, LNME, pp. 250–262, 2020.
https://doi.org/10.1007/978-3-030-46212-3_18

improve their activities to growth. Since more and more lifestyle support robots are recently introduced to practical use, it follows that the communication technology such as the interaction between human and robot comes to be indispensable.

b. To take advantage of utilizing the SME's expertise in the industries: SME in general are not good at conducting R&D work due to their various tendencies. The elemental technologies of special parts and components as in sensing, actuation for interaction used for robot are usually the areas of specialties for SME and they could do their best there. SME can utilize their existing own skills and technologies for their R&D work in such an elemental parts and modules in robotics field. Especially lifestyle support robot is consumer-oriented product which still have wide area to be developed and customized. The perspective is also expected to help reinforcing R&D work of SME and enhancing the company structure.

c. To change from competing the performance of hardware into promoting the cooperation between hardware and service performance: The emergence of newly industrialized countries brought the global competition which has resulted in making mass-production items such as bulk electronic appliances just ordinary "commodity". Therefore, it is important to seek the value in network quality rather than in mechanical property. The perspective suggests to change the SME situation from this commoditization, as in only hardware production, into adding service performance to hardware with digital technology.

2 Background

Recent engineering innovation in industries has decreased the overall value of fabricating process in manufacturing field, which has resulted in structural change in total manufacturing process. Fabrication is one of the major manufacturing parts and is located in the middle of value chain process. Fabrication used to be a value-added process but the progress of modularization technology has made it much simple to be handled easier and quicker. At the same time, the rise of emerging countries has decreased the value of fabrication part both in time and in cost. Afterwards this progress of fabrication technologies encourage the emergence of newly industrialized countries such as Asian countries and areas to play an increasingly important role in manufacturing industries, and eventually bulk electronic products primarily became just ordinary "commodity", not special and custom-made product. This commoditization rapidly expanded to various products such as many types of appliances, TV, Personal computer, mobile phone and even automobile, which resulted in battered Japanese SME industries. This trend of the lower value in fabrication part in manufacturing industries is called "smile-curve effect" which means, in the graph of added value vs process parts from design thorough sales, the added value on both ends of the process are relatively higher than the middle of the process, that is, the value of design part (left end) and marketing and sales part (right end) are always higher than fabrication part (middle). This shape looks like the mouth when smiling.

Let's trace a quick path through the situation of SME in Japan. They have played a great role in economic growth in Japan after World War 2. In the years of high-growth economy, SME have a function as Sub-Contractors in vertically integrated organization of production. It's a pyramid structure and top of the pyramid is big company, which strongly controlled whole activities and they also educated and trained SME people on the other hand. The pyramid structure has been transformed into more sophisticated style, namely KEIRETSU, which is more strategic and systematic grouping to share common interest. However, the structural change of manufacturing industries in Japan has induced a transformation of the relation between dominant big companies and SMEs and then disrupted the ties among them. Currently still KEIRETSU is active but sometimes it is often unable to function for such a huge organization to move quickly against significant changes in business such as ICT field. Therefore, in the near future, current manufacturing structure should be changed to be more flexible style, which means SME should change into more specialized style and the partnership such as divisional cooperation should be more important factor for industry in general. It could be a good opportunity for SME to grow and expand their business. Knowledge and experiences in this innovative change has possibility to create new digital-divide issues. It would not be an overstatement to mention that the proper action to this change can ensure to keep a company and its technology alive, especially in case of SME.

In the meantime, a digital technology has made significant progress in revolutionizing manufacturing processes. The digitalization continues ever after by creating a new designing system known as PLM (Product Lifecycle Management) which handles each step of the process virtually by using digital data, namely in computer system. This digital manufacturing process is simply based on the utilization of key characteristics of digital technology which has made a significant contribution to product innovation. In other words, digital manufacturing creates to obtain the right product at the right time, i.e. to get the product which matched the needs of the market. This trend is called Mass Customization which is examined and discussed in the later in this study.

3 Findings

The study reveals to apply "value linkage Concept" to the robotics field in possible future development schemes. The value linkage is key concept in the study to consider about the possibilities of future products and services. The concept indicates how to take action against new stream of change showing mainly mass customization movement, that is, how to manufacture as process strategies and what to manufacture as product strategies. In the future perspective from recent IOT trend, it is presented to the discussion about the importance to consider "value" which is linked to the relationship of "product-service-market" coordination and its enhancement, which is called Value Linkage in the previous study. The recent digital technology is powerful tool in order to ensure effective utilization of the Value Linkage concept, which consists of three types of Value, namely Service Value with Hardware, Diversified Value by Mass Customization and Market Value with Service. These three types of value handling are discussed as case study to apply Value Linkage Concept to the development work of robotics. Also, these three types of value are mutually linked with one another to interact more effectively.

3.1 Service Value with Hardware

In the first connection in Value Linkage Concept, the value signifies adding service value to hardware itself. In this case the key factor of the product is not its independent performance but its collaborative capability through other connected hardware and/or services. Current trend in market is from "simple hardware production" toward "hardware with support and aftercare (services)", in other words, from "selling out hardware only" toward "selling something with some service", which naturally places higher value on the existing product in market. Adding service value creates new product with service, which cannot be performed by original hardware alone. When we apply this service value concept to robotics field, we will find that its direction is not manufacturing robot itself as hardware but create some service by robot, such as life support type, by focusing on its collaborative capability between robot and human behavior. In another words, we should not simply look at the mechanism of robot but, as a matter of high priority, consider to generate some service by robot for human using adding service value concept.

3.2 Diversified Value to Mass Customization

In the second connection in Value Linkage Concept, the value signifies adding various type of functional value to simple hardware. In this case the key factor of the product is not in its routine job performance but in offering a wide range of services by its diversified capabilities for human. These values are based on Mass-customization purposes. General meaning of Value Linkage Concept in the second connection is the hardware which is capable to perform a variety of tasks depending on the demand, which is basically the opposite of the one for mass production. Mass production is a conventional system in industry and this system is suited for manufacturing standardized products in high volume. Another category, "high-mix low-volume manufacturing" is also widely accepted for creating greater variation of products, but its each volume is usually small. Ideal status of the second connection in Value Linkage Concept matched to mass customization is "high-mix high-volume manufacturing". Currently 3D printer displays great ability to showing variation in building objects. However high-volume manufacturing by 3D printer is still not be enough under present circumstances and this is one of the points we expect to be improved by digital technology towards the future high-mix high-volume manufacturing.

In case of the situation applied for robotics, this concept can be described as the robot with diversified value. This robot could respond and react differently and flexibly toward various request not only by the same person but also by a certain number of people who have different personalities among each other [6, 11, 12]. Recent sensor technologies including face and voice recognition functions is able to allow robot to individuate its response to react differently to the same person as well as to the different people by using collected data by various sensors, that is, interaction and interface tools. The reactions by robot are mass customized differently depending on response and behavior by human [5]. Figure 1 shows three possible factors to approach mass customized product and service are "multi and continuous system", "local production for local consumption" and "upper process dominance" and more in detail is shown in discussion part [8].

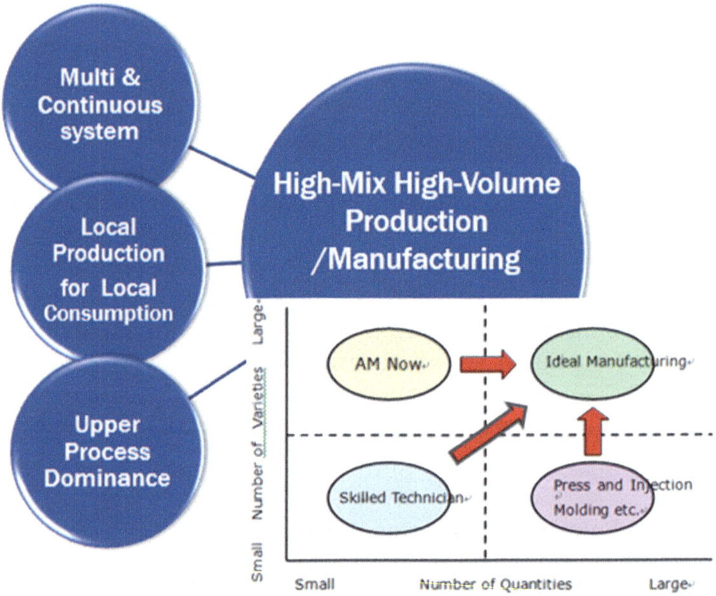

Fig. 1. Key factors to approach mass customized production

3.3 Market Value with Service

In the third connection in Value Linkage Concept, the Value can be in the data obtained from the market and the content of the existing service should be modified on demand and/or response to that collected data. In other words, the collected data is considered to be reference of market demand by which the existing service is updated through analysis and estimation of that data to be the most updated performance. What we call "big data" is a type of accumulated data through internet service or IOT, which usually has influence as the form of collected and analyzed data to the service or activities to be updated. This analyzed or modified data or information makes original service more flexible to the market trend and, in this manner, existing routine service can be modified to be new service in accordance to the market demand.

When we apply this third value concept to robotics field, we will find that the performance of the robot reflects the data collected from outside circumstances. Big data is one of the examples of the data from outside circumstances. Case studies are shown in discussion part.

4 Discussions

After reviewing recent development in robotics briefly, discussions show the case studies and the analysis which confirms relevance of the Value Linkage Concept applied to robotics field. The order of description is in the same way as in Findings part, that is, Service Value, Diversified Value and Market Value. Last part of discussion shows that Value Linkage is also effective for the development work by SME.

4.1 Recent Development in Robotics Field

The development of robotics is one of the encouraging fields to apply the Value Linkage concept. Industrial robots have been matters of mutual concern in manufacturing field for a long time in the past. Those robots are the machines which is mainly operated in the simple fixed action patterns by instructions from human. The working range of these industrial robots mostly have good distance from activity area of human and they are usually surrounded by physical wall or in fenced area. In the meantime, some types of robot have been developed to work more and more closely to people. Those are called service robot, which includes communication robot, assistant robot, guidance robot etc. They are working closely in the area of people's life. This expand of working territory have finally reached almost just to human body. Those are, what they call, muscle suite type of system such as HAL by CYBERDINE Inc [3]. The muscle suite is a cyborg-type wearable system for assisting and enhancing body function for such people that work in hospital, health care facility, logistic business etc. In this way robot itself have become more accessible hardware for individuals, therefore safety and favorable interaction with robot for the purpose of coexistence and cooperation are important and major concern for recent development aspect [15, 20].

4.2 Value Linkage to Robotics–Service Value

The typical case study on adding service value in manufacturing field is KOMTRAX, which is machine tracking system for 300 thousand dump trucks and construction machines etc. sold already in the world by Komatsu company in Japan [14]. Each machine has network sensors which send all operation data to remote office via satellite and mobile phone station. Those data are analyzed for customer for the purpose of planning efficient operation, reducing maintenance cost etc. as well as for Komatsu themselves for advance maintenance suggestion, sales forecast and so on. This is a good example to learn about "Hardware+Service" concept which means that they not only sell hardware only but sell them with some backup service via network [9]. This implies that they had turned mere hardware-only production around toward the hardware with some other service for human life. Their service function apparently promotes product sales itself. This case simply indicates what digital service featuring IOT is.

Adding service value to robotics field in the way as above allows you to take advantage of Value Linkage Concept. The recent sensor and network technology enable various type of interface for the communication between robot and human. It follows that adding various services to hardware, i.e. robot itself is much easier than before and more and more lifestyle support robots have recently come to the actual daily life [5, 21].

Guidance robot provides the service which supports disabled and elderly person (as operator) to lead to move around as intended direction in a hospital or public space. Operator can put a hand on the robot and push the grip toward intended direction, then internal sensor transfers its intention to instruct the robot to move as operator's intended direction. Operator can set the intended destination to reach, then the robot leads the operator to the intended destination by its own program with leaser and location sensors navigating obstacles on the way. LIGHTBOT manufactured by NSK Ltd. is the example of guidance robot [13, 16, 17].

Value Linkage Concept suggests us to pay attention to the importance of this service concept when we develop and design hardware (i.e. robot itself) in robotics field. V model development is effective tool for thinking about the service concept behind hardware. V model is basic and standard operating procedure in mainly software development, but we can apply this model to hardware production in robotics also, which is shown in Fig. 2. The design procedure usually starts from requirement definition down to integration positioned at the bottom and from integration the checking procedure goes up to operation validation. Whole procedure in this case, both design side (left) and checking side (right), handles mechanical related (or engineering and hardware related) phase and no human related phase there. Since we are discussing to pay attention to the relation between robot and human, it is important to consider about upper process before requirement definition because this upper process deeply related to human activities and its environment as in society and lifestyle. We should call this upper process as society & lifestyle phase, which contains the background, needs, cooperation, analysis, vision, assessment etc. and those are higher concept to drive the intension of human to expect its total system design [10, 18, 19].

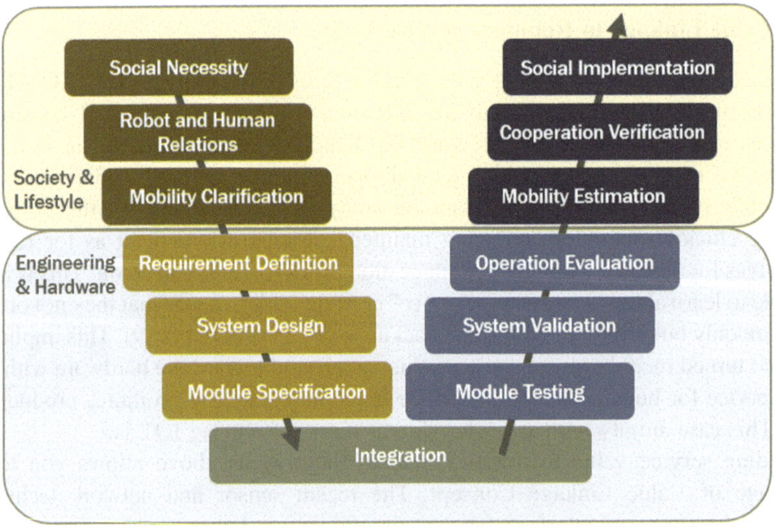

Fig. 2. V-Model in robotics development with value linkage concept

4.3 Value Linkage to Robotics–Diversified Value

When we closely consider the reality of the recent consumer market, we can witness the transformation in consumer's behavior toward the diversification of their preference patterns. In old days it was a time of shortage and people most likely would like to have something the same as what other people had. Manufacturers' responsibility at that time was to try to supply standardized commodities as much quantity as possible to clear the supply shortage and people satisfied those ready-made articles. Once the shortage has gone, people began to want different things. Presumably they would like to have different color, shapes, something different from what their neighbors had, in other words, variant products were the target for manufacturers at that time. What happened next is that people started to think that the commodity should be suitable for owner's character, preference, personality……, namely the time of diversity. Consequently, more and more mass customized product (bespoke product, that is, product with diversified value) has been in the marketplace for customers. What we may see next in the future when we further deepen and develop this personal direction would be something like the product featuring uniqueness for each individual.

Currently mass customized production (the high-mix and high-volume production) is not so efficient and effective in comparison with standardized mass production, because adding huge number of diversified values easily to a large amount of product still remains to be established yet. However, three ways shown in Fig. 1 as mentioned previously are possible key factors to approach mass customized production. The three key factors in Fig. 1 include "Multi and Continuous machine system" in hardware, "Local Production for Local Consumption" and "Upper Process Dominance" in digital manufacturing process as well as digital service production process. First key factor, what we call Multi and Continuous system, is connected and/or assembled process by combination of different methods. This is also called Multi-Systemization in which multi ways of production method is connected mutually to produce customized products efficiently [9]. The research work by Fraunhofer ILT in Germany shows an example of this method featuring 3D printer for a bio-fabrication system for artificial tissue engineering [4]. This multi and continuous process is the one connected with some different methods such as Inkjet, SLA, MPP and ES depending on its cell structure.

Local Production for Local Consumption, the second key factor, means that the production has been changing from a big fixed factory to local production firm by using network technologies such as IOT [1]. No matter where the production place is, they could deliver the design data through network. They execute production at the necessary point of consumption by using digital manufacturing system including 3D printer, which results in less logistics of finished products but more logistics of materials. Quite simply, for example, in case of space station far from the earth, they deliver materials and build something needed on site in the space station. The Local Manufacturing for Local Consumption helps producing bespoke products depending on its various necessity on site.

The last key factor for high-mix high-volume production is Upper Process Dominance, which means that the progress of PLM method makes upper part of the whole process much more dominant in terms with total optimization. Once digital manufacturing process completely covers the flow, we could control final product at rather beginning part of the total flow and which results in an easier way to mass customized products.

As noted as above, future direction of production would be mass customization in accordance with evolving customer preferences, that is, adding diversified value to existing product. Eventually production process should be changed towards high-mix high-volume production. Digital technologies and an equipment such as the 3D printer, as a leading-edge example, have encouraged sophisticated and diversified public demands in various industries. The typical and simple case example for the above stream of change, namely mass customization, is hearing aid. Hearing aid is a small medical equipment to support hearing-impaired individuals. The equipment should be fit precisely to ear canal. If not, the body of hearing aid is coming loose because one's ear canal itself moves during one is speaking or eating. Each person has a different inner shape of ear canal, therefore manufacturing process of hearing aid had been fully handmade and time-consuming product for shaping its outer shell depending on customer's ear canal. However recent digital manufacturing method made it possible to create full customized outer shell built by 3D printer. Digital technology can make it as precisely as customized demand.

Robotics is the possible fields also to apply the adding diversified Value from Value Linkage concept. Adding diversified value to hardware (i.e. robot) suggests us to create new service or function to match each customer's various type of desire, demand or necessity. Among recent digital technologies, 3D printer has good capability to create bespoke product which meets various need as explained. 3D printer as cooking robot is the second case study for adding diversified value category. This application of 3D printer as cooking robot is prepared mainly for patients in hospital and its building object in this case is various type of care food for them. Healthy person can eat everything as far as it is food, but patient needs properly prepared food depending on sickness or body conditions, not only in terms of nutrition but also easiness of chewing or swallowing. Usually in the kitchen at the hospital they cook food suitable for each person for each day, which is so complexed and sensitive work, therefore it requires proper arrangement with less care. Biozoon in Germany implemented and made it fit for practical applications by using 3D printing technology [2].

Recent development of digital and sensor technologies has begun to guide robots to behave in a similar or in much the same way as human do. Such a felicitous action can be achieved by recent interaction technologies as in using voice and facial recognition technologies appealing to our sense. These technologies allow robot to response timely and well-chosen action toward human side. Another possible and desirable feature of lifestyle support robot is to individuate it to act differently depending on different person by using some sensing technology such as facial recognition system, which verifies adding customized value from Value Linkage concept. Network technology also give them some flexible response feature from the market such as various websites, which shows market value in Value Linkage concept.

4.4 Value Linkage to Robotics–Market Value

Market value from Value Linkage Concept means the data collected, analyzed and modified from the circumstance of the target (i.e. normally existing service) which we would like to apply the value [9]. The collected data is considered to be the possible market demand and analyzed and modified data have some influence on the existing service. This analyzed or modified data or information makes original existing service more flexible to the market trend and, in this manner, existing routine service can be updated to be better service in accordance with the market trend and demand. The base matrix of the collected data is usually called big data. The matrix data is not necessarily big data and the information around the target (existing service) is good enough to handle.

Digital data is handled and utilized not only in manufacturing field but also in various area of our real society where our activities are. Those activities could be captured through network in the form of digital data, for example, text, image, sound, picture, video at the time when using SNS, purchasing commodities and services, transporting and so on. Those digital data are accumulated, analyzed and estimated and then its result is transformed into the data as a form of added value. The value-added data are given back to real society using application in mobile phone, cloud service, SNS etc. The data recirculation between our real activities and virtual world in our society is effective and rather easy to handle because of nature of digital data as in unlimited replication which enables high productivity and error-less transmission by which we can communicate signals exactly as it is without any change or noise.

Simple case study is a various type of communication robots which have been achieved a substantial improvement by interaction and interface technologies such as voice and facial recognition system put into commercial realities recently. Smart speaker is a typical communication robot which has widely gained in popularity. Network technology also give them some flexible response feature from the market such as various web-sites, which shows market value in Value Linkage concept. All three aspects of value connection in Value Linkage Concept is shown in Fig. 3. Nursing care system and robot in hospital and public institutions and facilities are another example of utilizing network system and interaction technology [7].

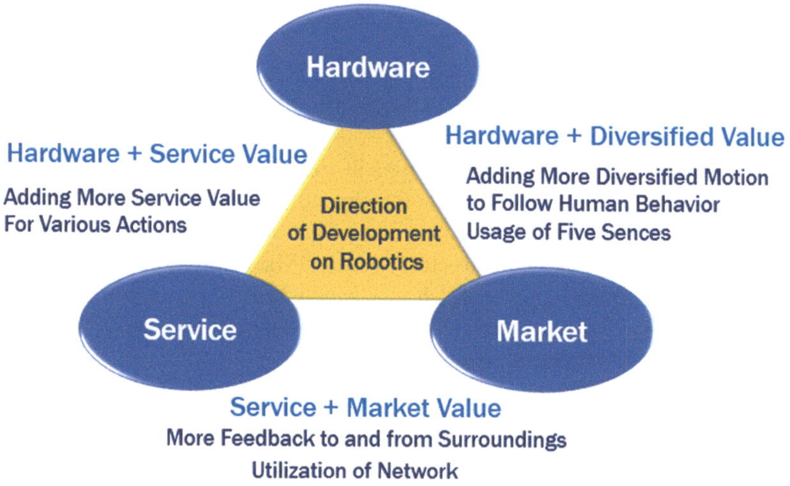

Fig. 3. Value linkage in robotics development

4.5 Digital Technologies for SME

Industrial automation and digital manufacturing including NC machinery and PLM have facilitate the development and growth of Japanese economy after damage and loss of WW2. Efficient and capable efforts performed by SME have contributed its rapid growth mainly as sub-contractors. PLM usually consists of Designing, 3D Modeling, CAE, 3D Prototyping, Knowledge Engineering, CAT etc. In those process digital data plays a prominent role and is commonly utilized among each steps of the process. This virtual system has been called Digital Manufacturing and widely applied for years mainly in the production of automobile, consumer appliances and other major manufacturing industries. Recently network technology has even greater role, for example, mobile phone and wireless LAN have brought IoT system.

SME is usually defined as the company with employee 250 and less and turnover 50 million EUR and less. SME in a comparison with countries indicates that Japan and Germany have some similar factors as in very high SME percentage in industry, R&D framework performed by public institute and long-term employment system. However more different aspects we should pay attention to these 2 countries is that they still try to take advantage in manufacturing industry in the world. That's Industry4.0 and Smart Society in Japan and the policy has come from trying to utilize the abilities and power of SME in each country. The comparison also implies that industry in Japan has rather vertical integration-oriented structure but, on the other hand, Germany is more individualistic and prefer differentiation. Two countries have common factors having come from its own background and different natural tendencies. Serbia, in comparison with the above two countries, shows quite similar figure of very high percentage of SME in industry, which is approximately 99.7%. Since SME in general plays a great role in economic growth, this figure means each of these three countries has strong industrial potential in society.

5 Conclusions-Robotics, Advantageous Field for SME

IT and digital technologies are mainly used in robotics industries. As mentioned in introduction, the effective way to overcome the suffered situation among SME is to utilize recent digital technology. Because one of the key characteristics of the digital technology is its inexpensive infrastructure, which SME could afford to install in their facilities. Most IT infrastructure is much less expensive compared with other ones in conventional industries as in steel making, automobile and so on. IT investment is relatively easier for SME and digital facility could be powerful weapon for them.

Secondly the Robotics field could be better suited for handling by SME due to the facts that robot generally consists of the technologies including sensing, actuation. and intelligence and the first two skill are fit for SME to develop since those are the elemental technology which is generally the area of specialties by SME.

The market of lifestyle support robot is also suite for SME because it is basically consumer-oriented product and still have wide area to be developed. The market includes health care, medical, daily life, those of which needs wide range of customization depending on customer's life situations, conditions and preferences.

Furthermore, if you make efforts to apply Value Linkage Concept to the development of robotics and achieve some products or services in a certain field, it could be invaluable knowledge and experience as SME since research and development is the weakness area of SME. Also, we will see the business would shift from mass production to value-oriented structure. The concept could strengthen and enhanced the company structure and its business quality.

References

1. Anderson, C.: Makers: The New Industrial Revolution. Crown Pub Inc., New York (2014)
2. Biozoon Food Innovations GmbH, Products/Seneopro. http://biozoon.de/en/products/seneopro/. Accessed Jan 2019
3. CYBERDYNE Inc., Products/HAL. https://www.cyberdyne.jp/english/products/HAL/. Accessed Dec 2018
4. Fraunhofer ILT, ArtiVasc 3D, Project Detail. https://www.artivasc.eu/en/projectdetails.html. Accessed Oct 2017
5. FUJISOFT Inc.: Communication robot PALRO. https://palro.jp/. Accessed Dec 2018
6. Hagita, N., Miyashita, T.: Networked co-creation system platforms facilitating human-robot learning. J. Rob. Soc. Japan **29**(10), 871–874 (2011)
7. Hirukawa, H.: Robotic devices for nursing care project. J. Rob. Soc. Japan **34**(4), 228–231 (2016)
8. Hohnoki, H., Tezuka, M., Ishikawa, T.: 3D printing technology: its effects and applications to manufacturing technology, New Technology Association of Japan (2014)
9. Hohnoki, H., Tezuka, M., Ishikawa, T.: New trend of digital manufacturing and its influences to SME in Japan, New Technology Association of Japan (2015)
10. Hohnoki, H., Tezuka, M., Ishikawa, T.: Recent development of interaction technologies for coexistence and cooperation with service robots, New Technology Association of Japan (2017)
11. Ishiguro, H., Miyake, N.: Toward a collaboratively creative society through human-robot symbiosis. J. Rob. Soc. Japan **29**(10), 868–870 (2011)
12. Ishiguro, H., Matsumoto, Y., Yoshikawa, M., et al.: An android synchronizing with human towards a good listener. J. Rob. Soc. Japan **29**(10), 879–882 (2011)
13. Kanagawa Prefectural Government, Robot Town Sagami: Future of man and machine, Introduction of Guidance robot by NSK Ltd. http://sagamirobot.pref.kanagawa.jp/products/product11/. Accessed Oct 2017
14. Komatsu Ltd., 22 June 2015. Press Release No. 028(2799). https://home.komatsu/en/press/2015/others/1194792_1810.html. http://smartconstruction.komatsu/whats.html. Accessed June 2015
15. NEDO (New Energy and Industrial Technology Development Organization): Project for practical applications of service robots, Research and Development of safety level verification technology (2014)
16. NSK Ltd., 26 November 2015. Press Release. http://www.nsk.com/jp/company/news/2015/press1126a.html. Accessed Dec 2018
17. NSK Ltd., 30 March 2017. Press Release. http://www.nsk.com/jp/company/news/2017/press0330b. Accessed Dec 2018
18. Ohba, K.: Robot Innovation, CAIST (Research Center for Advanced Information Science and Technology, The University of Aizu) Symposium (2017)

19. Ohba, K.: Safety level verification technology for service robot. J. Japan Soc. Precis. Eng. **81** (1), 3–4 (2015)
20. Ohba, K.: How to implement the system considering its safety. J. Japanese Soc. Eng. Educ. **63**(5), 23–27 (2015)
21. Okada, M.: Human-dependent weak robots for creating symbiotic relations with human. J. Rob. Soc. Japan **34**(5), 299–303 (2016)

Big Data Analysis as a Digital Service: Evidence Form Manufacturing Firms

Bojan Lalic, Ugljesa Marjanovic(✉), Slavko Rakic, Marko Pavlovic, Tanja Todorovic, and Nenad Medic

Faculty of Technical Sciences, University of Novi Sad, Novi Sad, Serbia
{blalic,umarjano}@uns.ac.rs

Abstract. Digital disruption is propelling manufacturers to move on towards digital transformation and deliver digital services based on predictive analytics. The literature agrees that digital technologies (i.e. big data) facilitate the service innovation of manufacturers by creating digital servitization. However, little research has specifically focused on the empirical data that analyze use of digital technologies in manufacturing firms in terms of technological intensity. The present study investigates the interaction between big data analysis, as a digital service, and firm characteristics (i.e. firm size and technological intensity). Our analysis used the Serbian dataset of 240 manufacturing firms from the European Manufacturing Survey conducted in 2018. The empirical results show that, in manufacturing firms, digital service based on predictive analytics is highly utilized in medium size firms. Furthermore, results indicate that high technology manufacturing firms in Serbia are not yet utilizing digital technologies to facilitate the service innovation in comparison to other innovation intensity characteristics.

Keywords: Digital servitization · Big data · Manufacturing firms

1 Introduction

Transition towards digitalization characterized by fast-paced technological advancements (i.e. artificial intelligence) is triggering complex challenges for micro, small and medium manufacturing companies [1–3]. The literature agrees that digital technologies (e.g. Internet of things, cloud computing and big data) facilitate the service innovation of manufacturers by creating digital servitization [4]. However, little research has specifically focused on the empirical data that analyze use of digital technologies in manufacturing firms in terms of technological intensity [5]. With this paper, we aim to guide researchers and practitioners with insights related to the use of big data analytics within the technological intensity framework.

In this paper, we complement the existing qualitative literature on big data as a digital service with a descriptive statistic. In addition, we provide a comprehensive overview of different firm size classes and different degrees of technological intensity using firm-level data covering 240 firms from Serbia.

L. Wang et al. (Eds.): AMP 2020, LNME, pp. 263–269, 2020.
https://doi.org/10.1007/978-3-030-46212-3_19

2 Theoretical Background

2.1 Big Data Analytics as a Digital Service

Research reports highlight that intelligent products, connectivity, cloud computing and big data analytics are expected to be disruptive for companies' business strategies and operational execution [4, 6]. Looking through history and humankind's relation to data and new knowledge, for the most part, data and knowledge were not recorded in any way [7]. Only through the invention of writing, data could be stored and preserved for future use, still activity of preservation of data and information was time consuming [7]. With the development and the increased use of modern information technologies tools we can record, store and analyze data on a new level. Due to digital revolution, huge amounts of digital data are generated by and collected from many different technical sources, like sensors, cameras, smart watches, social networks, mobile devices, Global Positioning System devices, Radio-Frequency Identification tags [8]. With this enormous quantity of data coming daily from billions of sources around the world, next step is to see how we can use it for a purpose, gain value from this inexhaustible source, and provide it as a service.

Big data are defined as dynamic information that is generated in complex systems with the characteristics of the three Vs: volume, velocity and variety [9]. Volume represents the amount of the data that is created, velocity represent the speed with which data is being created and variety represent the various types of data being gathered. Equipment and infrastructure necessary for setting up the big data systems are costly and

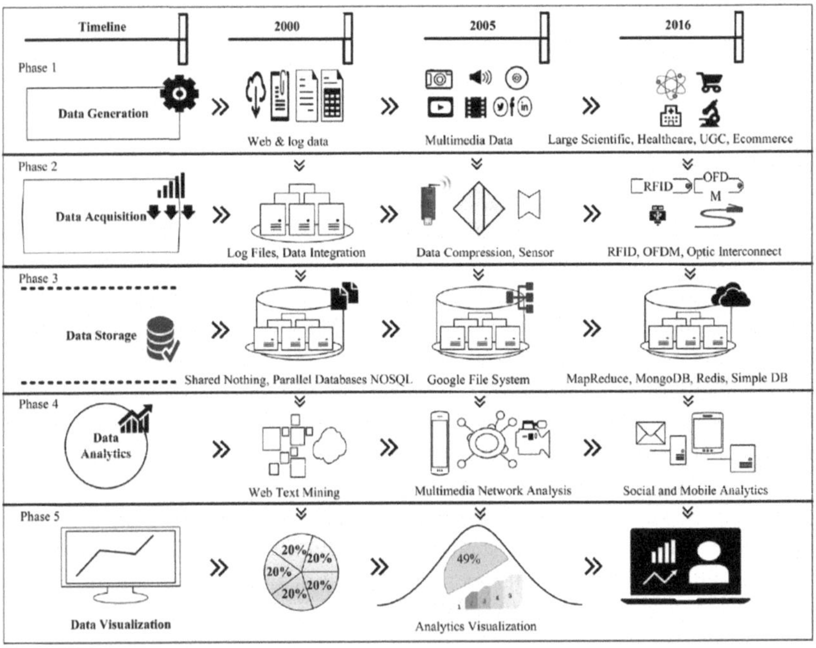

Fig. 1. Architecture of big data system [10].

needs lot of funds. In order to maintain the standards that are required for proper application of these technologies, it is essential to design intelligence systems that can efficiently gather, store and analyze real-time and historical data (see Fig. 1) [10].

In global market competition is ever growing. Companies are looking on how the generated data can help to achieve competitive advantage through smarter use of their data, especially data that through history were treated as side effect of business activity and regarded as no valued data [11]. In the past manufacturing companies used gathered data to solve issues related to product quality, organization of manufacture, storage, fault detection, maintenance etc. [12]. In the present big data has been established as an valuable tool for knowledge acquisition from the databases from manufacturing companies [13]. Development of the internet of things, cloud computing and predictive analytics makes possible the instalment of smart factories with linked devices and networks that are now capable to exchange information and communicate in real-time, thus improving operation performance [14].

2.2 Technological Intensity

Technological intensity is defined as the level of knowledge about manufacturing process incorporated in firms' products, and this indicator is typically measured relationship between research and development and firm's revenue [15]. The Organization for Economic Cooperation and Development and Eurostat are responsible for the classification of industrial sectors according to their level of technological intensity [16]. According to this classification, manufacturing firms are divided in four levels: high, medium-high, medium-low and low technological intensity. Moreover, technological intensity could be measured as the indicator of development and successful of manufacturing firms [15, 17, 18]. Furthermore, prior research shows that technological environment in which firms operate conditions the opening up of the innovation potential that determine largely the manufacturing firm's capacity to make success through their business models [19–21].

On the other side implications of the introduction of technological intensity in the application of innovative digital services, in the context of servitization in manufacturing firms have been neglected [1]. Thus, there are the need to study technological intensity in the field of digital manufacturing service with an eye on industry related factors.

3 Data and Methodology

For the purpose of our research descriptive survey research was employed, which is conducted under the international project European Manufacturing Survey (EMS). EMS is an international project coordinated by the Fraunhofer ISI Institute from Germany, which is oriented towards innovation in manufacturing companies considering all aspects of a manufacturing process in a standardized and systematized way [22–24]. The survey takes place every three years and considers manufacturing companies (NACE Rev 2 codes from 10 to 33) that have more than 20 employees. The dataset employed for the analysis in this research is built from 2018 data collection gathered from Serbian manufacturing companies. The dataset includes 240 companies of all manufacturing sectors. About 46% of the companies in the sample belong to the group of small companies having between 20 and 49 employees, additional 43% of the companies are medium-sized companies that have between 50 and 249 employees, and final 11% of the companies belong to the group of large companies having more than 250 employees.

Given our descriptive purpose, our data analysis relies on simple statistics. More specifically, we used descriptive statistics.

4 Results

In this research, we have analyzed the use of digital services based on big data analysis considering the size and the technological intensity of manufacturing companies.

The results of the use of digital services based on big data analysis by the size of companies are presented in Fig. 2. Most of the companies that use digital services based on big data analysis are medium sized companies with the share of 50%, followed by large companies with the share of 33% and small companies with the share of 17%.

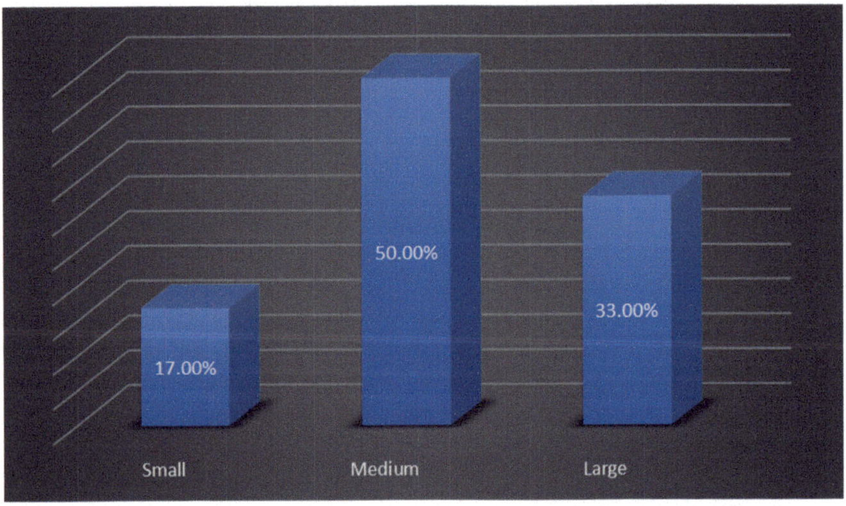

Fig. 2. The use of digital services based on big data analysis by the size of companies.

Figure 3 depicts classification of manufacturing firms in Serbia, which used digital services based on big data analysis according to technological intensity. There are no firms with high-technology intensity in Serbian manufacturing firms, which use digital services based on big data analysis. Moreover, there is the same number of manufacturing firms in the high-medium, low-medium and low technology intensity which use digital services based on big data analysis.

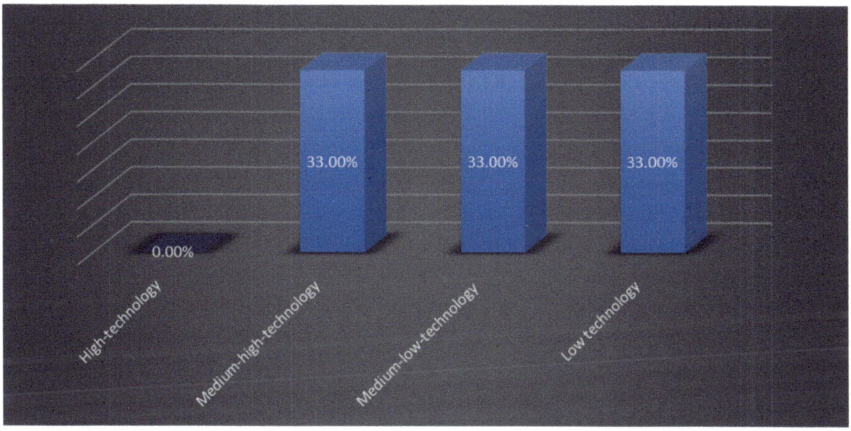

Fig. 3. Classification of manufacturing firms according to technological intensity.

5 Discussion

The results presented in this study contribute to the existing literature in multiple ways. First, the data obtained through a large, multisectoral survey allowed insights into digital servitization in transitional countries beyond evidence from case studies. Second, the analysis offers new information regarding the relationship between servitization in manufacturing firms and their size, as well the relationship between technological intensity of a manufacturing firm and the application of innovative digital services. Since there are no high-technology firms in Serbia that use digital services based on big data analysis, and there is an even distribution among the medium-high-technology firms, medium-low-technology firms and low-technology firms when it comes to operationalization of the above mentioned digital services, it could be concluded that there are other factors influencing this relationship. These results are not in line with previous studies on servitization and innovation intensity [25]. Having in mind the transitional character of the country of Serbia where the survey was performed, the lack of resources could be one of the significant contributors to the potentially insufficient utilization of the advanced services.

On the other hand, the classification according to the size of the companies, offers a more insightful perspective, showing the prevalence of medium companies that utilize the advanced digital services. Since medium and small companies represent most of the sample, it could be anticipated that the medium companies are the group that shows the highest occurrence of digital services deployment. As small companies have smaller operating budget and limited resources, the higher costs and risk of encompassing the implementation of novel business practices puts them in an inherent inferior position. Since large companies can operate without the strict budget constraints, there is a high number of large firms that utilize the digital services based in big data analysis. The study provides an alternative explanation for the interplay between digitalization and firm size [5].

6 Conclusion

The present study set out to challenge the simple assumptions underlying the causal relationship between big data analysis, company size, and technological intensity. The empirical results indicate that medium size companies tend to use more digital technologies (i.e. big data) to provide service in comparison to small and large firms. Furthermore, our results indicate that high technology manufacturing firms in Serbia are not yet utilizing digital technologies to facilitate the service innovation.

As with every study, the current study is not without limitations. The present study uses a relatively small sample of Serbian manufacturing companies, which may limit generalizability. For further research, authors could make a comparative context, primarily focusing the differences among developed and transitional economies. Furthermore, as the present study tested the link between only one digital service and firm characteristics (i.e. firm size, technological intensity), the effects of digitalization should be further studied to a variety of additional digital technology variables, such as Internet of things and cloud computing as a service.

References

1. Marjanovic, U., Rakic, S., Lalic, B.: Digital servitization: the next 'big thing' in manufacturing industries. In: Advances in Production Management Systems. Production Management for the Factory of the Future, pp. 510–517 (2019)
2. Lalic, B., Todorovic, T., Cvetkovic, N., Tasic, N., Marjanovic, U.: Strategic outsourcing of SMEs in the context of industry 4.0: evidence from Serbia. In: 3rd International Conference on the Industry 4.0 Model for Advanced Manufacturing, pp. 139–145 (2018)
3. Grosso, C., Forza, C.: Users' social-interaction needs while shopping via online sales configurators. Int. J. Ind. Eng. Manag. **10**(2), 139–154 (2019)
4. Ardolino, M., Rapaccini, M., Saccani, N., Gaiardelli, P., Crespi, G., Ruggeri, C.: The role of digital technologies for the service transformation of industrial companies. Int. J. Prod. Res. **56**(6), 2116–2132 (2018)
5. Kohtamäki, M., Parida, V., Patel, P.C., Gebauer, H.: The relationship between digitalization and servitization: the role of servitization in capturing the financial potential of digitalization. Technol. Forecast. Soc. Chang. **151**(July), 2020 (2019)
6. Hoffmann-Walbeck, T.: Smart factory: JDF and XJDF. J. Graph. Eng. Des. **9**(1), 5–9 (2018)
7. Hoy, M.B.: Big data: an introduction for librarians. Med. Ref. Serv. Q. (2014)
8. Belcastro, L., Marozzo, F., Talia, D.: Programming models and systems for big data analysis. Int. J. Parallel Emergent Distrib. Syst. (2019)
9. Viktor Mayer-Schönberger, K.C.: A summary of 'big data: a revolution that will transform how we live, work, and think' by Viktor Mayer-Schonberger and Kenneth Cukier [blog post]. New Books Br. (2013)
10. Bendre, M.R., Thool, V.R.: Analytics, challenges and applications in big data environment: a survey. J. Manag. Analytics (2016)
11. Niebel, T., Rasel, F., Viete, S.: BIG data–BIG gains? Understanding the link between big data analytics and innovation. Econ. Innov. New Technol. (2019)
12. Jun, C., Lee, J.Y., Kim, B.H.: Cloud-based big data analytics platform using algorithm templates for the manufacturing industry. Int. J. Comput. Integr. Manuf. **32**, 723–738 (2019)

13. Liu, C., Li, H., Tang, Y., Lin, D., Liu, J.: Next generation integrated smart manufacturing based on big data analytics, reinforced learning, and optimal routes planning methods. Int. J. Comput. Integr. Manuf. **32**, 820–831 (2019)
14. Tan, K.H., Ji, G., Lim, C.P., Tseng, M.L.: Using big data to make better decisions in the digital economy. Int. J. Prod. Res. **55**, 4998–5000 (2017)
15. Zawislak, P.A., Fracasso, E.M., Tello-Gamarra, J.: Technological intensity and innovation capability in industrial firms. Innov. Manag. Rev. **15**(2), 189–207 (2018)
16. Eurostat: Eurostat indicators on High-tech industry and Knowledge – intensive services (2018)
17. Ivanisević, A., Lošonc, A., Morača, S., Vrgović, P.: Exploring the business planning practices in SMEs in a developing country. Int. J. Ind. Eng. Manag. **10**(1), 105–114 (2019)
18. Iazzolino, G., Migliano, G., Dattilo, M.I.: The impact of intellectual capital on firms' characteristics: an empirical analysis on european listed manufacturing companies. Int. J. Ind. Eng. Manag. **10**(3), 219–237 (2019)
19. Lalic, B., Rakic, S., Marjanovic, U.: Use of industry 4.0 and organisational innovation concepts in the Serbian textile and apparel industry. Fibres Text. East. Eur. **27**(3), 10–18 (2019)
20. Segarra-Ciprés, M., Bou-Llusar, J.C., Roca-Puig, V.: Exploring and exploiting external knowledge: the effect of sector and firm technological intensity. Innov. Manag. Policy Pract. **14**(2), 203–217 (2012)
21. Berić, D., Stefanović, D., Lalić, B., Ćosić, I.: The implementation of ERP and MES Systems as a support to industrial management systems. Int. J. Ind. Eng. Manag. **9**(2), 77–86 (2018)
22. Lalić, B., Medić, N., Delić, M., Tasić, N., Marjanović, U.: Open innovation in developing regions: an empirical analysis across manufacturing companies. Int. J. Ind. Eng. Manag. **8**(3), 111–120 (2017)
23. Medic, N., Marjanovic, U., Zivlak, N., Anisic, Z., Lalic, B.: Hybrid fuzzy MCDM method for selection of organizational innovations in manufacturing companies. In: International Symposium on Innovation and Entrepreneurship (ISIE), pp. 109–116 (2018)
24. Marjanovic, U., Lalic, B., Majstorovic, V., Medic, N., Prester, J., Iztok, P.: How to increase share of product-related services in revenue? Strategy towards servitization. In: Advances in Production Management Systems. Smart Manufacturing for Industry 4.0, vol. 536, pp. 57–64 (2018)
25. Dachs, B., Biege, S., Borowiecki, M., Lay, G., Jäger, A., Schartinger, D.: Servitisation of European manufacturing: empirical evidence from a large-scale database. Serv. Ind. J. **34**(1), 5–23 (2014)

Abrasive Flow Machining of 3D Printed Metal Parts – A Scientific Review with Extension on Industrial Needs

Luka Kastelic(✉), Davorin Kramar, and Franci Pušavec

Faculty of Mechanical Engineering, Ljubljana, Slovenia
luka.kastelic@fs.uni-lj.si

Abstract. Nowadays, the use of additive technologies is increasing. Despite the exceptional capability to produce complex geometries, additive technologies are unable to produce components or functional surfaces within tight tolerances and integrity demands. With other words, "as build" surface qualities are poorer in comparison to a conventional machined surfaces. Therefore, components have to be post-machined. Lately, abrasive flow machining (AFM) is offering improvements in such cases. However, many process parameters, e.g. abrasive media property, temperature, velocity, etc. influencing the performance of AFM and their understanding is crucial for successful implementation of AFM into the industrial applications. This paper presents critical scientific review of AFM, with an emphasis on post-machining of 3D printed metal parts.

Keywords: Abrasive flow machining · Complex geometry · Additive manufacturing · Surface quality

1 Introduction

Additive technologies or 3D printing methods make it possible to produce products of complex shapes which are difficult or impossible to process with conventional ones. In this area, the Abrasive Flow Machining offers advantages especially in the machining of internal channels, such as cooling channels in tool inserts. Due to this advantage, the process is currently of particular interest in the molding tool industry. The paper will present the basic characteristics of the AFM process, as well as the current research on the processing of 3D printed products with the help of AFM and other processes.

2 Abrasive Flow Machining

The Abrasive Flow Machining (AFM) is a polishing process where polymeric medium with added abrasive is used as a tool. Deformability and non-Newtonian behavior of the abrasive medium (shear-dependent behavior) allows to process parts with complex geometries which processing by conventional methods is not possible [1].

L. Wang et al. (Eds.): AMP 2020, LNME, pp. 270–278, 2020.
https://doi.org/10.1007/978-3-030-46212-3_20

2.1 AFM System

The AFM machining system consists generally of an AFM machine, workpiece and abrasive medium [2]. Depending on the direction of movement of the abrasive medium, the AFM devices are divided into [3]:

- one-way AFM: in this process, the abrasive medium is pushed only in one direction, as shown in Fig. 1a,
- two-way AFM: In this process, the abrasive medium is alternately pushed first in one direction and then in the other, as shown in Fig. 1b,
- orbital AFM: in addition to the flow of the abrasive medium, the orbital vibrations are also present in the process, as shown in Fig. 1c.

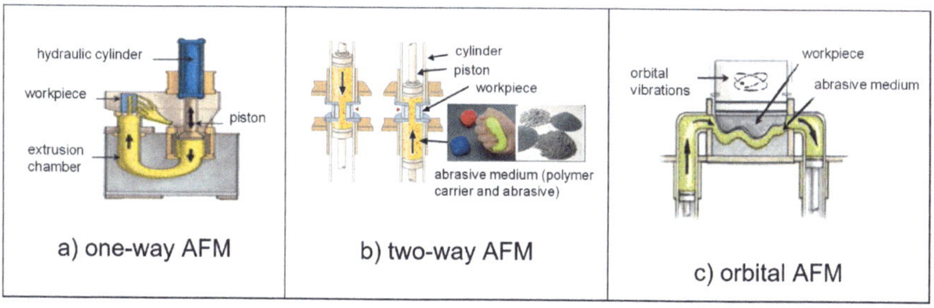

Fig. 1. Types of abrasive flow machine [3]

As the authors of the contributions [4–6] wrote, the main influencing parameters affecting AFM are divided into:

- the setting parameters to be set on the machine,
- on the properties of abrasive media,
- using the auxiliary tools/structures/mechanisms and
- the properties of the workpiece.

The selected parameters which has the biggest influence on the AFM are shown in the Table 1.

Table 1. Main influencing parameters on AFM [3]

Configuration parameters	Media properties	Workpiece properties
Extrusion pressure Number of cycles	Type of carrier medium Type of abrasive Rheological properties of the abrasive medium Abrasive mesh size	Chemical composition Hardness Geometry
	Concentration of the abrasive in the carrier medium additives	Initial surface texture and roughness

An AFM is a completely mechanical machining with cutting based on undefined cutting geometry. The abrasive medium consist of a viscoelastic medium and abrasive, which allow it to machine the surface. The main material removal mechanisms that occur in AFM treatment are: elastic deformation, plastic deformation (ploughing of the abrasive grains on the surface) and micro-cutting of the material [3].

3 Review of the Former Researches

In the field of AFM machining of conventional produced metal materials, a lot of different studies have been done, but there is a lack of studies on 3D printed metal materials. The reason for this is probably in a fairly new 3D printing technology that is still being developed and introduced in the industry.

Peng et al. [7], were investigating a surface improvement of the 3D printed AlSi10Mg aluminum alloy which was post-processed with a two-way AFM process. The efficiency of the processing was measured by drilling a hole into the surface of the 3D printed sample, which was then used as a reference for the determination of the height of the material removed in the measurements with the confocal laser microscope VK-X250. In Fig. 2, a surface topology after 15, 90 and 390 AFM cycles is displayed. They stated that after 15 cycles, only some of the particles that were adhered to the surface were removed, after 90 cycles, these larger particles were mostly removed, but craters remained. The craters were almost completely removed after 390 cycles. The surface roughness was reduced from the initial Sa = 14 μm to Sa = 1.8 μm.

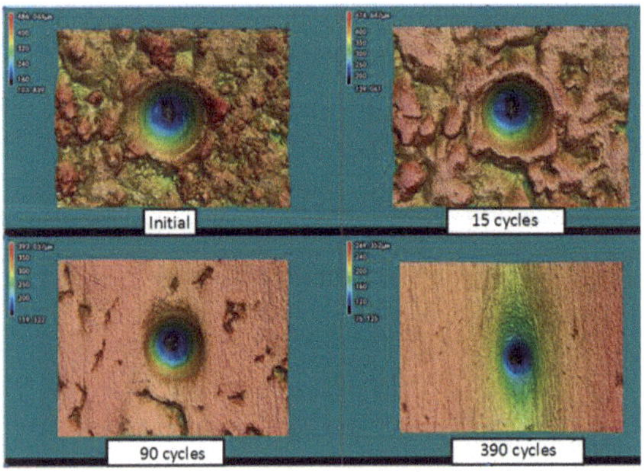

Fig. 2. Surface topology during AFM process at the reference hole [7]

Regarding the material removal, they stated that in the initial cycles it is mostly due to the removal of adhered particles on the material, and in subsequent cycles, the removal of the base material becomes more apparent. The residual stresses were also measured, where they found that there are tensile stresses on the surface of the printed product, but after AFM machining, compressive residual stress is induced on the surface.

Tina Bremstein et al. [8] measured the wear of abrasive grains before and after machining of an austenitic stainless steel X2CrNiMo17, which was produced by the Selective Laser Melting process (SLM). They found that worn-out abrasives reduce the efficiency of the process, and the roughness of the treated surface increases, as the plating and rolling of the abrasive particles along the workpiece surface increases. Due to the higher concentration of abrasive particles in the abrasive medium, this becomes more solid and more elastic, which forces the abrasive particles deeper into the surface, resulting in a more rough finish. It was found that the use of worn abrasive media leads to up to 20% lower quality of the surface and up to 30% lower machining efficiency compared to the machining where a new abrasive is used.

The die-casting or plastic injection molding tools, are heavily stressed due to temperature extensions and shrinkage. The roughness of the surface and the residual tensions on the surface thus significantly affect their lifetime. Duval-Chaneac et al. [9], therefore found that the viscosity of the polymer and the concentration of the abrasive influence the roughness and residual stresses of the heat treated and non-heat treated samples from the maraging 300 steel that were produced by the SLM process. Two-way AFM system was used, during the experiments the normal force on the sample wall, the temperature of the sample and the abrasive medium were controlled, so the constant conditions were ensured. After 25, 50, 75, 150 and 200 cycles, the surface roughness Sa and Sq was measured, and it was found that the heat treatment had no significant effect on the roughness of the machined surface, but the viscosity of the carrier medium, the abrasive concentration and the size of abrasive grains do. The samples on which the machining was carried out with higher viscosity mediums had a lower surface roughness Sa and Sq. The higher concentration of abrasive grains also affects faster processing, and the lower roughness of the machined surface after the same number of cycles performed. When measuring the residual stress on the surface of the sample, it was found that in the case of untreated samples, the compressive residual stresses after AFM machining occur in the longitudinal direction (0°) and in the direction perpendicular (90°) to the abrasive flow. In the case of heat-treated samples, the compressive residual stresses, were induced in the longitudinal direction (0°), but in direction perpendicular (90°) to the abrasive flow, no significant residual stress was measured. These results indicate that in the case of untreated samples there is a greater effect of plowing or lateral material flow than in the heat-treated samples, which is the reason for higher residual compressive stress on the surface of the non-heat treated samples.

Uhlmann et al. [10] were working on CFD (Computational Fluid Dynamics) simulations of the AFM process, simulating the flow of the abrasive medium at the treatment of the sample piece and the turbine blade. The properties of the abrasive medium were obtained by means of experiments, and then they were compared with Maxwell's model. The viscosity dependence of the shear rate was introduced into the CFD simulation using the Cox-Merz rule. The simulations were performed on the basis of solving the Navier-Stokes equations in a stable state. As a result of the simulation, the removed material was not monitored, as this would result in much more complex simulations. Instead of that, they compared the level of shear on the surface of the workpiece with the final finished surface, where both results showed that the height of the shear is the highest and therefore the most effective machining at the center of the surface. In addition, an important conclusion is also, that the rounding of the edges is inevitable as it is a consequence of the flow of the abrasive medium, which, due to the damping pressure, removes the material at the edge. They also concluded that the flow of the abrasive medium can be made as uniform as possible with appropriate obstructions or directors of the abrasive medium which can be placed along the workpiece.

On the combination of abrasive and chemical machining, Mohammadian et al. [11] worked. The samples were made of the Inconel 625, manufactured using the SLM method. Since the chemical and AFM treatment were combined, the viscoelastic polymer was not used for the carrier medium of abrasive particles, as it usually is at AFM, but an acid with abrasive added. They found that the combination of abrasive and chemical flow machining reduces the polishing time by as much as two-thirds compared to the use of individual procedures. In addition, it has been found that increasing the fluid velocity increases the polishing depth.

William Gilmore, in his master's thesis [12], compared the surface of the samples, which were processed by ultrasonic peening and abrasive flow machining. The sample material was 316L stainless steel which was manufactured using the SLM process. With experiments he found that both processes have their advantages and disadvantages, so none is universal for machining of 3D printed products. When machining with AFM, a lower roughness of the treated surface can be achieved, but the restriction is characterized by rapid changes in the cross section, which makes it impossible to reach evenly treated surfaces. AFM is a really good process for channels processing which are adapted for the flow of the fluid. The advantage of ultrasonic surface treatment is that the surface is hardened and the change of the workpiece size is minimal, since the material is not removed. However, the roughness of the treated surface is worse than at AFM machining, and the treatment of internal surfaces is limited or strongly dependent on their geometry.

4 Industrial Example: Polishing of Plastic Injection Molding Tool

Abrasive flow machining (AFM) is also usefull in practice, as demonstrated by the project in which the AFM process was introduced for polishing the multi-nozzle plastic injection molding tool [13]. For plastic products manufactured by injection molding, a high quality surface of the finished product is often required. In order to achieve this, it is necessary to produce tool or tool inserts which functional surfaces are polished. This can be achieved by manual polishing, which is very time-consuming, monotonous, and tiring for the worker. For this purpose, the polishing of tool inserts with the abrasive flow machining (AFM) process was introduced in this project. The 3D model of the tool insert is shown in the cross section in Fig. 3, where the polished surfaces are marked with blue color.

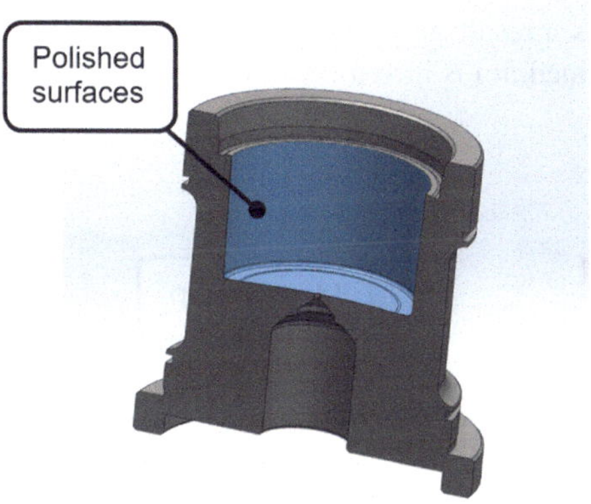

Fig. 3. Cross section of the tool insert.

Together with Extrude Hone Corporation in Germany, a clamp preparation was made to fit the tool inserts onto the AFM machine and perform experiments. Polishing with AFM, as compared to manual polishing, has provide a 12-times shorter polishing cycle. This is particularly important for multiple nesting tools where time and cost savings are significant, and the customer gets the ordered products quicker, which increases competitiveness on the market. The quality of AFM polished surfaces did not achieve high gloss finish, such as manual polishing, but sufficient according to customer requirements. However, improvements would be needed, in particular, on the construction of the clamping device of the cartridge, and in the parameters of the AFM process, in order to achieve a better and more uniform quality of the polished surface.

5 Innovative AFM for Increasing Supervision and Efficiency of the Process

In the field of abrasive flow machining, the current development trend focuses on increasing efficiency and control over the process itself. The latter was analyzed by the Laboratory for Machining (UNI-LJ) [14]. The results demonstrate the positive effect of AFM machining on roughness and residual stress on the surface of AISI D2 tool steel after EDM (Electrical Discharge Machining). Depending on the findings, AFM treatment reduces unwanted tensile residual stresses left after EDM treatment, and reduces the roughness of the treated surface.

Additionally, an Abrasive Flow Machining with movable mandrel (AFMmm) [15] was developed, using a special pin with which it is possible to increase and locally control the speed profile of the abrasive medium. Figure 4 shows the velocity profile inside the workpiece as a result of the Finite Element Model (FEM), where the speed profile of the abrasive medium is increased at the point of narrowing the slit due to the fixed/movable pin.

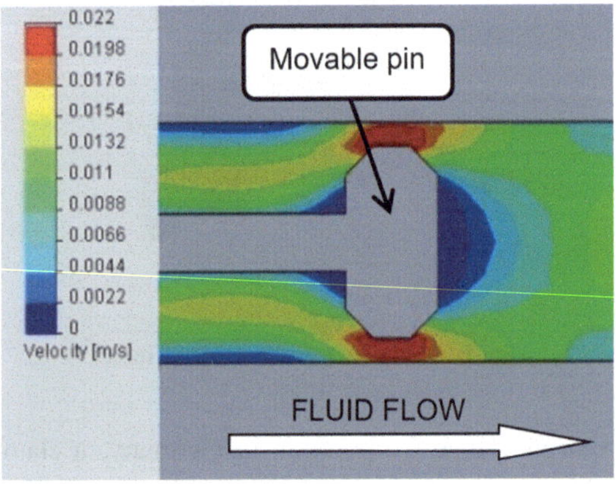

Fig. 4. The abrasive media speed profile of the FEM analysis (AFMmm) [15]

With this system, it is possible to increase the control over the topography of the machined surface and to shorten the processing time which also affects on the lower energy consumption of almost 20%. The innovation (Fig. 5) is in the patenting procedure [16].

In Fig. 5, a moving pin (ball) is shown on sketch 1, which can be used for polishing non-linear channels; in sketch 2, a pin is fixed to the machine piston, the pin on the sketch 3 is used for polishing channels where it is necessary to polish on several different diameters. Moving pin on the sketch 4 can be moved independently with respect to the machine piston or also rotate with the concept on the sketch 5. This area requires the implementation of moving pins on the machine itself, as the technical implementation of the movable pins is quite demanding, and currently standard AFM machining machines are not yet adapted for processing with AFMmm.

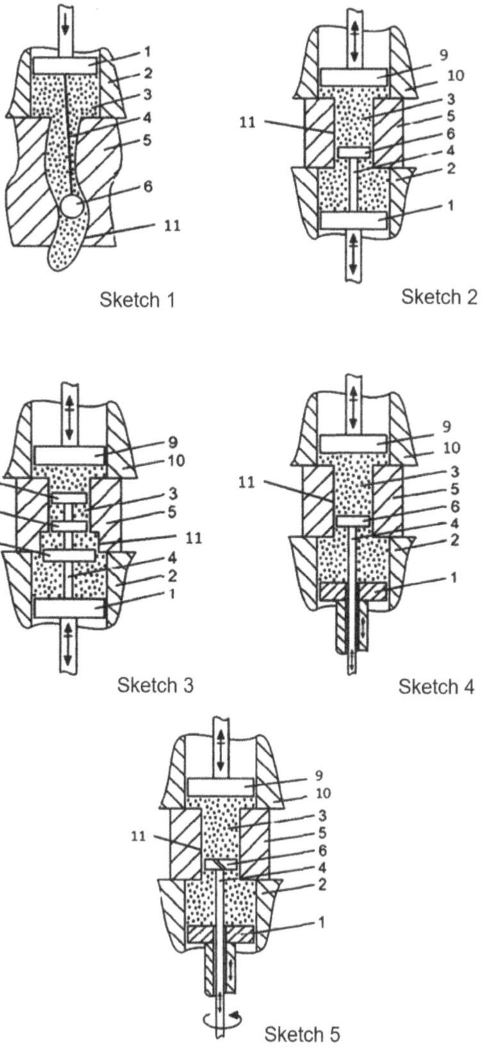

Fig. 5. Various versions of movable pins [16]

6 Conclusions

According to literature review and previous research, it has been found that AFM is mostly used in tooling, for polishing cooling channels and/or effective surfaces of molds, and is also used to process the surfaces of turbine blades, nozzles, etc. The researches were focused primarily on the determination of the influence of the process parameters on the roughness of the treated surface, the residual stress on the surface and the efficiency of the machining.

References

1. Williams, R.E., Melton, V.L.: Abrasive flow finishing of stereolithography prototypes. Rapid Prototyping J. **4**(2), 56–67 (1998)
2. Rhoades, L.J.: Abrasive flow machining: a case study. J. Mater. Process. Technol. **28**, 107–116 (1991)
3. Sankar, M.R., Jain, V.K., Ramkumar, J.: Abrasive flow machining (AFM): an overview. In: Intelligent Machining Systems and Multi-scale Manufacturing, December 2008
4. Williams, R.E.: Investigation of the abrasive flow machining process and development of a monitoring strategy using acoustic emission. Dissertation, University of Nebraska, Lincoln (1993)
5. Loveless, T.R., Williams, R.E., Rajurkar, K.P.: A study of the effects of abrasive flow machining on various machined surfaces. J. Mater. Process. Technol. **47**, 133–151 (1994)
6. Gorana, V.K., Jain, V.K., Lal, V.K.: Forces prediction during material deformation in abrasive flow machining. Wear **260**, 128–139 (2006)
7. Peng, C., Fu, Y., Wei, H., Li, S., Wang, X., Gao, H.: Study on improvement of surface roughness and induced residual stress for additively manufactured metal parts by abrasive flow machining. Procedia CIRP **71**, 386–389 (2018)
8. Bremerstein, T., Potthoff, A., Michaelis, A., Schmiedel, C., Uhlmann, E., Blug, B., Amann, T.: Wear of abrasive media and its effect on abrasive flow machining results. Wear **342–343**, 44–51 (2015)
9. Duval-Chaneac, M.S., Han, S., Claudin, C., Salvatore, F., Bajolet, J., Rech, J.: Experimental study on finishing of internal laser melting (SLM) surface with abrasive flow machining (AFM). Precis. Eng. **54**, 1–6 (2018)
10. Uhlmann, E., Schmiedel, C., Wendler, J.: CFD simulation of the abrasive flow machining process. Procedia CIRP **31**, 209–214 (2015)
11. Mohammadian, N., Turenne, S., Brailovski, V.: Surface finish control of additively-manufactured Inconel 625 components using combined chemical-abrasive flow polishing. J. Mater. Process. Technol. (2017)
12. Gilmore, W.: An evaluation of ultrasonic shot peening and abrasive flow machining as surface finishing processes for selective laser melted 316L. Master thesis, California Polytechnic State University (2018)
13. Stanonik, B.: Application of abrasive fluid machining. Bachelor thesis, Faculty of Mechanical Engineering, University of Ljubljana (2019)
14. Kenda, J., Pušavec, F., Kermouche, G., Kopač, J.: Surface integrity in abrasive flow machining of hardened tool steel AISI D2. Procedia Eng. **17**, 172–177 (2011)
15. Pušavec, F., Kenda, J., Kopač, J.: Modeling and energy efficiency of abrasive flow machining on tooling industry case study. Procedia CIRP **13**, 13–18 (2014)
16. Kenda, J., Pušavec, F., Kopač, J.: Arrangements and methods for abrasive flow machining. Ministry of Economic Development and Technology, Patent (2014)

Rejuvenation of Business Management Tools in Industry 4.0

Dragan Đuričin(✉) and Iva Vuksanović Herceg

Faculty of Economics, University of Belgrade, Belgrade, Serbia
dragan.djuricin@ses.org.rs, ivav@ekof.bg.ac.rs

Abstract. The world of business economics (and management) traditionally has been viewed as relatively linear. In such context, competitive dynamics depends on contingency between structural factors and contextual factors, as well as characteristics of a representative company. But, the context has been changed under the impact of Industry 4.0. By synthesizing the breakthroughs from cyber and physical (and/or biological) worlds, it gave rise to an almost endless stream of combinatorial innovations. There are two major consequences of the previous transition. First, universal connectivity as the new free good enables that the world of engineering reaches the levels of complexity and dynamism typical for non-linear systems. Second, emerging amalgams of cyber and physical breakthroughs trigger in business management transformation of linear value chain into exponential value chain (or platform), actually a non-linear system. Mentioned structural changes lead to convergence of the engineering and business management in conceptual terms. In this paper we explore the ways in which Industry 4.0 can offer a powerful and consistent platform for implementation of conventional business management tools. We have been inspired by two achievements. First, to map out the impact of Industry 4.0 on double paradigm change, both in macro and micro (or business) management. Second, to explore, with key details, the impact of the paradigm change in business management on effectiveness of conventional management tools. By doing this, we wish to promote the broader and systemic thinking, synthesizing micro and macro management perspectives into a single point of view that is actually based on the reversibility principle.

Keywords: Industry 4.0 · Paradigm change · Combinatorial innovation · Reversibility principle · Exponential value chain · Information value loop · Micro management tools

1 Introduction

The Great Recession of 2008 definitely confirms that the neoliberal model of growth and related economic policy platform do not lead to a sustainable and inclusive growth, both toward the people (full employment and decent jobs) and the nature (environmental conservation). When a complex system like economy grows within a materially finite context and with ignorance of negative external effects and adverse implications, some deviations from expectations like financial bubbles, pollutant gases bubble, income inequality, and environmental degradation in particular, could only be explained as

L. Wang et al. (Eds.): AMP 2020, LNME, pp. 279–293, 2020.
https://doi.org/10.1007/978-3-030-46212-3_21

consequences of the model's premises. The last crisis has reminded us that adherence to the current economic system represents a betrayal of future generations. No doubt, the market forces cannot stop the negative external effects and stagnation trap.

Behavior of business organizations and economy as a whole should provide better balance between the society and the nature. The previous perspective has been addressed in many discussions dedicated to the new economy rules, particularly in the Stockholm Statement [2]. In contrast to the neoliberal model of growth based on market fundamentalism and the Washington Consensus [22] as related policy platform, the new consensus illuminates that the market on its own is not capable of managing serial structural transformations inspired by Industry 4.0. So, new interest around mission-driven industrial policies is growing.

The impact of Industry 4.0 is ambivalent, it holds both promises and perils. If not managed properly, it will exacerbate existing structural imbalances from the past, create the new ones, and slow down the progress towards climate crisis resolution. Business organizations and economy as a whole can no longer continue to operate under the old rules. Rewriting the rules, in fact, means a paradigm change in management. The reversibility principle (or feedback loop) as a basic principle of functioning in physical systems is the foundation of a double paradigm change in economics (and macro management) and business economics (and micro management).

Implementation of this principle in macro management leads to the growth model inspired by the idea of circular (regenerative or shared) economy and heterodox economic policy platform [6], both combining economic progress with environmental and social responsibility. Paradigm change in micro management triggers radical changes in business model, organizational structure, and strategy of business organizations. It enables proliferation of combinatorial innovations through economy, as a whole. The implementation of the same principle in new macro management paradigm, in fact, means broadening the existing development goals, introduction of new development initiatives focused on environmental sustainability and mission-oriented industrial policies for tradable sector combined with automatic stabilizers in core economic policies (monetary, fiscal, and competition).

Search for solutions of the legacy problems is also relevant. Circular economy is an alternative to linear production systems. Also, "green transition" needs coordination of the visible hand of the state and invisible hand of the market. So, heterodox policy platform provides at the same time verticalization of research and development within frontier technologies development and education improvements (long life education for reskilling and upskilling workforce) through the "visible hand" of the state along with horizontalization of innovative products and services through the market "invisible hand".

In Industry 4.0, creation and use of actionable information gives to the reversibility principle the role of a key transformation rule. Related performance improvements on a micro management level trigger the paradigm change on a macro management level in the same direction. Namely, feedback loop is another focal point which should be respected in the model of growth and related economic policy platform definition.

Universal connectivity is the ultimate free good in Industry 4.0. It orchestrated an almost endless stream of combinatorial innovations, by enabling a greater efficiency and superior value proposition. The explanations come from the fact that a deeper managerial visibility of the structure of component costs combined with better insights into the client needs triggers broadening of actionable information data base and, as a consequence, adequate decisions. Above all, paradigm change rejuvenates conventional micro management tools, making their implementation more effective and efficient.

The last stance is exactly what this paper tries to promote. Our intention is to present a comprehensive picture of the ways in which the business platform, as the key consequence of universal connectivity and combinatorial innovations, can offer rejuvenation of conventional micro management tools like quality control, activity-based costing, value-based management, manufacturing execution system, and enterprise resource management.

The structure of the paper follows the abovementioned. After the introduction, the second part deals with the impact of major contextual forces on paradigm change in micro management. The third and the fourth part explain the consequences of paradigm shift in micro management from two relevant angles: business model (and organization), and strategy. The fifth, and most important part, reveals rejuvenated applications of some conventional micro management tools. The last part presents some concluding remarks and thoughts.

2 Paradigm Shift in Business Management

Paradigm change in business management is in urgent need for updating, having come under the impact of universal connectivity and combinatorial innovations.

Despite great potential of Industry 4.0, gaps around spread of noise (or misinformation), cybercrime, and problems with algorithmic biases and big data are still large. Moreover, complexity of the business ecosystem grows faster than the system itself. Figure 1 indicates that the possible interconnections (or flows) in business ecosystem grow with the square of the number of participants (or nodes). Consequently, ability to use transaction data (find, classify, aggregate, and analyze) in order to get the so-called "actionable information" grows faster than the opportunity of using it for concrete decision making. Indeed, it is a significant threat not only for prosperity, but also for business continuity.

In all sectors actionable information is a prerequisite of a competitive advantage. New dynamic favors the pursuit of fast growth and incentivizes business organizations to expand from value chain to value network with the aim to gain control of critical infrastructure, data flows and actionable information. In the new context the reversibility principle is going to be a key rule for capitalization of actionable information.

Advanced (or additive) manufacturing, sometimes called smart automation, implemented on production phase of the linear value chain is a typical example of the reversibility principle. Advanced manufacturing is actually an amalgam of cognitive technologies, artificial intelligence, and robotics. Cognitive technologies *via* digital tweens of the innovative product communicate with the machinery in a unique way to tell it what to do.

Value creation based on information is a similar process with value creation from the physical value chain. The "Information Value Loop" concept developed by *M. Raynor* and *M. Cotteleer* [18] is the framework that allows a multiple feed-back

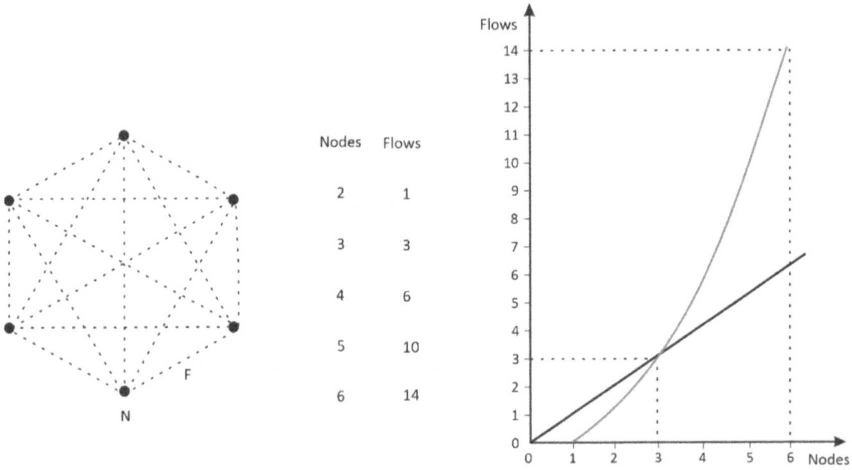

Fig. 1. Relations between nodes and flows

loops of information or flow of transaction data from physical to digital and back to physical content. It is the nexus of activities and related data that are successively created, classified, summarized, analyzed, and communicated in order to be transformed in actionable information (see Fig. 2).

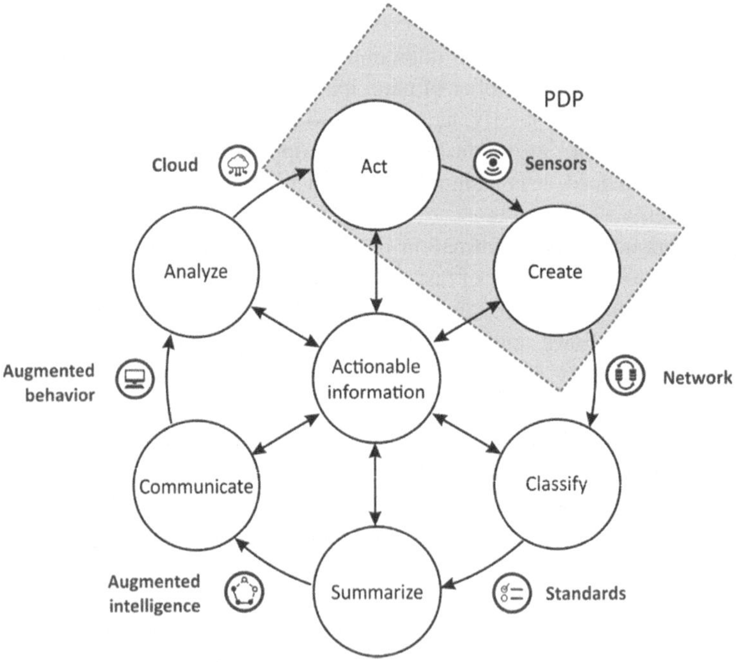

Fig. 2. The Information Value Loop, Source: Modification based on [18, p. 55]

Each loop consists of three activities: creation of digital record (transaction data) related to the physical activity, real time exchange of transaction relevant data between data bases with the aim of creating actionable information, and implementation of some algorithms to translate actionable information in concrete business actions (or transformation of digital context into physical context). Consequently, the concept incorporates physical-to-digital-to-physical loop (or PDP loop).

The previous example explains how reversibility principle is implemented in value chain by enabling creation of the feedback loop from physical back to digital, from digital back to physical, and from digital back to physical content. Each stage of the loop is supported by specific technologies, virtual and/or physical. For example, an activity, monitored by sensor technology, creates some transaction data. Analysis of transaction data are meant to explain all kinds of analytical support of data analytics to operations stage of the value chain. Artificial intelligence helps to complete the loop. It enables the automated and autonomous action of machinery to be implemented through actionable information.

3 Business Model (and Organizational Structure) Change

Domination of standardized technologies and/or products is one of the key characteristics of the previous stage of economic development. The related business model has been developed, more or less, as a reaction to a predictable demand pattern. Conventional business economics set of rules is based on behavior of a representative company. Under such proposition, a business organization was structured for efficiency/effectiveness. This orientation leads to division of labor and functional hierarchy. Unfortunately, functional silos restrict collaboration, limit knowledge sharing, as well as identification and annulation of a competence gap. They continuously decrease the ability to react adequately to frequent, interrelated, and radical changes.

Reactive business model and functional hierarchy do not work when symbiosis of different technologies is the main rule of competitive dynamics. Being in the intersection between the physical and virtual world, a modern business organization has started to make digital transformation, in terms of virtualization and sharing. The combinatorial innovation as a hallmark of Industry 4.0 goes hand in hand with a cognitive diversity. Empowered network of teams is infrastructure for this symbiosis. Namely, new organization provides a network of teams. Teams must be formed and disbanded rapidly and with minimal transaction costs.

Rapid advances in connectivity and industrial internet of things (IIoT) are becoming critical for the new business model. In mapping the future beyond the digital frontier, we see that singular technologies are ingredients in combinatorial innovations as well as a recipe for transformation. In the near future, the ways people interact with technology should be replaced with synchronous intelligent interfaces.

Industry 4.0 offers ongoing competitors huge and vigorous opportunities for differentiation based on advanced manufacturing and deeper client insights along with cost-cutting based on real time costing methodologies. Combinatorial innovations also create new competitors, threatening incumbents, reshaping conventional value chains and industries, as well as promoting new business models with transformation power for the economy and the society, as a whole. According to *Ch. Christensen* [4], combinatorial innovations are mainly disruptive.

After digital (and organizational) transformation, there are so many possible choices for a player of the competitive game regarding suppliers, buyers, technology vendors, communication protocol providers, and system integrators [3]. To bring together different resources and technologies and make them usable in an optimal way, business organizations need facilitators or some form of platform.

A platform is a physical and virtual space enabling participants to realize their intentions. Actually, the platform is a business model (or ecosystem) of business organizations in which multiple players are connected and attracted (see Fig. 3). Important functionality of the platform is pricing. Namely, the platform is a two-sided market space, in which one party affects the volume of transactions while balancing the price level paid by the other parties.

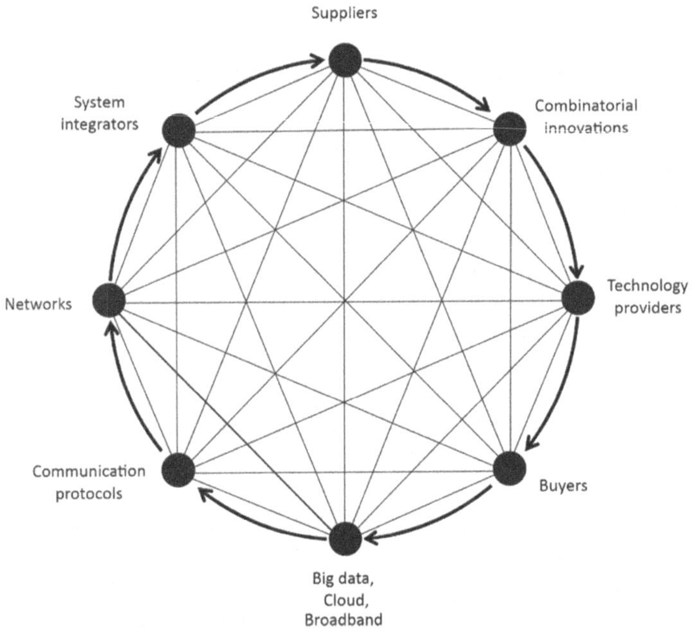

Fig. 3. Platform as a business model

The platform connects different technologies in a combinatorial way. Technologies, from sensors and communication protocols, to networks technologies like 5G, data analytics (big data, cloud computing, broad band, etc.), and cognitive tools and, their integration within IIoT, artificial intelligence, virtual reality, etc., are key enablers of growth. Members of the emerging ecosystem then use row technologies to create tailor-made solutions with the aim to reach the user's demand simultaneously with cost reduction. Technology suppliers provide alternative pathways by offering a possibility for different users to find relevant content, services, and solutions within a platform.

Indeed, the platform provides a trial-and-error mechanism for new combinatorial innovations. Companies frequently use prototyping to learn about the potentials of some combinatorial innovation before performing a large scale production. Thinking about big ideas and starting with small implementation is compatible with fast scaling. Along with diversification effect, the economy of scale effect is also possible, even for a niche player, based on the agglomeration effect on the global market.

In the face of identified opportunities generated by the platform, many companies are diving precipitately into digital transformation. To escape obstacles on this journey, related tools remain a valuable guide.

4 Broadening the Strategy Scope

Conventional business economics proposes that the perfect market structure leads to optimal resource allocation. In such case, beating rivals is the purpose of the strategy. Competitors which calculate higher profit margin on total cost and marginal costs are ready to decrease the margin keeping in mind that any competitor entering the industry will contribute in aggregate supply only if its price covers total costs per unit, at least. Fundamental defect of such line of reasoning is that such market structure is more or less static, as well as the positioning based on pricing strategy.

Static industry structure and static positioning are irrelevant when a continuous stream of combinatorial innovations influences dynamic competition, by making a new entry, substitution effect and, even more, disruption of incumbents. When "disrupt or be disrupted" is the name of the competitive game, without adequate strategy, the threat of being left behind the technological frontiers increases dramatically.

In *M. Porter's* strategy formulation framework, [14, 15], the key to success in the competitive game lies not in a low price with the aim of taking away the market share from the main competitor, or eventually from the whole market ("winner-takes-all"), but in ability to create a unique and value-based competitive advantage. When competition is based on actionable information, a better analogy for industry dynamics might be the win-win instead of the zero-sum-game.

The level of complexity, rapidity of change, uncertainty, mutual interactions, and the level of ambiguity that strategists need to deal with in a modern business ecosystem are going up. Indeed, the deep understanding of major forces of change helps to amplify their transformative power beyond *M. Porter's* framework, promoting cost-cutting, differentiation, and focusing as generic strategy options.

The paradigm shift in business management means that the focus of strategy covers not only cost reduction, but also, and predominantly, the value creation. Namely, cost-cutting and differentiation are not mutually exclusive alternatives. Robotics, smart automation, and cognitive technologies can lead simultaneously to cost reduction that is much more significant than the historical standards, while still allowing much higher consumer satisfaction.

Interaction between the borderless environments (both internal and external) with the fast development of frontier technologies causes transformation from linear to exponential value chain. Respect toward mentioned requirements needs broadening of the strategy scope. In *M. Porter's* terminology the value chain is perceived to be linear. Connectivity makes the linear value chain augmented by transforming it into exponential, dynamic value chain with circular feedback loops of resources, money, and information.

In the new setting the collaboration dominates over competition. Collaboration enables platform participants to optimize their global footprint, by building strategic alliances across the platform. We live in the era when the technology change is facilitating the formation of strategic alliances and partnership with external parties which can deliver different material components and intangibles in the value chain of single participants. In the new context, the linear value chain of one industrial organization needs to be understood as a part of the exponential value chain or network of independent value chains of suppliers, customers, competitors, innovative start-ups, online sellers, off-line sellers, and other stakeholders (regulators, platform providers, cloud providers, big data providers, etc.). Figure 4 provides a schematic view of the exponential value chain.

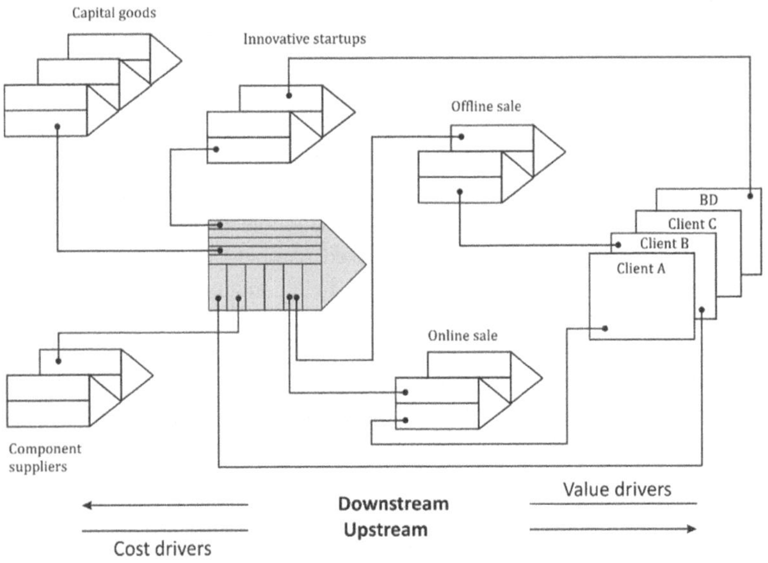

Fig. 4. Exponential value chain

The exponential value chain is the weapon to trounce rivals based on strategy, particularly keeping in mind that in an environment with endless combinatorial innovations multiple winners can thrive and coexist.

5 Deconstructing Rejuvenation of Micro Management Tools

Changes in the focus and the scope of strategy already discussed triggered rejuvenation of some tools and their alignment in strategy formulation and implementation. The following analysis exhibits certain evolutionary links between traditional framework and contemporary practice.

5.1 Quality Management

Industry 3.0 with ICT breakthroughs in the background of operations, actually, boosted the usability of the variety of quality control management techniques. Since the early 1980 s, inspired by a "zero defect", the adoption and improvement of the quality control management techniques such as Six Sigma [13] and Total Quality Management [1] was growing.

In Industry 4.0 the new wave of ICT breakthroughs, along with combinatorial innovations from the virtual and physical world, enable a continuous quality control (actually "controlling"), or the shift from intermittent quality control to strategic quality controlling.

The new concept is based on a triple feedback loop (see Fig. 5). Let us suppose that one of the strategic initiatives to improve the market position of a company producing machinery is to increase the life span of key components of its products. First activated feedback in the process is strategic learning. This feedback contributes to the formation of the digital twin of innovative products. Artificial intelligence uses data from data analytics, based on cognitive technologies. The second feedback considers making forecasted value based on resource allocation in advanced manufacturing. The third feedback is a traditional quality control feedback. Sensor technology can create information about the rotation, vibration and temperature of machinery communicating transaction data with the central server where they can be classified, aggregated and analyzed through artificial intelligence in conjunction with standards and clients' expectations identified by cognitive technologies. In doing so, a business organization can create predictive model of failure of the key parts, taking actions on maintenance only when failure is likely. Such system of strategic quality controlling would create value in the form of extended pre-maintenance life time of the machinery and reduced maintenance costs.

5.2 Cost Management

Activity-based costing (ABC) is the tool developed in Industry 3.0 with the aim to solve deficiencies of standard costing method by covering all activities that drive costs.

The method is causing an organization to manage activities not costs, recognizing the cost as simply an outcome of undertaken activities. This method has been inspired by cost optimization through identification of specific drivers for direct costs and overhead costs (secondary and tertiary) depending on the activity from the value chain.

ABC assumes two steps. First, resource costs are tied to activities in the value chain, using various resource drivers. Second, activities are tied to cost objects (products, parts, services, etc.), using various activity drivers. Cost optimization is

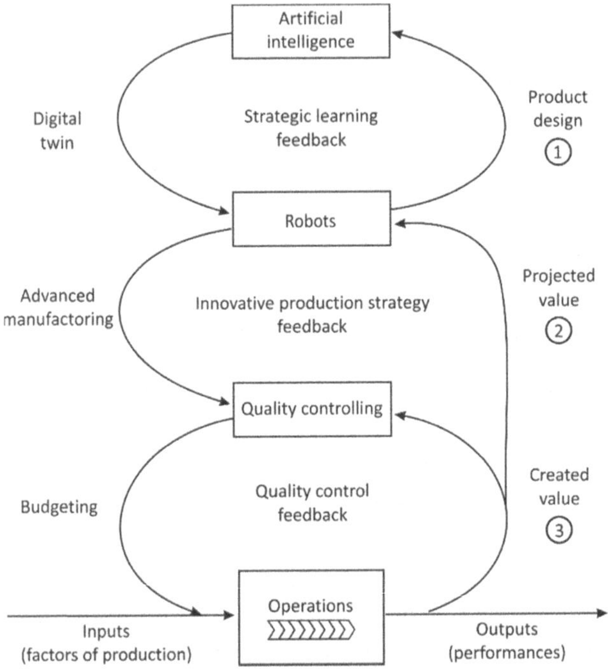

Fig. 5. Strategic quality controlling

based on linking modules (or activity pools) by using resource and activity drivers which show the relationship between the sources of costs and destination and allocation of activity cost pool to product/services.

In Industry 4.0 we see dramatic changes in cost structure, making ABC a more attractive costing method. Overhead expands significantly, particularly inspired by digital transformation, emerging as a major component cost. Also, variability of overhead is a matter of fact. Overhead variability is driven in accordance with the range and complexity of the products, customers and selling channels. When overhead dominates direct cost and variability of overheads is a matter of fact, ABC is a more relevant costing method than direct costing.

Initially, the concept proved highly useful during the mass automated production [7, 12]. In this way, ABC led to efficiency improvement and performance measurement improvement. Also, it provides better base for optimization of the product mix [5].

Also, ICT breakthroughs offer new possibilities regarding costing methods, since they allow for real time data acquisition and the shift toward real time ABC [19, 21]. Namely, the data about resource drivers and activity drivers are collected in real time using sensor technologies (ID readers, RFID, etc.), and other virtual technological breakthroughs like IIoT, BD, cloud computing, broadband, etc. New technologies rejuvenate conventional ABC which is, in some sense, time consuming and costly, difficult to scale and related with the granularity of data problem. Also, ICT breakthroughs embedded the use of real-time ABC in the wider context as a decision support system that provides a robust basis for business analytics.

5.3 Performance Management

In Industry 3.0 cash flow became a key metrics in performance measurement system ("cash is a fact, profit is an illusion"). The related performance management system is based on the creation of the value through identification, measurement, and the use of broader base of value drivers as factors increasing cash inflows and cost drivers as factors influencing cash outflow increase. There are various techniques evolving from *A. Marshal's* concept of economic profit to economic value added [16, 17]. Balanced Scorecard (BSC) developed by *R. Kaplan* and *D. Norton* [9] as a truly holistic tool actually integrating leading indicators (or cash flow based) and lagging (or profit based) indicators of the company success.

Mentioned tools provide ground breaking advance in assuming more strategic approach to performance measurement system [8]. The overall comprehensive platform integrating a number of interrelated techniques aimed at maximization of the client/customer' life time value with shareholders' value is known as Value Based Management or VBM [23].

5.4 Strategic Management

In Industry 3.0, Enterprise Resource Planning (ERP) performs as the backbone of management information system. Traditional ERP is capable of supplying strategists predominantly with cost data. However, this is only one way of supplying the actionable information, integrating the data about standardized costs. Also, traditional ERP system provides the information on an aggregate level, while real time records about working hours, utilization of machines, material loss, and the like can hardly be provided.

Manufacturing Execution System (MES) is a hallmark of operations management. MES will make an optimal production plan by considering how to arrange the advanced manufacturing in accordance with formulated strategy [10]. In the operations stage of the value chain, all resources are tracked and their real time status data is displayed in the MES. Namely, the production line can be broken down into individual machines to collect the data (quantity of raw material, machine time, manpower time, etc.). Also, the collection of the quality control data can be used to achieve quality management.

We are living in a time when innovative products and processes fundamentally determine the strategic vision. An innovative production strategy is a way to reach strategic objectives based on frontier technologies, enabling the implementation at various points of the value chain, particularly in front stages (design, construction and digital twin) and operations. Figure 6 represents a simple abstraction of two building blocks of the strategic management process including micro management tools like ABC, BSC, VBM, and ERP as hallmarks. The flow diagram is used to simplify complex relations, decision-making points, and feedback loops that lie in the background.

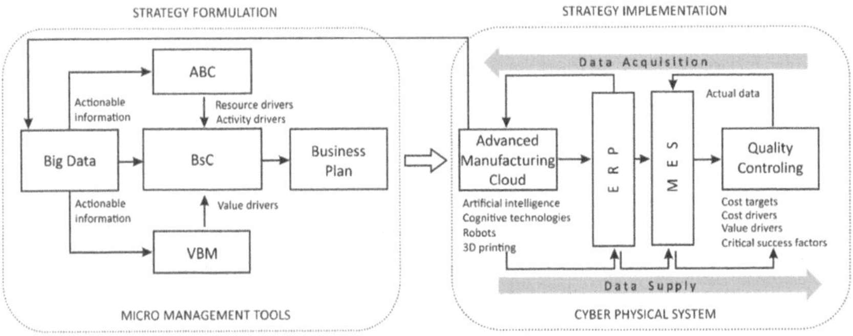

Fig. 6. Strategic management tools: integrated view

Executing strategy is about organization focused on strategy. In fact, it is about managing performance toward a predetermined direction. Big Data combined with Business Intelligence is going to be the core of emerging real time ABC, BSC and VBM based on them. Precisely, the real time ABC provides actionable information about resource drivers and activity drivers and VBM provides break-even and value drivers, all playing the role of inputs for strategy formulation based on BSC.

VBM is also a complementary method with strategic management based on BSC. More precisely, VBM aligns management processes of the quality control and strategic management with the value creation [24]. VBM plays across several areas like formulation and implementation of the strategy with the highest potential for the long-term value creation, identification of the key performance indicators and their correlations with value drivers on a company, business unit, product, brand, or customer level.

The very essence of the operational part of the VBM is the identification of the specific performance variables or "value drivers" that lead to value creation given the business strategy. Industry 4.0 impacts the value drivers in two ways. First, in the digital environment there are new value drivers, such as strategic quality controlling (or digital quality management) in the quality control area, real time supply chain optimization in the inventory management area, human-robot collaboration and digital performance management in the operations area, etc. On the other hand, improved MES allows real time data acquisition about value drivers. This way the information about the value drivers plays the role of the lead performance indicator.

Big data, cloud computing technologies, and broadband technologies can be used for real time decision making. Namely, after real time data acquisition, a machine-to-machine data feedback allows for the advanced manufacturing. The entire cyber-physical system is used to integrate the data to automatically manage and control production processes in real time, as well as to measure the operating performance [20]. As a result, decisions about innovative products, optimal product mix, equipment layout, and production protocols can be made to achieve the defined value proposition.

Thanks to the ERP-MES link, operations management database turns transaction data into actionable information. During the strategy implementation, all resources used are tracked with real time ABC.

These days, ERP is only one way of supplying standard costing. The strategy formulation is derived from the immense quantity and quality of information used for identification of the demand level, resource drivers, activity drivers, and value drivers. When translating strategy defined in BSC format into a business plan and investment projects, the cost drivers, the cost targets and the value drivers simultaneously play the role of the critical success factors, operating goals and the performance measures. Before the operations take place, MES database integrates the information about unit quantities for material, labor, and overhead, and the value drivers turn them into actionable information (production process mapping) thanks to the ERP-MES link [10]. Namely, in advanced manufacturing environment, VBM, BSC and ABC are embedded in ERP system and, then, connected to MES.

6 Conclusion

In Industry 4.0 framework business management and engineering are viewed as non-linear systems. In both cases the reversibility principle is going to be a silver lining of systems being managed. Considering previous, in this paper we briefly present snap-shots of recommendations based on views of economists supporting circular economy new deal and heterodox economic policy platform.

Summarizing the emerging contours on the new paradigm in business economics (and management), we see that business organization of the future should be con-centrated not only on further cost reduction, but first and foremost, on combinatorial innovation and value creation, not violating circular economy proposals. Harmonizing contradictory requirements of different stakeholders with the sustainable and inclusive vision of future development, the company of the future is going to be the "symphonic company" operating in the new space and with the new way of competing.

The symphonic company will change the strategy focus and broaden the strategy scope. Key consequences of these changes are combinatorial innovations as a propo-sition of competitive dynamics and exponential value chain (or platform) as infras-tructure. Both changes provide rejuvenation of conventional business management tools and their improvement with new functionalities. Tools like strategic quality controlling, real time ABC, advanced BSC, VBM, and new releases of MES/ERP are in focus. Along with IIoT, BD, cloud computing, broadband, 5G network, and other components of the digital infrastructure, these tools define the new way how data are being acquired on a real time basis and transformed into actionable information through the Information Value Loop to create values which are environmentally not damaging.

References

1. Ahire, S.L., Golhar, D.Y., Waller, M.A.: Development and validation of TQM implementation constructs. Decis. Sci. **27**(1), 23–56 (1996)
2. Alkire, S., Bardhan, P., Basu, K., Bhorat, H., Bourguignon, F., Deshpande, A., Lin, J.Y., Moene, K., Platteau, J.P., Saavedra, J., Stiglitz, J.E.: Stockholm statement: towards a consensus on the principles of policymaking for the contemporary world. (2016). https://scholar.google.com/scholar?hl=sr&as_sdt=0%2C5&q=Alkire%2C+S.%2C+Bardhan%2C+P.%2C+Basu%2C+K.%2C+Bhorat%2C+H.%2C+Bourguignon%2C+F.%2C+Deshpande%2C+A.%2C+Lin%2C+J.Y.%2C+Moene%2C+K.%2C+Platteau%2C+J.P.%2C+Saavedra%2C+J.+and+Stiglitz%2C+J.E.%2C+2016.+Stockholm+Statement%3A+Towards+a+Consensus+on+the+Principles+of+Policymaking+for+the+Contemporary+World.&btnG=
3. Ardito, L., Petruzzelli, A.M., Panniello, U., Garavelli, A.C.: Towards industry 4.0. Bus. Process Manag. J. **25**(2), 323–346 (2019)
4. Christensen, C.M.: The Innovator's Dilemma: When New Technologies Cause Great Firms to Fail. Harvard Business Press, New York (2013)
5. Cooper, R., Kaplan, R.S.: Measure costs right: make the right decisions. Harvard Bus. Rev. **66**(5), 96–103 (1988)
6. Đuričin, D., Vuksanović-Herceg, I.: Industry 4.0 and paradigm change in economics and business management. In: Jun, N., Majstorovic, V.D., Djurdjanovic, D. (eds.) Proceedings of the 3rd International Conference on the Industry 4.0 Model for Advanced Manufacturing, pp. 37–56. Springer, Heidelberg (2018)
7. Hooper, M.J., Steeple, D., Winters, C.N.: Costing customer value: an approach for the agile enterprise. Int. J. Oper. Prod. Manag. **21**(5/6), 630–644 (2001)
8. Ittner, C.D., Larcker, D.F.: Are nonfinancial measures leading indicators of financial performance? An analysis of customer satisfaction. J. Account. Res. **36**, 1–35 (1998)
9. Kaplan, R.S., Norton, D.P.: Putting the balanced scorecard to work. Econ. Impact Knowl. **71**(5), 315–324 (1998)
10. Kletti, J.: Manufacturing Execution System-MES, pp. 61–78. Springer, Heidelberg (2007)
11. Knight, J.A.: Value-Based Management: Developing a Systematic Approach to Creating Shareholder Value. McGraw Hill, New York (1998)
12. Koltai, T., Lozano, S., Guerrero, F., Onieva, L.: A flexible costing system for flexible manufacturing systems using activity based costing. Int. J. Prod. Res. **38**(7), 1615–1630 (2000)
13. Pande, P.S., Holpp, L.: What is Six Sigma? McGraw-Hill, New York (2001)
14. Porter, M.E.: Competitive Strategy: Techniques for Analysing Industries and Competitors. Free Press, New York (2000)
15. Porter, M.E., Millar, V.E.: How information gives you competitive advantage. Harvard Bus. Rev. **63**, 149–160 (1985)
16. Rappaport, A.: Creating Shareholder Value: The New Standard for Business Performance. Free Press, New York (1986)
17. Rappaport, A.: Creating Shareholder Value: A Guide for Managers and Investors. Simon and Schuster, New York (1999)
18. Raynor, M.E., Cotteleer, M.J.: The more things change: value creation, value capture, and the internet of things. Deloitte Rev. **17**, 51–65 (2015)
19. Terkaj, W., Tolio, T.: The Italian flagship project: factories of the future. In: Factories of the Future, pp. 3–35. Springer, Cham (2019)

20. Tsai, W.H., Lan, S.H., Huang, C.T.: Activity-based standard costing product-mix decision in the future digital era: green recycling steel-scrap material for steel industry. Sustainability **11** (3), 889 (2019)
21. Wagner, T., Herrmann, C., Thiede, S.: Industry 40 impacts on lean production systems. Procedia CIRP **63**, 125–131 (2017)
22. Williamson, J.: Democracy and the 'Washington Consensus'. World Dev. **21**(8), 1329–1336 (1993)
23. Young, D.: Economic value added: a primer for European managers. Eur. Manag. J. **15**(4), 335–343 (1997)
24. Young, S.D., Byrne, S.F.: EVA and Value-Based Management: A Practical Guide to Implementation. McGraw-Hill Professional Publishing, New York (2001)

Influence of the Orientation of Steel Parts Produced by DMLS on the Fatigue Behaviour

Nebojša Bogojević[1], Snežana Ćirić-Kostić[1](✉), Aleksandar Vranić[1], Giorgio Olmi[2], and Dario Croccolo[2]

[1] Faculty of Mechanical and Civil Engineering in Kraljevo, University of Kragujevac, Kraljevo, Serbia
cirickostic.s@mfkv.kg.ac.rs
[2] Department of Industrial Engineering (DIN), University of Bologna, Bologna, Italy

Abstract. The goal of this paper is to present studies of the influence of orientation of steel samples during additive manufacturing to their fatigue behaviour. The samples were produced from maraging steel EOS MS1 and stainless steel EOS PH1 using direct laser metal sintering technology. Three sets of samples were manufactured for each of the materials, with slopes of longitudinal axis of the samples being 0° (horizontal), 45° (slanted) and 90° (vertical) with respect to the horizontal building plane. All the samples were post-processed by heat treatment, shot-peening and machining, and tested according to the ISO 1143 standard. The curves for finite life domain were calculated using ISO 12107, and an estimation of the fatigue limit was made by Dixon-Mood method. The obtained results show that the building orientation has no significant influence on fatigue strength of maraging steel samples, while the stainless steel samples with slanted orientation of the axis have fatigue strength of up to 20% higher than the samples with horizontal or vertical orientation of the axis.

Keywords: Fatigue behaviour · Fatigue limit · S-N curve · Additive manufacturing · DMLS · Build orientation

1 Introduction

The paper is focused on studying of dependence of the fatigue strength on the orientation of steel samples during the process of direct metal laser sintering (DMLS). The study is a part of a research program, carried out within the framework of the Horizon 2020 project A_MADAM, that aims to improve knowledge about the dynamic behavior of additive manufacturing products [1].

Additive Manufacturing (AM) technologies represent a family of manufacturing technologies that, unlike more conventional subtractive and forming technologies, build a part by addition of raw material. The most important advantages of AM are their ability to be used for manufacturing of products with complex shape and the short lead-in times due to the independence of the manufacturing equipment on product. These advantages make AM technologies the optimal choice for production of prototypes

L. Wang et al. (Eds.): AMP 2020, LNME, pp. 294–305, 2020.
https://doi.org/10.1007/978-3-030-46212-3_22

("rapid prototyping" applications) and small series of products ("rapid manufacturing" applications), but also leaves them as the only choice in numerous shape-integrated applications (lightweight products based on cellular design, tools with conformal cooling channels, highly efficient turbine blades and heat exchangers, etc.) [2].

All the current AM technologies have layerwise production principle, which means that a product is made by addition of successive parallel thin layers of material. Each layer represents a cross-section of the product, which is calculated by software for preparation of production on the basis of the CAD model. Regarding that the whole manufacturing process is controlled by the computer software, the AM technologies became available after "IT revolution" and massive production of low-cost computers by the end of XX century.

As the mechanical strength is an important characteristics of components of mechanical systems, the choice of AM technologies that are used for manufacturing of mechanical parts is limited to a narrow set of technologies that may process metals, alloys, high-performance polymers and composite materials. Such AM technologies are based on the principle of joining of the powder of material by high-energy beam, and they are called powder-bed-fusion technologies. The most common powder bed technologies are Direct Metal Laser Sintering, Selective Laser Melting and Electron Beam Melting that are used for processing of metals and alloys and the Selective Laser Sintering for processing of polymers and polymer-based composites [3–5].

Direct Metal Laser Sintering (DMLS) technology uses laser beam as the high-energy beam for melting of powders of iron, copper, nickel, aluminium and titanium alloys. The materials and their densities used in AM are comparable to metal alloys obtained by traditional technologies, which makes DMLS technology the most popular technology for AM production in automotive and aerospace industry, but also a preferable choice for production of advanced tools and cooling components. The key breakthrough in improving the DMLS technology was development of appropriate "scanning strategies", i.e. time order of the exposition of different parts of a layer to the laser-beam [6].

As all the other AM technologies, DMLS is still new, and the knowledge of the mechanical properties of the metal parts produced with AM technologies is still scarce. In the literature are mostly presented the results which describing the static characteristics and very few papers presents the fatigue testing of the parts produced with AM of the steel, titanium, aluminium and nickel alloys [7–19]. It is still not known if the calculation principles developed for traditional technologies may be applied to the parts manufactured by the DMLS technology, and not even if the parts manufactured by DMLS technology have deterministic behaviour regarding the fatigue strength. The questions of influence of the production process parameters and post-processing procedures to the dynamic behaviour of DMLS products are still open. On the other hand, the dynamic behaviour of products is critical in all automotive and aerospace applications, and this discrepancy between the existing and knowledge and needs was inspiration for the research and results presented in this paper.

2 Experiment

The conducted experimental study is based on ISO 1143 standard for fatigue testing by rotating bending [20]. This standard defines the testing procedure, the load scheme, and the sample geometry. The four-point bending load was selected as one of the possible loads defined by the standard, and the sample geometry for this type of testing is presented at the Fig. 1. The hourglass shape of the samples with 6 mm diameter at the gauge and 10 mm diameter at the head was chosen as the smallest shape recommended by the standard. With the selected testing strategy, the bending moment has constant value over the whole gauge length and stress is equally distributed within the gauge, as it is presented in Fig. 2.

The samples were tested at the Alma Mater Labs of University of Bologna/Italy/, on the machine for four-point bending load shown in Fig. 2. All the tests were performed under reverse bending load (stress ratio R = −1) at the frequency of 60 Hz. Before testing, all the samples were measured to check their diameters at the gauge and at the head. Roughness measurement and throughout were also checked against nominal values provided by ISO 1143 standard. The initial stages of testing of each sample set were aimed at determining fatigue behaviour in finite life domain(-N relation) and rough estimation of fatigue limit. The modified Dixon staircase method was used to obtain more accurate estimation of fatigue limit with related maximum likelihood band [21]. The processing of data in finite life domain has been performed according the ISO 12107 standard [22]. Stress and life cycles were linearly interpolated in logarithmic coordinates. The lower and upper limits of the -N curves were determined based on standard deviation with probabilities of failure of 10% and 90% respectively for 90% confidence level. The series of failure and non-failure tests outcomes have then been processed by Dixon method for a life duration of 10 million cycles, which was set as the run-out limit.

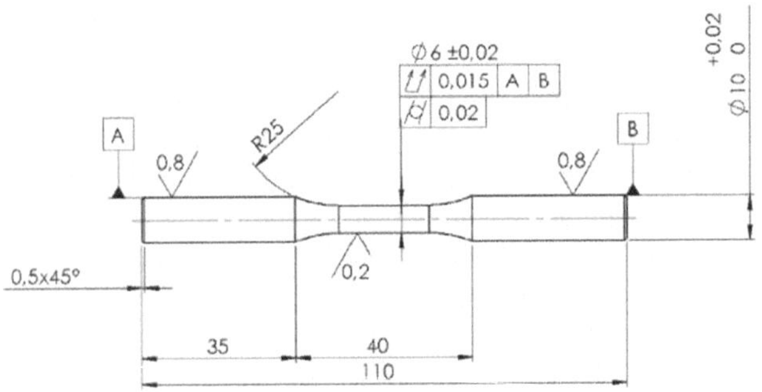

Fig. 1. Sample geometry [20].

Samples were produced by DMLS machine EOSINT M280 (EOS GmbH – Electro Optical Systems, Krailling-Munich/Germany/) in the "3D Impulse" laboratory of Faculty of Mechanical and Civil Engineering in Kraljevo/Serbia/. The EOSINT M280 machine is equipped with Ytterbium 200 W laser.

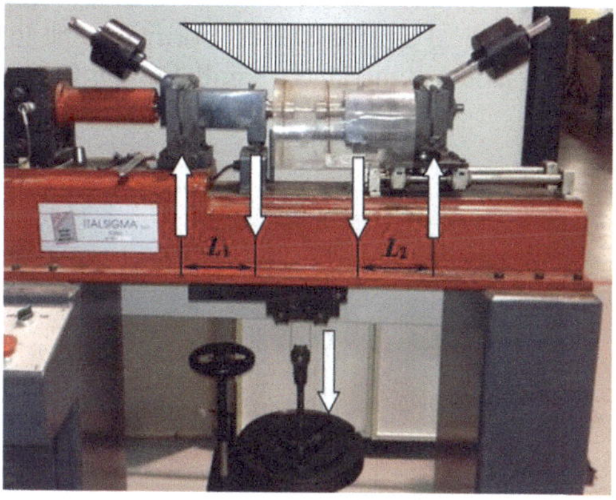

Fig. 2. Rotating bending machine with load distribution shematics.

Materials used for sample manufacturing are maraging steel EOS MS1 equivalent to DIN 1.2709 [23] and the stainless steel EOS PH1 equivalent to DIN 1.4540 [23].

For the selection of the production process parameters were used recommendations of the manufacturer of the machine for the presented materials. For MS1 maraging steel were applied process parameters defined according as EOS "Performance" set of the parameters, with the layer thickness set to 40 μm. For PH1 stainless steel was used the EOS "Surface" set of the parameters, with the layer thickness set to 20 μm.

During the manufacturing process, the samples were connected to the base plate with the support structures. These structures have double role, first to remove the heat from the manufacturing area, and second to keep the parts at fixed positions during the manufacturing process.

After the DMLS manufacturing process, the samples were first shot-peened by steel spheres with approximate diameter of 0.7 mm for the purposes of cleaning of residual powder and improvement of the surface quality. After the shot-peening, heat treatment was performed according to the EOS materials data sheet recommendations [23], which means that the MS1 samples were exposed to temperature of 490 °C for 6 h, while the PH1 samples were exposed to temperature of 482 °C for 3 h. Heat treatment considerably lowers the amount of residual stress accumulated in samples due to the temperature gradients that arise during the manufacturing process. Finally, after the heat treatment, the samples were removed from the building plate using wire electro discharge machine (EBM).

In order to study the influence of the samples orientation during the production process, six sample sets were manufactured, a three sets for each of the two materials. During the production, the samples of each of the three sample sets had different orientations of the longitudinal axis with respect to the building plane. The longitudinal axis of the samples of the first set were normal to the horizontal building plane (vertical axis-denoted by "V"), the longitudinal axis of the samples of the second set were parallel to the horizontal building plane (horizontal axis-denoted by "H"), while the longitudinal axis of the samples of the third set were inclined to the horizontal building plane by the angle of 45° (slanted axis-denoted by "S"), as it is shown in the Fig. 4. In this way, the samples with vertical axis had layers normal to the longitudinal axis, and the samples with horizontal axis had the layers parallel to the longitudinal axis (Fig. 3).

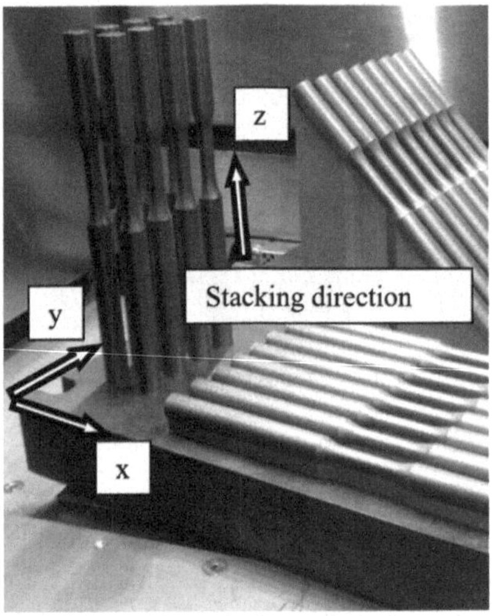

Fig. 3. Samples orientation on the base plate

All the produced samples have the diameters increased by 1 mm for MS1 and 2 mm for PH1 samples. The increased diameters enable to achieve the surface quality required by the ISO 1143 standard by additional machining.

The complete list of the manufactured samples is presented in the Table 1.

Table 1. Produced number of the samples for testing

Orientation of the longitudinal axis	Material (thickness of the allowance for machining)	
	Maraging steel MS1 0.5 mm	Stainless steel PH1 1 mm
Vertical	MS1-V: 8 samples	PH1-V: 10 samples
Horizontal	MS1-H: 8 samples	PH1-H: 10 samples
Slanted	MS1-S: 8 samples	PH1-S: 10 samples

3 Results

The results of the fatigue testing are presented in Table 2 (MS1 and PH1 samples). In the table is given the information about the longitudinal axis orientation of the sample.

Table 2. Results of testing of samples made from MS1 and PH1

Vertical							
Sample	Stress [MPa]	Life [cycles]	Failure	Sample	Stress [MPa]	Life [cycles]	Failure
MS1-V1	699	2277295	Y	PH1-V1	651	4834809	Y
MS1-V2	665	3374203	Y	PH1-V2	711	1871476	Y
MS1-V3	596	6090458	Y	PH1-V3	590	108926	Y
MS1-V4	524	–	N	PH1-V4	590	68686	Y
MS1-V5	560	–	N	PH1-V5	470	–	N
MS1-V6	560	–	N	PH1-V6	560	43729	Y
MS1-V7	596	–	N	PH1-V7	530	–	N
				PH1-V8	560	2807208	Y
				PH1-V9	530	2564861	Y
				PH1-V10	500	5047111	Y
Horizontal							
Sample	Stress [MPa]	Life [cycles]	Failure	Sample	Stress [MPa]	Life [cycles]	Failure
MS1-H1	699	3780607	Y	PH1-H1	420	–	N
MS1-H2	665	4926903	Y	PH1-H2	550	144726	Y
MS1-H3	579	–	N	PH1-H3	524	167829	Y
MS1-H4	610	8225283	Y	PH1-H4	500	–	N
MS1-H5	579	1642162	NV	PH1-H5	500	728708	Y
MS1-H6	579	–	N	PH1-H6	475	8423284	Y
MS1-H7	610	9262114	Y	PH1-H7	651	47315	Y
Slanted							
Sample	Stress [MPa]	Life [cycles]	Failure	Sample	Stress [MPa]	Life [cycles]	Failure
MS1-S1	699	1368541	NV	PH1-S1	640	–	N
MS1-S2	665	1042346	NV	PH1-S2	670	8344160	Y
MS1-S3	550	–	N	PH1-S3	640	–	N
MS1-S4	579	8997765	Y	PH1-S4	670	–	N
MS1-S5	699	3582162	Y	PH1-S5	700	–	N
MS1-S6	665	4309539	Y	PH1-S6	880	573080	Y
MS1-S7	550	–	N	PH1-S7	730	9012402	Y
MS1-S8	579	–	N	PH1-S8	790	4974052	Y
				PH1-S9	820	1776278	Y
				PH1-S10	850	497854	Y

For each of the sample orientations, the results of testing of each of the samples from the sample set are described by the sample identifier (column "Sample"), the nominal stress value at the gauge (column "Stress"), the observed number of cycles ("Life") and by the final outcome of the test (column "Failure"). In the last column, the failure outcome is indicated by "Y", the run-out outcome is indicated by "N", while the "NV" indicates that the test is not valid because the break occurred at the head, instead at the gauge, of the sample (Fig. 4). The samples without indication of the testing outcome were not tested because they were broken during machining process (two samples from MS1 and three samples of PH1).

Since DMLS manufacturing process is expensive, at the initial experiment plan consisted of eight samples per set, as this was considered as the minimum number size of a set to obtain finite life domain curve and fatigue limit value. At the time of planning, there was no indication of potential problems that could arise during machining or testing process. After the problems appeared with the MS1 samples that were tested first, the number of the samples in the PH1 sets was increased to ten.

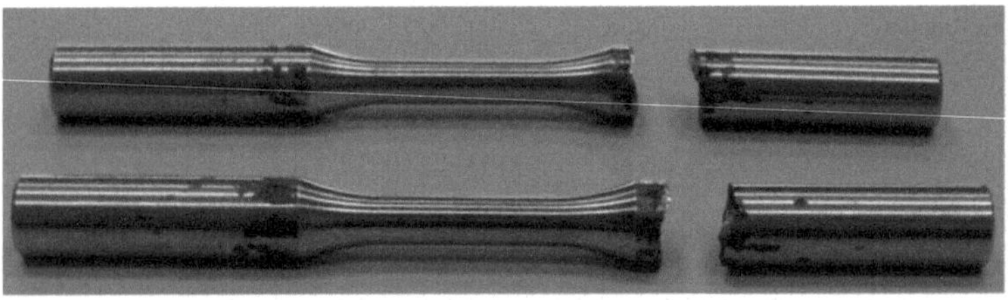

Fig. 4. Failure on sample head

The number of the samples that passed the planned test is given in Table 3.

Table 3. Final number of tested samples

Orientation of the longitudinal axis	Material and thickness allowance for machining	
	Maraging steel MS1 1 mm	Stainless steel PH1 2 mm
Vertical	MS1-V: 7 samples	PH1-V: 10 samples
Horizontal	MS1-H: 7 samples	PH1-H: 7 samples
Slanted	MS1-S: 8 samples	PH1-S: 10 samples

4 Discussion

4.1 σ-N curves

The results were processed according to the ISO 12107 standard to determine curves in finite life domain. The bending stresses and the corresponding number of cycles to failure were presented in log-log diagram, and the σ-N curves were retrieved using linear regression. Run-outs are indicated by arrows.

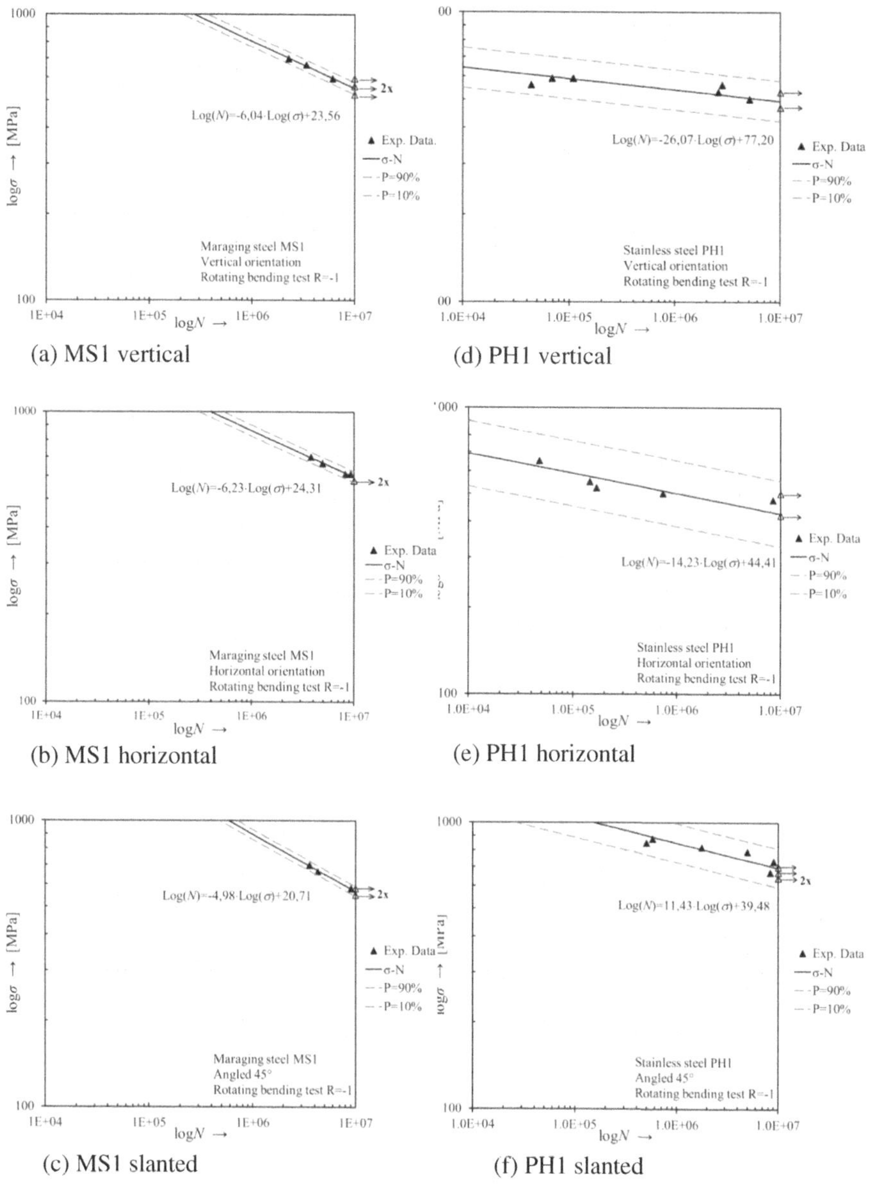

Fig. 5. σ-N curves for maraging steel MS1 and stainless steel PH1

If more than one run-out occurred at the same stress level, the number of run-outs is written at the end of the arrow (for example "2×" indicates 2 run-outs). Details about material type, sample orientations and type of load are also included in diagrams. Six plots are presented at figure Fig. 5. The three vertical plots on the left side present the trends of the σ-N curves derived from testing of the MS1 samples for with vertical, horizontal and slanted axis ((a), (b) and (c) respectively). The three vertical plots on the right side are trends of the σ-N curves derived from testing the PH1 samples with vertical, horizontal and slanted axis ((d), (e) and (f) respectively). This three by two array of plots makes easy to compare the influence of the building orientation (by columns) and the results of different materials (by rows).

By comparison of the graphs (a), (b) and (c) at the figure Fig. 5, it can be notices that the slopes of the graphs are very similar for all three MS1 sample orientations. It can also be noticed that their run-outs are at similar stress levels and that the probability bands are narrow (close to σ-N curve).

The results of the PH1 samples tests are strikingly different. The most important difference is that the slope of the σ-N curve for samples with slanted axis is not similar to the slopes of the σ-N curves of the samples with vertical and horizontal samples. The obtained results suggest that the PH1 samples with horizontal and vertical axis seem to be more sensitive to dynamic loads than PH1 samples with slanted axis. Further difference in comparison with MS1 samples is that the run-outs of PH1 samples with slanted axis occured at higher stress values than run-outs of the PH1 samples with horizontal or vertical direction (and MS1 samples for that matter). The third difference in comparison with MS1 samples is that the probability bands for all three PH1 sample orientations are wider than for the MS1 samples. Finally, the results of the PH1 sample tests in finite life domain show larger data scattering than the results of the MS1 samples.

4.2 Fatigue Limit (FL)

Without of the post-processing procedure (in the "as-built" state) the ultimate tensile strength (UTS) was 1100 MPa for MS1 samples and 1050 MPa for PH1 samples [23]. After the post-processing by age hardening the UTS os MS1 and PH1 become 1930 MPa and 1310 MPa, respectively [23]. It may be noticed that, while the UTS of the materials differ by about only 5% in the as-built state, the age hardening causes substantial difference between the UTS of the materials, raising it to 47% in favour of MS1 [23].

The fatigue limit values were calculated with 95.5% confidence level using the modified Dixon staircase method for all six sample sets. Comparative diagram of FL for MS1 samples with the three axis orientations is given at Fig. 6. The FL ispresented with the bar graphs with appropriate confidence bands considering twice standard deviation. It can be noticed that all three sample sets have close values of FL.

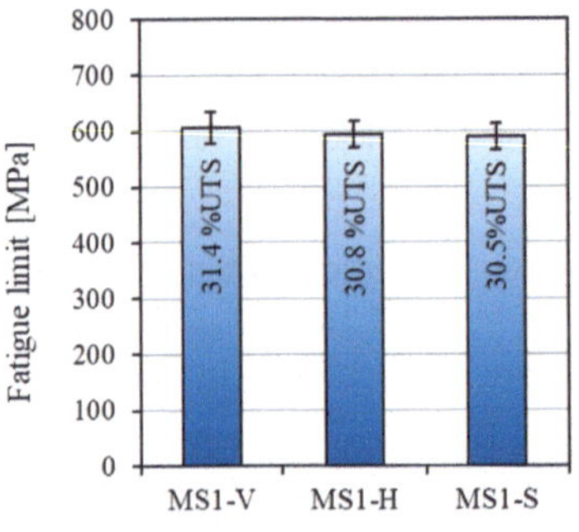

Fig. 6. Fatigue limits for MS1

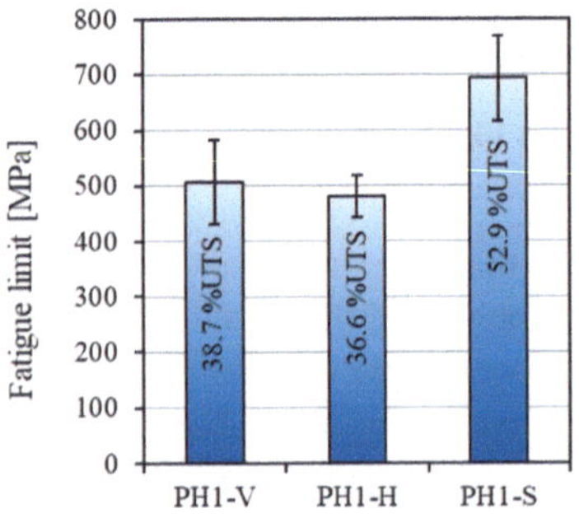

Fig. 7. Fatigue limits for PH1

While MS1 samples show consistent FL around 30% of UTS regardless the orientation of the longitudinal axis during production, the PH1 samples show different behavior. The samples with vertical and horizontal axis have similar estimated FLs of 507.4 MPa (38.7% of UTS) for vertical orientation and 479.85 MPa (36.6% of UTS) for horizontal orientation. The PH1 samples with slanted axis, on the other hand, have estimated FL of 692.5 MPa, which is 52.9% of UTS. It is unexpected that even if UTS of PH1 is lower than for MS1, the PH1 samples with slanted axis have higher FL value than MS1 sample sets. The fatigue limits of PH1 samples with all three orientations of the axis are compared in the bar graph on Fig. 7, with their confidence intervals considering the twice standard deviation. One may notice that for all the samples except the PH1 samples with slanted orientation have FL/UTS ratios much lower than the commonly accepted 50% value for metallic materials [24, 25].

5 Conclusion

The paper presents results of analysis of influence of build orientation to the fatigue strength of parts manufactured by direct metal laser sintering (DMLS). Six sample sets, three per material type, were manufactured on EOSINT M280 DMLS machine and all were machined to final dimension according ISO 1143 standard for rotary bending testing. The obtained results of testing were sufficient to construct and process σ-N curves with their confidence bands (for 10% and 90% failure probability and 90% confidence level) in finite life domains and fatigue limits for all six sample sets involving 49 samples. The fatigue behaviour of the DMLS produced samples shows deterministic nature which leads that the standard methodologies for calculation of the fatigue strength may be applied.

The statistically processed results for maraging steel MS1 have indicated that part orientation has no significant influence on fatigue strength in finite or infinite life domain. FL were estimated to be close to 30% of UTS for this material. Regarding the stainless steel PH1, the processed results show higher FL/UTS ratio close to 40% for the samples with horizontal and vertical axis. These FL are considerably lower than FL of MS1 samples with corresponding orientations, which is in accordance to lower values of UTS for PH1 samples than for MS1 samples. These results suggest that, in general, the DMLS production process leads to products with lower fatigue resistance than traditional technologies. The most probable reason for this is presence of increased amount of material defects, porosities and irregularities in microstructure of the DMLS material [18, 19]. The most intriguing results, however, were obtained for PH1 samples with slanted axis. With fatigue limit of 692.5 MPa, which is around 53% of UTS, these samples showed the highest fatigue strength of all studied sample sets. The obtained result suggests that the proper selection of the building orientation of the parts can improve their fatigue resistance even if the basic material has lower UTS. Therefore, the DMLS may have even some positive effects to fatigue strength of the products. The most probable explanation of the observed effect is that boundaries between layers prevent or extend crack propagation [17–19].

Further research is needed to better understand the effects of DMLS microstructure, residual stresses, the optimal post-processing methodologies and studies of other materials.

Acknowledgments. The research presented in this paper has received funding from the European Union's Horizon 2020 research and innovation programme under the Marie Skłodowska-Curie grant agreement No. 734455.

References

1. Advanced design rules for optimal dynamic properties of additive manufacturing products. Horizon 2020 project No. 734455. www.a-madam.eu
2. Vranić, A., Bogojević, N., Ćirić-Kostić, S., Croccolo, D., Olmi, G.: Advantages and Drawbacks of Additive Manufacturing, IMK-14 – Research & Development in Heavy Machinery, vol. **21**, 2/2017, pp. EN57-62, Kruševac (2017)
3. Bourella, D.L., Beaman, J.J., Leu, M.C., Rosen, D.W.: A brief history of additive manufacturing and the 2009 roadmap for additive manufacturing: looking back and looking ahead. In: Proceedings of RapidTech, pp. 24–25 (2009)
4. Aliakbari, M.: Additive manufacturing: state-of-the-art, capabilities, and sample applications with cost analysis. Master Science thesis, Royal Institute of Technology, Sweden (2012)
5. Herderick, E.: Additive manufacturing of metals: a review. Mater. Sci. Technol. **2**, 1413–1425 (2011)
6. http://www.e-manufacturing.it/downloads/e-manufacturing_distinctive_features_MET.pdf. Accessed 29 June 2019
7. Meneghetti, G., Rigon, D., Cozzi, D., Waldhause, W., Dabalà, M.: Influence of build orientation on static and axial fatigue properties of maraging steel specimens produced by additive manufacturing. Procedia Struct. Integr. **7**, 149–157 (2017)
8. Naiju, C.D., Adithan, M., Radhakrishnan, P.: An investigation of process variables influencing fatigue properties of components produced by direct metal laser sintering. CIRP J. Manufact. Sci. Technol. **4**(1), 60–66 (2011)
9. Kempen, K., Yasa, E., Thijs, L., Kruth, J.P., Van Humbeeck, J.: Microstructure and mechanical properties of selective laser melted 18Ni-300 steel. Phys. Procedia **12**, 255–263 (2011)
10. Bača, A., Konečná, R., Nicoletto, G., Kunz, L.: Influence of build direction on the fatigue behaviour of Ti6Al4 V alloy produced by direct metal laser sintering. Mater. Today Proc. **3**(4), 921–924 (2016)
11. De Vree, W.K.: On the influence of build orientation on the mechanical properties of direct metal laser sintered (DMLS) Ti-6Al-4V flexures. Master thesis, Faculty of Mechanical Engineering, Technical University of Delft, The Netherlands (2016). http://resolver.tudelft.nl/uuid:e775454d-b4c8-4b59-b126-c6b508201afa. Accessed 25 Dec 2019
12. Brandl, E., Heckenberger, U., Holzinger, V., Buchbinder, D.: Additive manufactured AlSi10Mg samples using selective laser melting (SLM): microstructure, high cycle fatigue and fracture behavior. Mater. Des. **34**, 159–169 (2012)
13. Edwards, P., Ramulu, M.: Fatigue performance evaluation of selective laser melted Ti–6Al–4V. Mater. Sci. Eng. A **598**, 327–337 (2014)

14. Scott-Emuakpor, O., Schwartz, J., George, T., Holycross, C., Slater, J.: Bending fatigue life comparison between DMLS and cold-rolled nickel alloy 718. In: MFPT 2014-The Prognostics and Health Management Solutions Conference, Virginia Beach, Virginia, pp. 20–22 (2014)
15. Nicoletto, G.: Directional and notch effects on the fatigue behavior of as-built DMLS Ti6Al4V. Int. J. Fatigue **106**, 124–131 (2018)
16. Nicoletto, G.: Efficient determination of influence factors in fatigue of additive manufactured metals. Procedia Struct. Integr. **8**, 184–191 (2018)
17. Croccolo, D., De Agostinis, M., Fini, S., Olmi, G., Vranic, A., Ciric-Kostic, S.: Influence of the build orientation on the fatigue strength of EOS maraging steel produced by additive metal machine. Fatigue Fract. Eng. Mater. Struct. **39**(5), 637–647 (2016)
18. Croccolo, D., De Agostinis, M., Fini, S., Olmi, G., Bogojevic, N., Ciric-Kostic, S.: Effects of build orientation and thickness of allowance on the fatigue behaviour of 15–5 PH stainless steel manufactured by DMLS. Fatigue Fract. Eng. Mater. Struct. **41**(4), 900–916 (2018)
19. Croccolo, D., De Agostinis, M., Fini, S., Olmi, G., Robusto, F., Ciric-Kostic, S., Vranic, A., Bogojević, N.: Fatigue response of as-built DMLS maraging steel and effects of aging, machining, and peening treatments. Metals **8**(7), 505 (2018). (1–21)
20. International Organization for Standardization, ISO 1143:2010 Standard - Metallic materials – Rotating bar bending fatigue testing, International Organization for Standardization (ISO) Geneva Switzerland (2010)
21. Dixon, W.J., Massey, F.J.: Introduction to Statistical Analysis, vol. 344. McGraw-Hill, New York (1969)
22. International Organization for Standardization, ISO 12107:2003. Metallic Materials – Fatigue Testing – Statistical Planning and Analysis of Data, International Organization for Standardization (ISO) Geneva Switzerland (2003)
23. https://www.eos.info/material-m. Accessed 10 May 2019
24. Niemann, G., Winter, H., Hohn, B.R.: Maschinenelemente. Springer, Berlin (2005)
25. Stoffregen, H.A., Butterweck, K., Abele, E.: Fatigue analysis in selective laser melting: review and investigation of thin-walled actuator housings. In: 25th Solid Freeform Fabrication Symposium, pp. 635–650 (2014)

New ISO Geometrical Product Specification Standards as a Response to Industry 4.0 Needs

Zbigniew Humienny(✉)

Institute of Machine Design Fundamentals, Warsaw University of Technology, Narbutta 84, 02-524 Warsaw, Poland
zbigniew.humienny@pw.edu.pl

Abstract. The task of international standardization to provide design, manufacturing and quality assurance teams the specification tools based on clear and unambiguous rules is discussed. The current state in implementation of the ISO/TC 213 new approach to develop complete set of rule based standards in the field of geometrical tolerancing is analysed. It is mentioned that fourth edition of the ISO 1101 standard that establish fundamentals for geometrical tolerancing is still case based standard without directly listed rules that make this standard difficult for digital utilization. Proposal of rewriting clause from the ISO 1101 in the rule based way is given. The currently available ISO GPS standards that are rule based are listed and shortly reviewed.

Keywords: Geometrical tolerancing · GPS rules · GPS tools · ISO 1101

1 Introduction

Term Digital Twin that is associated with Industry 4.0 idea has a number of definitions. Regarding geometrical product specification (GPS) the following definition *The Digital Twin is a set of virtual information constructs that fully describes a potential or actual physical manufactured product from the micro atomic level to the macro geometrical level. At its optimum, any information that could be obtained from inspecting a physical manufactured product can be obtained from its Digital Twin* shall be recalled [1]. It is stated in [2] that the application of fully automated techniques within planning processes is not yet common practice. That opinion is still valid. Deficits are observed in the course of the use of a fully automated data acquisition of the underlying process data, a key element of Industry 4.0. The aim of this paper is to point the deficits reasons regarding currently available tools for geometrical product specification as well as to discuss new specification tools given in recently publish standards and trends in the development of new standards that aim to fully incorporate geometrical tolerances into product digital model.

The ISO Technical Committee ISO/TC 213 *Dimensional and geometrical product specifications and verification* is responsible for international standards relating to the tolerancing specification and verification of mechanical components. The brief analysis of new set of customer-facing ISO GPS tolerancing standards is shown in [3]. The authors present set of tools that enable a designer precise description of his intend and specification how far the actual product may be away from its nominal model and still

L. Wang et al. (Eds.): AMP 2020, LNME, pp. 306–312, 2020.
https://doi.org/10.1007/978-3-030-46212-3_23

work properly. Particularly twenty two ISO GPS modifiers for tolerancing of linear and angular size given in the ISO 14405-1,3 are enumerated and new specification tools from recently published ISO 1101:2017 standard are listed in the tables with some comments and examples of application. The new modifiers are allocated in following categories: four dispersion parameters (P, V, T, Q); six objective functions and constraints for association operations (C, G, X, N, E, I) and five symbols for association operations (Ⓒ, Ⓖ, Ⓣ, Ⓧ, Ⓝ). Recalled new parameters were introduced to give a designer tools that enable more precise specification of characteristics to determine functional performance of a component and take advantage of possibilities of their verification offered by coordinate measuring systems that are not available in traditional metrology based on measuring equipment such as callipers, micrometres, measuring plates, dial indicators and hard gauges.

The GPS standards are only mentioned once in [4], but analysis of 49 references provide a reader excellent overview of other aspects of the geometrical variations management in the context of Industry 4.0.

2 Geometrical Tolerancing Standards – Sets of Examples or Sets of Rules

The awareness that plus/minus tolerancing is not sufficient for unique definition of the parts geometry (Fig. 1) raised up in industry in the middle of XX century. Finally in 1969 recommendation ISO/R 1101-1: *Tolerances of form and of position – Part 1 Generalities, symbols, indications on drawings* was published. It was replaced in 1983 by first edition of the standard ISO 1101 *Technical drawings – Geometrical tolerancing – Tolerancing of form, orientation, location and run-out – Generalities, definitions, symbols, indications on drawings*. Similar needs in the USA drove to publish standard USASI Y14.5-1966 *Geometric dimensioning and tolerancing* that was preceded by three editions of military standard MIL-STD-8. Currently the 4th edition of International Standard ISO 1101:2017 and 6th edition of American Standard ASME Y14.5-2018 are valid.

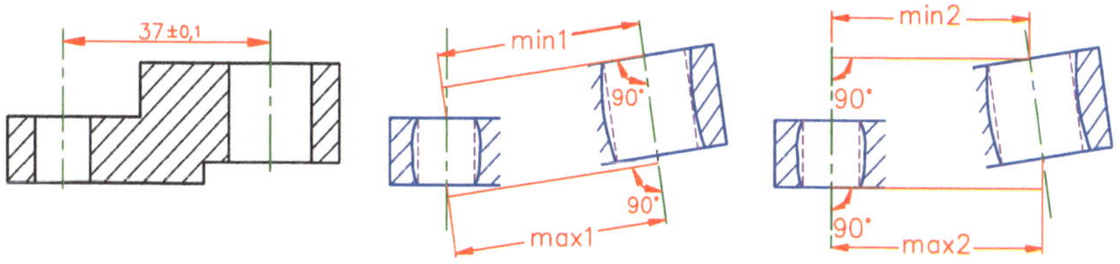

Fig. 1. Plus/minus tolerancing does not provide the clear specification for verification of the distance between the hole axes. Even if holes form deviations are neglected, at least four measurement results are possible. Specification is ambiguous because datums and geometrical tolerances are not used.

The both standards were significantly improved and expanded from their first editions. However there are still published as set of examples of various application of fourteen geometrical tolerancing symbols for selected typical cases. Each drawing is accompanied by explanation. A user of the standard shall interpolate/extrapolate given examples to specify tolerance for his particular component with intended function. This is pretty simple for typical cases similar to presented in the standards however may be error prone for complex requirements and highly depends on a designer skills. Therefore, successive editions of the ISO 1101 [3, 5] were more and more extensive, with more and more drawings with new examples being given in the standard.

To overcome various shortcomings in the traditional tools for specifying geometrical requirements of components the ISO Technical Committee ISO/TC 213 decided to develop unambiguous standards stating clear rules and propose adequate tools for drawing indication and specification of all necessary conditions that might affect any measurement result [6]. In the ISO/TC 213 works two aims may be distinguish:

- creation of new tools (symbols) that enable precise translation of particular more sophisticated functional requirements to specification;
- transformation and systematization of current standard statements that are worldwide verified in industrial practice into set of rules.

The goal to publish all standards related to geometrical tolerancing with explicitly formulated rules is ambitious and difficult. The ISO 1101 standard represents the initial basis and describes the required fundamentals for geometrical tolerancing. It is stated in the ISO 1101 scope: *This document defines the symbol language for geometrical specification of workpieces and the rules for its interpretation.* More over the Note 1 in the scope starts from the statement: *This document gives rules for explicit and direct indications of geometrical specifications.* So a user that start to study this standard may have impression that he can find exhaustive set of rules that enable him tolerancing of form, orientation, location and run-out according any of his needs. The ISO 1101 has not been developed in this way. The traditional way in which the standard is written is user friendly and makes it more understandable for greenhorn designers or programmers for coordinate measuring systems (CMS). On the other hand lack of directly formulated rules is a drawback during attempt of development interfaces between 3D drawings and CNC machines software (for manufacturing) or interfaces between 3D drawings and CMS software (for components verification). Such links are necessary to realise the concept of digital twins.

The GPS system more advanced user may extract a number of statements from the ISO 1101 text and rewrite them as rules that shall be applied for unique specification of geometrical tolerances (Fig. 2 and Fig. 3). For example based on the Clause 6 *Geometrical features* the following rules may be formulated:

- Rule #X1: A geometrical specification applies to a single complete feature unless specifically indicated otherwise (repetition of feature principle from ISO 8015);
- Rule #X2: When the geometrical specification refers to the integral feature, the geometrical specification indication shall be connected to the toleranced feature by a reference line and a leader line terminated on …;

- Rule #X3: When the geometrical specification refers to a derived feature (a median point, a median line, or a median surface), it shall be indicated by a reference line and a leader line terminated by an arrow on the extension of the dimension line of a feature of size …;
- Next Rules ….

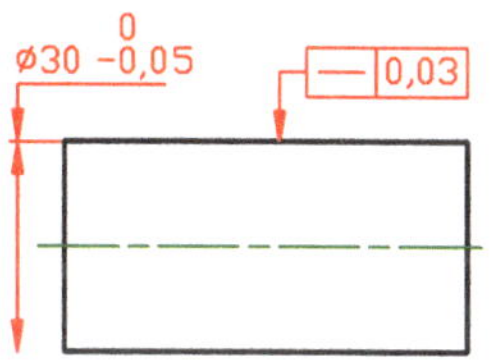

Fig. 2. Straightness of the generating line – example of the Rule #X1 and Rule #X2 implementation.

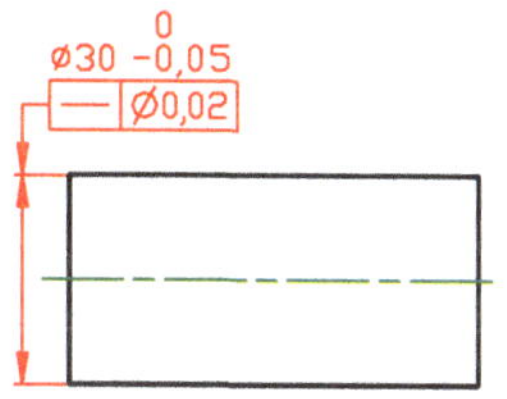

Fig. 3. Straightness of the axis – example of the Rule #X1 and Rule #X3 implementation.

Above may be also done by CNC or CMS software developers. The given example is simple and obvious. It is almost sure that any vendor will create rules with identical meaning, but the crucial question is: Will it work for all necessary rules? Unfortunately rather not, because if it will be so simple the last edition of the ISO 1101:2017 prepared by the ISO/TC 213 experts would be developed as a set of rules. So it is very risky to let CNC or CMS software developers to formulate relevant rules that they will use to develop appropriate reliable interfaces for digital processing of geometrical specifications that is crucial for realizing the digital twin.

The first standard in which the ISO/TC 213 implemented new rule based approach was the ISO 2692 *GPS – Geometrical tolerancing – Maximum material requirement (MMR), least material requirement (LMR) and reciprocity requirement (RPR)*. It was published in 2006 with 14 rules.

Currently in the following ISO GPS standards sets of rules that shall be applied to achieve unique specifications are given:

- ISO 1660:2017 *GPS – Geometrical tolerancing– Profile tolerancing and combined geometrical specifications*; includes 12 rules;
- ISO 2692:2014 *GPS – Geometrical tolerancing – Maximum material requirement (MMR), least material requirement (LMR) and reciprocity requirement (RPR)*; includes 14 rules (3rd edition, but 2nd edition with implemented rule based approach);
- ISO 5458:2018 *GPS – Geometrical tolerancing – Pattern and combined geometrical specifications*; includes 5 rules;
- ISO 5459:2011 *GPS – Geometrical tolerancing – Datums and datum systems*; includes 10 rules.

The ISO 8015:2011 *GPS – Fundamentals – Concepts, principles and rules* contains 13 principles that due to their content may be considered as general rules supplemented by rules for indication of default specification operators and rules for indication of special specification operators as well as rule for statements in parentheses.

The above summary shows new trend in the ISO/TC 213 policy – in majority of recently published standards concerning geometrical tolerancing the rule based approach is applied. The advantage of the rule base approach implemented in the ISO 5458:2018 is shown below. The drawings in Fig. 4, 5 and 6 are made with application of: Rule B: *constraints,* Rule C: *indication of single indicator pattern specification* and Rule D: *indication of a multiple indicator pattern specification.*

For specification in Fig. 4: (*) the axis of each of the two cylindrical position tolerance zones is perpendicular to datum A and is within a radius 20 mm with respect to datum B; (**) the symmetry plane of each pair of parallel planes that establish two symmetry tolerance zones is perpendicular to datum A and passes through datum B.

For specification in Fig. 5: (*) two cylindrical tolerance zones create combined zone; their axes are situated on one plane perpendicular to datum A that passes through datum B and are within a radius 20 mm with respect to datum B; (**) two tolerance zones established by pairs of parallel planes create combined zone; their symmetry planes are situated on one plane perpendicular to datum A that passes through datum B.

For specification in Fig. 6: (*) axes/symmetry planes of tolerance zones are situated on two mutually perpendicular planes that are also perpendicular to datum A and pass through datum B; (**) axes of two cylindrical tolerance zones are situated within a radius 20 mm with respect to datum B.

Combination of individual geometrical specifications, grouping tolerance zones together was always a bit tricky and complicated task. Due to rules listed in the ISO 5458 securing of functional requirements on drawings has become easy and unique.

Beside standards prepared by the ISO/TC 213 standard ISO 16792:2015 *Technical product documentation – Digital data practices* that refers to the ISO GPS standards shall be recalled. The requirements for preparation, revision and presentation of digital product definition data are given in this standard. The ways to indicate geometrical tolerances in 3D models (Fig. 7) created in CAD systems are shown in the ISO 16792.

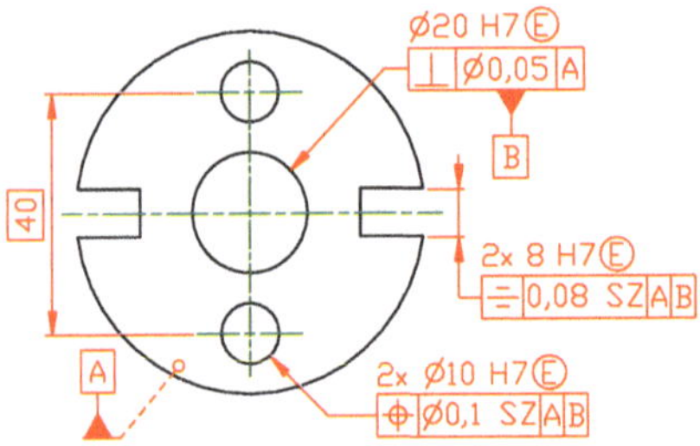

Fig. 4. Each feature is toleranced independently. There are no constraints for angle between holes, grooves. holes and grooves.

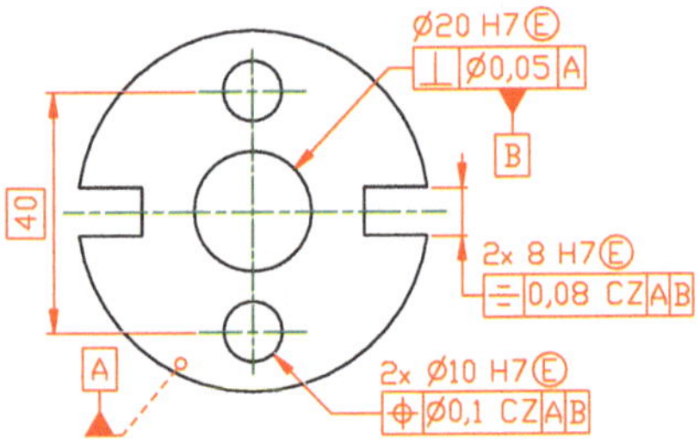

Fig. 5. Pattern of two holes and pattern of two grooves are toleranced independently. There are no constraints for angle between patterns.

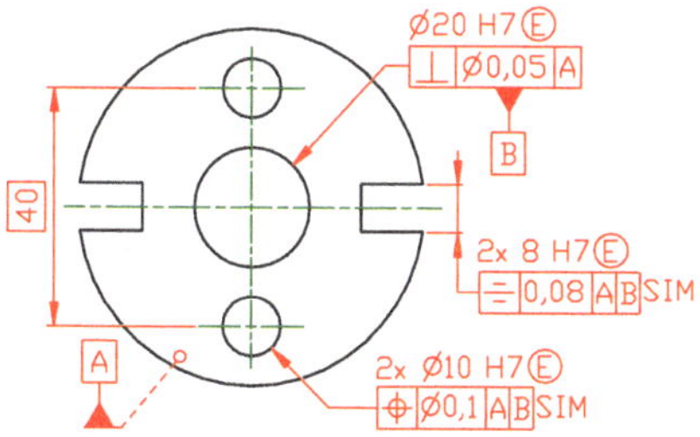

Fig. 6. Four tolerance zones: two cylindrical tolerance zones for position tolerance and two zones between two pairs of parallel planes for symmetry tolerance are locked together. The four tolerance zones have internal constraints (location and orientation) to each other and external constraints to datums A and B (location and orientation).

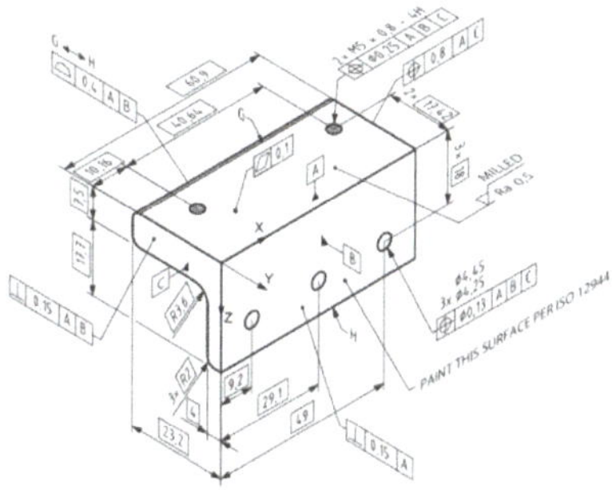

Fig. 7. Model with displayed tolerance indications [ISO 16792]. Current CMS software is not able to extract data for geometrical tolerancing from CAD model. Only nominal geometry can be processed.

3 Conclusions

In recent years the ISO/TC 213 in new GPS standards implemented number of new tools that enable precise specification of geometrical requirements and determination how far actual component may be away from its nominal geometry and still work according a designer intend. More over many statements in standards were formulated more precise due rule based approach when standards were revised.

The components in Fig. 2, 3, 4, 5 and 6 are intentionally presented on 2D drawings. The latest editions of the ISO 1101 and the ISO 1660 contain 3D drawings with geometrical tolerances. However when components are specified in 3D drawings only their nominal geometry may be imported by CMS software. So the attempts towards incorporation of geometrical tolerances in digital processing of component models are made, but the software developers have to work out and implement suitable algorithms.

References

1. Grieves, M., Vickers, J.: Digital twin: mitigating unpredictable, undesirable emergent behavior in complex systems. In: Kahlen, F.-J., Flumerfelt, S., Alves, A. (eds.) Trans-Disciplinary Perspectives on Complex System, pp. 85–114. Springer, Switzerland (2016)
2. Uhlemann, T.H.J., Lehmann, Ch., Steinhilper, R.: The digital twin: realizing the cyber-physical production system for Industry 4.0. Procedia CIRP **61**, 335–340 (2017)
3. Morse, E.P., Shakarji, C.M., Srinivasan, V.: A brief analysis of recent ISO tolerancing standards and their potential impact on digitization of manufacturing. Procedia CIRP. **75**, 11–18 (2018)
4. Schleicha, B., Wärmefjord, K., Söderberg, R., Wartzack, S.: Geometrical variations management 4.0: towards next generation geometry assurance. Procedia CIRP **75**, 3–10 (2018)
5. Humienny, Z.: State of art in standardization in GPS area. CIRP J. Manuf. Sci. Technol. **2**, 1–7 (2009)
6. ISO/TC 213 Business Plan. https://www.iso.org/committee/54924.html. Accessed Dec 2019

Multi-modal Multi-agent Path Finding with Optimal Resource Utilization

Aysu Bogatarkan[1], Esra Erdem[1], Alexander Kleiner[2], and Volkan Patoglu[1(✉)]

[1] Faculty of Engineering and Natural Sciences, Sabancı University, İstanbul, Turkey
{aysubogatarkan, esraerdem, vpatoglu}@sabanciuniv.edu
[2] Robert Bosch GmbH, Corporate Research, Stuttgart, Germany
alexander.kleiner@de.bosch.com

Abstract. The multi-agent path finding (MAPF) problem is a combinatorial search problem that aims at finding paths for multiple agents (e.g., robots) in an environment (e.g., an autonomous warehouse) such that no two agents collide with each other. We study a general version of MAPF, called mMAPF, that involves further challenges, such as multi-modal transportation modes, a set of waypoints to visit for each agent, and consumption of different types of resources. We introduce a declarative method to solve mMAPF, using answer set programming that provides a flexible formal framework to address all these challenges while optimizing multiple objectives.

Keywords: Multi-agent Path Finding · Answer Set Programming · Declarative problem solving · Autonomous warehouses

1 Introduction

Autonomous robot teams are increasingly deployed for industrial applications in warehouses and production. Commonly tasks are within intralogistics in which the goal of the team is to efficiently transport crates and pallets between stationary locations such as packing stations and conveyer entry points. For example, in Amazon's Kiva system or Ali Baba's smart warehouses, there exist multiple robots picking and delivering relevant shelves with products to human workers so that orders can be completed efficiently in time. While these systems heavily depend on engineered infrastructures, i.e., warehouses build from scratch with movable shelves, a substantially larger fraction of intralogistics problems are arising from scenarios with existing infrastructure in which the aforementioned solutions generally do not apply. Challenges for robots in conventional environments are robust methods for Simultaneous Localization and Mapping (SLAM) and Multi-agent Path Finding (**MAPF**). Whereas lidar-based approaches are successfully deployed for solving the SLAM problem today, only inflexible baseline approaches have been deployed for **MAPF** so far.

MAPF problem aims to find a plan for multiple agents to reach their destinations in a certain environment with static obstacles, subject to some constraints on the maximum or the total plan length. Every agent can be considered as a dynamic obstacle for

L. Wang et al. (Eds.): AMP 2020, LNME, pp. 313–324, 2020.
https://doi.org/10.1007/978-3-030-46212-3_24

other agents. Therefore, obstacles and agents are leading to some constraints for the executability of the plan: agents cannot pass through obstacles, and agents cannot collide with each other. While single-agent shortest pathfinding can be solved in polynomial time [7], **MAPF** (with constraints on the plan length) is an intractable problem [19] due to the latter constraint that no two agents can be in the same location at the same time.

MAPF has been investigated in Artificial Intelligence by utilizing various heuristic search algorithms. Some of these studies address time efficiency and minimize the maximum plan length of an agent. Some of them address energy efficiency and minimize the total sum of distance traveled by the agents. However, many realistic conditions observed in warehouses have not been considered in these studies. For instance, the robots' battery levels change as they travel around, and it may be necessary for them to be charged to complete their tasks. Furthermore, some parts of the warehouses, for instance with human occupants or tight passages, may necessitate robots to move slowly to ensure safety.

Along these lines, to handle more realistic autonomous warehouse scenarios, a mathematical model general enough to handle multi-objective optimizations and multimodal transportation conditions is needed. Furthermore, the computational framework is required to be flexible such that a large set of variations of **MAPF** problems can be addressed.

Motivated by these challenges, we mathematically model a general version of **MAPF** (called **mMAPF** – multi-modal **MAPF** with resources) as a rich graph problem and introduce a flexible method to solve **mMAPF** declaratively.

Our method relies on the declarative programming paradigm Answer Set Programming (ASP) [2, 3, 15, 17, 18]. By utilizing an expressive formal language and efficient solver of ASP, our method can handle the following variations of **MAPF**:

- **Multi-objective optimization:** Our method can find priority-based optimal solutions with respect to time optimization (the minimum plan length), energy optimization (the sum of distances) and their combinations.
- **Waypoints:** Our method can handle scenarios where robots can visit several locations (to pick and deliver different items) on their way to the goal. Note that the determination of the order of items to be collected requires further decision making while computing optimal solutions.
- **Resource constraints:** In addition to time constraints and total energy consumption, our method also considers individual resource such as battery consumption of each robot. As robots travel, their battery levels decrease. There exist charging stations scattered around the warehouses such that robots can charge their batteries. Therefore, while finding optimal solutions, our method also considers battery consumption and the possibility of including charging stations in their itineraries.
- **Multi-modal transportation:** Our method considers different transportation modes, for instance, for regions where robots should move slow or where they are allowed to move fast, while computing optimal solutions.

In the following, once we define our mathematical model for **mMAPF** as a graph problem, we describe how to solve **mMAPF** using ASP. We illustrate an application of our method with an interesting scenario, emphasizing the advantages listed above.

2 Related Work

There are mainly two kinds of **MAPF** solvers: some of them use search-based problem solving (mostly based on a variant of A* search), and some of them use declarative problem solving.

For instance, Silver [21] introduces an incremental method where the paths of agents are computed one by one with A* [13]; once a path is found for an agent, it is considered as an obstacle for other agents. Luna and Bekris [16] propose to compute the paths of agents independently, and then resolve the conflicts (i.e., when two agents collide with each other) with respect to some push-and-swap rules (e.g., there should be at least two free vertices in the graph). Chouhan and Niyogi [5, 6] propose a similar solution where the paths are computed independently; but the conflicts are resolved differently by assigning priorities to agents. Other search-based algorithms, like [8, 14, 25], also compute paths independently; in case a collision occurs, it is resolved by replanning one of the conflicting agents' route. Sharon et al. [20] propose a different method that performs a search on a tree based on the conflicts between agents.

Declarative methods reduce **MAPF** to formal frameworks (e.g., ILP, SAT, ASP) and use general problem solvers to find plans. Yu and Lavalle [26] model **MAPF** as a network flow problem and use an ILP solver to optimize the makespan (the time when the last robot reaches its goal) or the total distance traveled by all robots. Surynek et al. [23, 25] reduce **MAPF** to SAT and use a SAT solver to optimize the makespan or the sum of costs. Erdem et al. [9] model **MAPF** as a logic program and use an ASP solver to optimize the makespan or the distance.

None of the earlier works is applicable to multi-modal transportation and considers utilization of different resources. Our flexible method for **mMAPF** generalizes Erdem et al.'s ASP-based solution for **MAPF** [9] by including different transportation modes that allow priority-based optimization of multiple resource utilizations, and deciding the order of waypoints/charging stations visited by the agents on their ways to the goals.

3 mMAPF: Multi-modal MAPF with Optimal Resource Utilization

mMAPF can be viewed as a generalization of **MAPF** to enable multiple transportation modes and to take resource consumptions of the robots into account.

Let us first introduce some concepts and notation before we define **mMAPF**.

A traversal *f of a path* $P = \langle w_1, w_2, \ldots, w_n \rangle$ in a graph G, where every $w_l \in V$ and every $\langle w_l, w_{l+1} \rangle \in E$ within some time $t \in \mathbb{Z}^+$, is an onto function that maps every nonnegative integer less than or equal to t to a vertex in P or to *intransit*, such that, for every w_l and w_{l+1} in P and for every $x < t$.

- if $mode(\langle w_l, w_{l+1}\rangle) = normal$ and $f(x) = w_l$, then $f(x+1) = w_l$ or $f(x+1) = w_{l+1}$.
- if $mode(\langle w_l, w_{l+1}\rangle) = slow$ and $f(x) = w_l$, then $f(x+1) = w_l$, or $f(x+1) = intransit$ and $f(x+2) = w_{l+1}$.

We denote by $f(P)$ a traversal f of a path P (within time t).

Let f_i and f_j be traversals of two different paths P_i and P_j by agents a_i and a_j respectively, in a graph G within some time t. We say that the traversals f_i and f_j *do not collide with each other* within time t if the following three cases hold:

Case 1. For every time $x, x' \leq t$ such that $f_i(x) \neq intransit$, $f_j(x') \neq intransit$ the following holds: if $f_i(x) = f_j(x')$ then $x \neq x'$. That is, if the same vertex is visited by agents a_i and a_j then it should be visited at different times. Intuitively, no two agents can be at the same location at the same time. The type of collisions eliminated in this case are illustrated in Fig. 1.

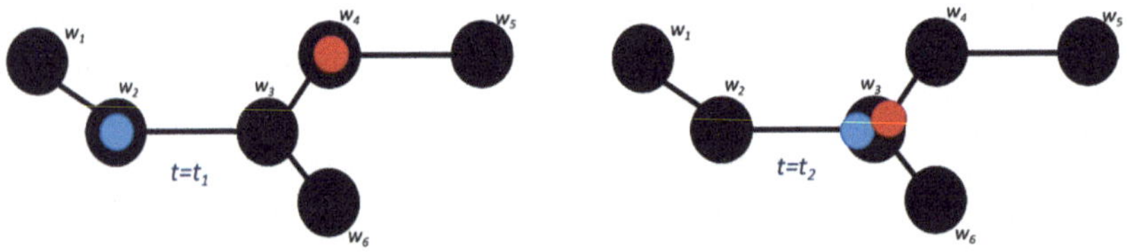

Fig. 1. Case 1. A collision occurs when two agents are at the same location at the same time. In this figure, either $mode(\langle w_2, w_3\rangle) = normal$ and $mode(\langle w_3, w_4\rangle) = normal$, or $mode(\langle w_2, w_3\rangle) = slow$ and $mode(\langle w_3, w_4\rangle) = slow$. In the former case, the collision occurs at time $t_2 = t_1 + 1$, after the agents start moving towards w_3 at time t_1. In the latter case, the collision occurs at time $t_2 = t_1 + 2$, after the agents start moving towards w_3 at time t_1.

Case 2. For every time $x < t$ such that $mode(\langle f_i(x), f_i(x+1)\rangle) = normal$, the following holds: if $f_i(x) = f_j(x+1)$ then $f_i(x+1) \neq f_j(x)$. That is, a normal edge cannot be visited by agents a_i and a_j in reverse directions at the same time. Intuitively, no two agents can swap their locations along a normal edge at the same time. The type of collisions eliminated in this case are illustrated in Fig. 2.

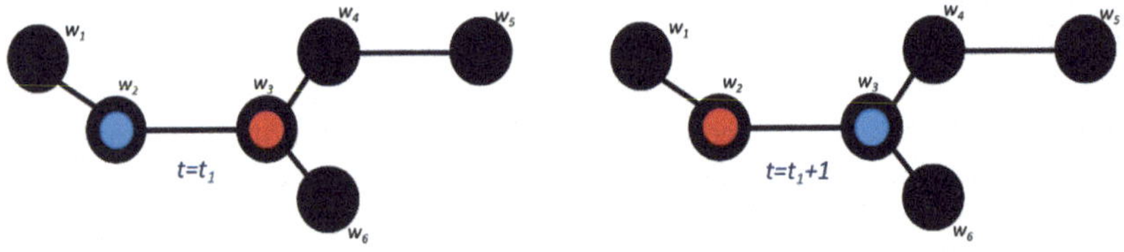

Fig. 2. Case 2. A collision occurs when two agents in the left figure move along the normal edge $\langle w_2, w_3\rangle$ at time t_1 towards each other, as they try to swap their places as in the right.

Case 3. For every time $0 < x \leq t-2$ and for every vertex $u, v \in V$, such that $mode(u, v) = slow$ and $f_i(x) = u, f_i(x+2) = v$,

- if $f_j(x-1) = v$ and $f_j(x) = intransit$, then $f_j(x+1) \neq u$; and
- if $f_j(x) = v$, then $f_j(x+2) \neq u$.

The first condition ensures that no two agents can swap their places along a slow edge if one of them is already in transit. The second condition ensures that no two agents can swap their places along a slow edge, if they are already located at the endpoints of the slow edge. The collisions eliminated in these cases are illustrated in Figs. 3 and 4.

Fig. 3. Case3(a). A collision occurs when two agents try to swap their places along a slow edge $\langle w_1, w_2 \rangle$ and one of them is a bit ahead in transit.

Fig. 4. Case3(b). A collision occurs when two agents located at the endpoints of a slow edge $\langle w_1, w_2 \rangle$ try to swap their places.

Now we can define **mMAPF** as a computational problem, in terms of its input and output, as shown in Fig. 5. Intuitively, graph G characterizes the warehouse where the agents move around, set C describes where charging stations are located in the warehouse, set S describe where agents can be located initially and in the end, set O denotes the parts of the environment covered by the static obstacles, set M denotes transportation modes, function *mode* denotes the parts of the corridors where the agents should travel slowly or where they are allowed to go faster, positive integer n denotes the number of agents, set A denotes the set of n agents, functions *init* and *goal* describe initial locations and goal locations of agents, set B describes the battery levels, function *init_battery* describes the initial battery levels of agents, set W_{ai} describes the set of waypoints for each agent a_i, and positive integer τ is an upper bound on plan lengths.

Given these input, **mMAPF** asks for, for each agent a_i, a path P_i in G from $init(a_i)$ to $goal(a_i)$, a traversal f_i of this path within time $u \leq \tau$, and a battery level function b_i showing how the agent's battery level changes during the traversal. **mMAPF** ensures about P_i that all the waypoints W_{ai} are visited by the agent a_i without colliding any static obstacles O. **mMAPF** ensures about f_i that the agents do not collide with each other while traversing their paths. **mMAPF** ensures about b_i that the agents' batteries have sufficient amount of energy (by charging at stations C, when needed) so that the agents can complete their plans.

mMAPF is an intractable problem: unless P $\neq$ NP, there does not exist a polynomial time algorithm to solve this problem.

Input:
- a graph $G = (V, E)$, where V is the set of vertices and E is the set of edges,
- a set $C \subset V$ of charging nodes,
- a set $S \subset V$ of station nodes,
- a set $O \subseteq V$ of obstacles,
- a set $M = \{slow, normal\}$ of edge modes,
- a function $mode : E \rightarrow M$,
- a set $A = \{a_1, \ldots, a_n\}$ of agents, $(n > 0)$,
- a function $init : A \rightarrow S$,
- a function $goal : A \rightarrow S$,
- a set $B = \{1, .., maxBatteryLevel\}$ of battery levels,
- a function $init\_battery : A \rightarrow B$,
- a set $W_{a_i} \subseteq V$ of waypoints for each agent $a_i \in A$,
- a positive integer τ for maximum plan length.

Output: For every agent $a_i \in A$, for some positive integer $u \leq \tau$,
- a path $P_i = \langle w_{i,1}, \ldots, w_{i,n_i} \rangle$ of length n_i $(n_i \leq u)$,
  - that the agent a_i will follow to reach its goal location from its initial location (i.e., $w_{i,1}$=$init(a_i)$ and w_{i,n_i}=$goal(a_i)$),
  - every waypoint $w_i \in W_{a_i}$ is included in the path P_i,
  - without colliding with any obstacles (i.e., $w_{i,j} \in V \setminus O$)
- a traversal f_i of the path P_i within time u, such that
  - for every other agent $a_j \in A$ with a path P_j and its traversal f_j within u, $f_i(P_i)$ and $f_j(P_j)$ do not collide with each other.
- a battery level function $b_i : \{1, 2, \ldots, u\} \rightarrow B$ where, for every $x < u$,
  - if $f_i(x) \in C$, then $b_i(x+1) = max(B)$,
  - if $f_i(x) \notin C$, then $b_i(x+1) = b_i(x) - 1$.

Fig. 5. Problem definition

4 Solving MMAPF Using ASP

Answer Set Programming (ASP) [2, 3, 15, 17, 18] is a knowledge representation and reasoning paradigm that is oriented towards combinatorial search problems as well as knowledge-intensive applications. The idea of ASP is to represent a problem and relevant knowledge as a "program" and to reason about the program by computing its models (called "answer sets" [11, 12]). These models characterize solutions of the problem, and can be computed by "ASP solvers" such as CLINGO [10].

We solve **mMAPF** using ASP by (i) representing it as a program in an ASP language (in this case, the input language of the ASP solver CLINGO), (ii) using an ASP solver (in this case, CLINGO) to find the answer sets for the program, and (iii) extracting the solutions from the answer sets, if there is an answer set.

Let us describe how we represent **mMAPF** in the input language of CLINGO, so that the interested readers can use it directly to experiment with it.

Describing the input and the output. In the following, suppose that `t` is the maximum possible length for a plan (i.e., τ) and `b` is the maximum battery level (i.e., *maxBatteryLevel*). We represent the vertices of G by atoms of the form `vertex(X)` and edges by atoms of the form `edge(X,Y)`. The agents can be located at vertices but also may be in transit, so we define locations explicitly:

```
location(Y) :- vertex(Y).
location(intransit).
```

The transportation modes are described by atoms of the forms `mode(X,Y,n)` or `mode(X,Y,s)` for normal and slow modes of edges respectively. The vertices covered by static obstacles are described by atoms of the form `obstacle(X)`. Agents are described by atoms of the form `agent(A)`. The initial and goal vertices for each agent are defined by atoms of the form `init(A,X)` and `goal(A,Y)` respectively. The initial battery levels are described by atoms of the form `init_battery(A,B)`.

We describe the output by atoms of the forms `plan(A,T,X)` (agent `A` is at location `X` at time `T`) and `batteryLevel(A,T,B)` (agent `A` has battery level `B` at time `T`).

Generating the paths and their traversals. We generate plans of agents recursively. Every agent A starts its plan at time step `0` at its initial location `X`.

```
plan(A,0,X) :- init(A,X), agent(A).
```

Consider any agent `A` who visits location `X` at time `T`. Agent `A` can wait at its current location `X` (if `X` denotes a vertex but not `intransit`) until the next time step `T + 1`:

```
{plan(A,T + 1,X)}1 :- plan(A,T,X), vertex(X), time(T), T < t.
```

Alternatively, the agent can move to the adjacent vertex `Y`. Then agent `A` will be at `Y` at the next time step, if `X` and `Y` have a normal edge between them.

```
{plan(A,T + 1,Y)}1 :- plan(A,T,X), edge(X,Y), mode(X,Y,n),
time(T), T < t.
```

If there is a slow edge between `X` and `Y`, the agent moves to `Y` in two time steps: in the first step, it becomes `intransit` state; in the second step, it becomes at `Y`.

```
{plan(A,T + 1,intransit)}1 :- plan(A,T,X), edge(X,Y),
mode(X,Y,s), time(T), T < t-1.
1{plan(A,T + 2,Y): edge(X,Y), mode(X,Y,s)}1 :-
plan(A,T + 1,intransit), plan(A,T,X), time(T), T < t-1.
```

Validity of paths and their traversals. The second rule above utilizes cardinality expressions [22]. The paths generated recursively above should satisfy the existence and uniqueness constraints: every agent should be at some location at each time step; every agent cannot be at two different locations at the same time.

```
:- {plan(A,T,Y): location(Y)}0, agent(A), time(T).
:- 2{plan(A,T,Y): location(Y)}, agent(A), time(T).
```

Note that these constraints ensure that there are not forks in a path and the path is connected without any gaps.

Every agent should visit its goal as well as the waypoints.

```
:- goal(A,X), not visit(A,X).
:- waypoint(A,X), not visit(A,X).
```

Here `visit(A,X)` describes which vertices are visited by each agent.

```
visit(A,X) :- plan(A,T,X).
```

If there is an obstacle on vertex `X`, no agent visits it.

```
:- plan(A,T,X), obstacle(X), agent(A), time(T).
```

Collision constraints. No two agents are at the same place at the same time, except when they are both in transit.

```
:- plan(A1,T,X), plan(A2,T,X), agent(A1;A2), A1 < A2,
X! = intransit.
```

This constraint eliminates the types of collisions described in Case 1 (Fig. 1).

Swapping is not allowed along a normal edge.

```
:- plan(A1,T,X), plan(A1,T + 1,Y), plan(A2,T,Y), A1 < A2,
plan(A2,T + 1,X), agent(A1;A2), mode(X,Y,n), T < t.
```

This constraint eliminates the types of collisions described in Case 2 (Fig. 2).

Swapping is not allowed along a slow edge, either. For that, we first define the transition of an agent along a slow edge `(X,Y)` starting at time step `T`:

```
slow(A,T,X,Y) :- plan(A,T,X), plan(A,T + 1,intransit),
plan(A,T + 2,Y), mode(X,Y,s), T < t-1.
```

Then, we ensure that swapping is not allowed on a slow edge.

```
:- slow(A1,T,X,Y), slow(A2,T-1,Y,X), T > 0, T < t-1, A1! = A2.
:- slow(A1,T,X,Y), slow(A2,T,Y,X), T < t-1, A1 < A2.
```

These constraints eliminate the types of collisions described in Case 3 (Figs. 3–4).

Battery levels should remain positive. We define the battery level of an agent recursively starting from its initial battery level.

```
batteryLevel(A,0,B) :- init_battery(A,B), agent(A).
```

At each step `T`, if the agent is not at a charging station, its battery level reduces by 1.

```
batteryLevel(A,T + 1,B1-1) :- batteryLevel(A,T,B1),
plan(A,T,X), not charging(X), agent(A), time(T), T < L,
planLength(A,L).
```

If the agent is at a charging location, its battery level may quickly get to the maximum level or the agent can move forward without charging its battery.

```
1{batteryLevel(A,T + 1,b); batteryLevel(A,T + 1,B1-1)}1 :-
plan(A,T,X), batteryLevel(A,T,B1), charging(X),
agent(A), time(T), T < L, planLength(A,L).
```

Then we ensure that the battery level cannot be less than the minimum level 1.

```
:- batteryLevel(A,T,B), B < 1.
```

Optimizing the plan length. We identify, for each agent, when it reaches the goal (i.e., the time step L) ensuring that it visits all of its waypoints on the way.

```
planLength(A,L) :- #max{T: plan(A,T,X), goal(A,X)} = L,
agent(A).
:- plan(A,T,X), waypoint(A,X), planLength(A,L), L < T.
```

Note that L denotes the plan length for each agent.

Then we identify the maximum of the plan lengths for all agents.

```
maxPlanLength(M) :- #max{T: planLength(A,T)} = M.
```

and ask CLINGO to minimize it by the following weak constraint [4]:

```
: ~ maxPlanLength(M). [M@1]
```

Further optimizations. We can ask CLINGO to minimize the total plan lengths to reduce the total energy consumption:

```
: ~ planLength(A,L). [L@1,A]
```

or we can also ask CLINGO to minimize the total number of charging the batteries:

```
: ~ batteryLevel(A,T,b). [1@1,A,T]
```

For multi-objective optimization, suppose that we want to minimize the maximum plan length first since we want the tasks to be completed by a given time. Next, we want to minimize the total energy consumption and ensure that robots do not wander around redundantly. Finally, we want to ensure that robots charge their batteries only when needed, by minimizing the number of times they charge. We can express these multiple optimizations by setting the priorities of weak constraints accordingly.

Solving a problem instance using CLINGO. Once the input is described by a set of facts, we can compute an answer set for the program above using the ASP solver CLINGO. After that, we can extract the atoms of the forms plan(A,T,X) and batteryLevel(A,T,B), to identify the traversals of agents and how their battery levels change along the way. An example scenario is illustrated in the next section.

5 An Example MMAPF Scenario

Let us illustrate how our method can be applied to solve an example **mMAPF** scenario, illustrated in Fig. 6. In this example, the warehouse consists of three shelf units denoted as obstacles (black cells). The charging stations are located at cells 24 and 27 highlighted by yellow, and the corridor where the agents should go slowly, covers cells 3–8 highlighted by red. There are two robots located at opposite corners of the warehouse. Robot *A1* start with an initial battery level of 10, while robot *A2* has an initial battery level of 8. The maximum battery level is set to 10. Each robot wants to collect some items on its way (denoted by stars that match the color of the relevant robot) and to deliver these items to the opposite corner of the warehouse. Therefore, eventually, they want to swap their places.

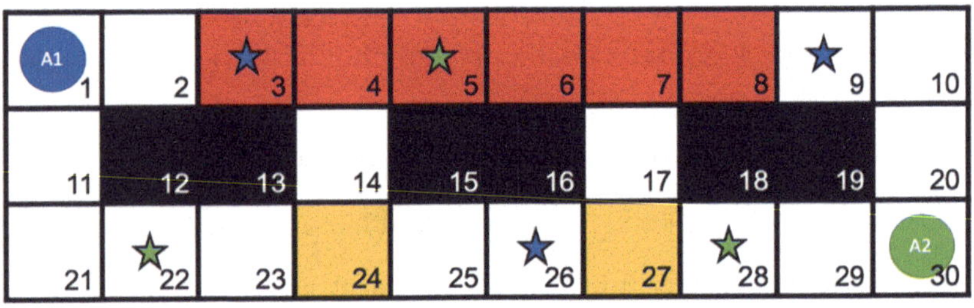

Fig. 6. A **mMAPF** instance. Initially, robot *A1* is located at cell 1 and robot *A2* is located at cell 30. The goal of *A1* is to deliver some items to cell 30, while the goal of *A2* is to deliver some items to cell 1. Each robot's waypoints are shown by stars with the same color as that robot.

A solution for this **mMAPF** instance is computed by CLINGO. Colored paths in Fig. 7 denote paths followed by the robots. Note that each robot visits its waypoints on the way to the goal. The traversals of these paths are shown in Table 1. The traversal of each slow edge (e.g., *A2* moving from cell 6 to 5) takes two time steps. The battery levels of the robots are also shown in this table. The battery level decreases at each time step, unless a robot is at a charging station. The battery level gets to its maximum when robots decide to charge while at a charging station. For instance, *A2*'s battery level increases to 10 when the charging station at cell 27 is visited.

Table 1. The table presents (i) the traversals of the blue and green paths shown in Fig. 7, by the blue robot *A1* and the green robot *A2*, respectively, from time step 0 to 18, and (ii) how the battery levels of these two robots change during these traversals.

Time	0	1	2	3	4	5	6	7	8	9	10	11	12	13	14	15	16	17	18
A1 location	1	2	3	t	4	14	24	25	26	27	17	7	t	8	9	10	20	30	–
A1 battery	10	9	8	7	6	5	4	3	2	1	10	9	8	7	6	5	4	3	–
A2 location	30	29	28	27	17	7	t	6	t	5	t	4	14	24	23	22	21	11	1
A2 battery	8	7	6	5	10	9	8	7	6	5	4	3	2	1	10	9	8	7	6

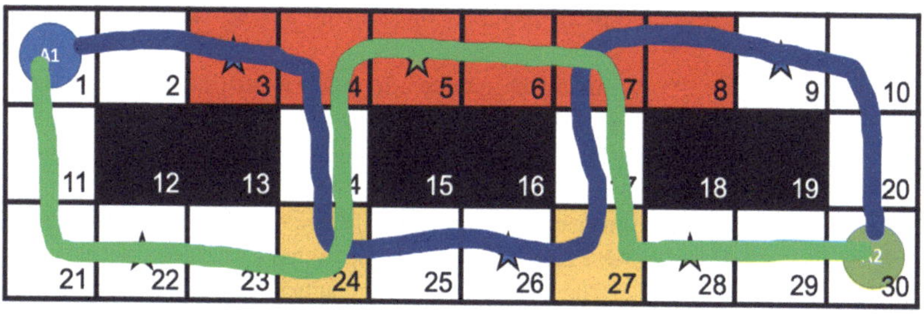

Fig. 7. A solution for the **mMAPF** instance shown in Fig. 6. *A1* follows the blue path while *A2* follows the green path to reach their goals, ensuring they collect items at their waypoints.

6 Conclusion

We introduced a general mathematical model for **mMAPF** problems as a rich graph problem that allows different transportation modes for parts of the environment, priority-based optimization of multiple resource utilizations, and agents to visit waypoints to pick/deliver some items on the way to the goal. Based on this model, we introduced a method to declaratively solve **mMAPF** using the expressive formalism and efficient solvers of ASP. The generality of our model and its declarativeness provide a formal yet flexible framework to investigate alternative solutions in autonomous warehouses. Particularly the flexibility of our framework allows to tailor solutions for highly diverse scenarios comprising various constraints, as they can arise in industrial domains.

Based on the theoretical setup provided in this paper, we plan to extend our studies with a comprehensive experimental evaluation to better understand the scalability of our method, and by considering changes in the environment as investigated in our earlier studies [1].

References

1. Bogatarkan, A., Patoglu, V., Erdem, E.: A declarative method for dynamic multiagent path finding. In: Proceedings of the 5th Global Conference on Artificial Intelligence, pp. 54–67 (2019)
2. Brewka, G., Eiter, T., Truszczynski, M.: Answer set programming at a glance. ACM Commun. **54**(12), 92–103 (2011)
3. Brewka, G., Eiter, T., Truszczynski, M.: Answer set programming: an introduction to the special issue. AI Mag. **37**(3), 5–6 (2016)
4. Buccafurri, F., Leone, N., Rullo, P.: Enhancing disjunctive Datalog by constraints. IEEE Trans. Knowl. Data Eng. **12**(5), 845–860 (2000)
5. Chouhan, S.S., Niyogi, R.: DMAPP: a distributed multi-agent path planning algorithm. In: Proceedings of AI, pp. 123–135 (2015)
6. Chouhan, S.S., Niyogi, R.: DiMPP: a complete distributed algorithm for multi-agent path planning. J. Exp. Theor. Artif. Intell. **29**(6), 1129–1148 (2017)

7. Dijkstra, E.W.: A note on two problems in connexion with graphs. Numer. Math. **1**(1), 269–271 (1959)
8. Dresner, K.M., Stone, P.: A multiagent approach to autonomous intersection management. J. Artif. Intell. Res. (JAIR) **31**, 591–695 (2008)
9. Erdem, E., Kisa, D.G., Oztok, U., Schueller, P.: A general formal framework for pathfinding problems with multiple agents. In Proceedings of AAAI (2013)
10. Gebser, M., Kaminski, R., Kaufmann, B., Schaub, T.: Clingo = ASP + control: Preliminary report. In Proceedings of ICLP (Technical Communications) (2014)
11. Gelfond, M., Lifschitz, V.: The stable model semantics for logic programming. In: Proceedings of International Logic Programming Conference and Symposium, pp. 1070–1080 (1988)
12. Gelfond, M., Lifschitz, V.: Classical negation in logic programs and disjunctive databases. New Generat. Comput. **9**, 365–385 (1991)
13. Hart, P.E., Nilsson, N.J., Raphael, B.: Correction to "a formal basis for the heuristic determination of minimum cost paths". SIGART Newslett. **37**, 28–29 (1972)
14. Jansen, R., Sturtevant, N.: A new approach to cooperative pathfinding. In: Proceedings of AAMAS, pp. 1401–1404 (2008)
15. Lifschitz, V.: Answer set programming and plan generation. Artif. Intell. **138**, 39–54 (2002)
16. Luna, R., Bekris, K.E.: Efficient and complete centralized multi-robot path planning. In Proceedings of IROS, pp. 3268–3275 (2011)
17. Marek, V., Truszczynski, M.: Stable models and an alternative logic programming paradigm. In: The Logic Programming Paradigm: A 25-Year Perspective, pp. 375–398. Springer, Heidelberg (1999)
18. Niemela, I.: Logic programs with stable model semantics as a constraint programming paradigm. Ann. Math. Artif. Intell. **25**, 241–273 (1999)
19. Ratner, D., Warmuth, M.K.: Finding a shortest solution for the n × n extension of the 15-puzzle is intractable. In: Proceedings of AAAI, pp. 168–172 (1986)
20. Sharon, G., Stern, R., Felner, A., Sturtevant, N.R.: Conflict-based search for optimal multi-agent pathfinding. Artif. Intell. **219**, 40–66 (2015)
21. Silver, D.: Cooperative pathfinding. In: Proceedings of AIIDE, pp. 117–122 (2005)
22. Simons, P., Niemelae, I., Soininen, T.: Extending and implementing the stable model semantics. Artif. Intell. **138**(1), 181–234 (2002)
23. Surynek, P.: On propositional encodings of cooperative path-finding. In: Proceedings of ICTAI, pp. 524–531 (2012)
24. Surynek, P., Felner, A., Stern, R., Boyarski, E.: Efficient SAT approach to multiagent path finding under the sum of costs objective. In: Proceedings of ECAI, pp. 810–818 (2016)
25. Wang, K.-H.C., Botea, A.: Fast and memory-efficient multi-agent pathfinding. In: Proceedings of ICAPS, pp. 380–387 (2008)
26. Yu, J., LaValle, S.M.: Planning optimal paths for multiple robots on graphs. In: Proceedings of ICRA, pp. 3612–3617 (2013)

The Effects of Milling and Drilling Process Parameters and Different Tool Path Strategies on the Quality of the CFRP Composites

Grigore Marian Pop(✉), Emilia Campean, Liviu Adrian Crisan, and Mihai Tripa

Technical University of Cluj-Napoca, Muncii Blvd., no. 103-105, Cluj-Napoca, Cluj, Romania
grigore.pop@muri.utcluj.ro

Abstract. Carbon fiber reinforced plastic (CFRP) composites are being used at a greater scale which is increasing the demands on automated production to improve productivity. These types of materials can be stronger than steel, about 40% lighter than aluminium, up to 80% lighter than steel and as stiff as titanium. Composite materials represent a good alternative to engineering materials, providing several important advantages in comparison to conventional materials, such as: light weight, mechanical and chemical resistance, low maintenance costs, high specific strength, higher stiffness and temperature stability, allowance of free forms modelling and specific design. Surface roughness evaluation is very important for many fundamental problems such as friction, contact deformation, heat and electric current conduction, tightness of contact joints and positional accuracy. For this reason, surface roughness, has been the subject of experimental and theoretical investigations for many decades. Also, surface roughness imposes one of the most critical constrains for the selection of machines and cutting parameters in process planning. In order to economically machine these materials with high part qualities, improvements in machining strategies must be made. This chapter presents the researches regarding different milling and drilling strategies in order to determine the best quality of the finished product. Flatness, parallelism, cylindricity, roughness and dimensional tolerances were measured using 3D CMM and the results were analysed using the Design Expert software.

Keywords: CFRP · Machining strategy · Design of experiments · Roughness · Infrared thermography · Education

1 Introduction

Composite materials represent a good alternative to conventional materials, providing several important advantages, such as: light weight, mechanical and chemical resistance, low maintenance costs, high specific strength, higher stiffness and temperature stability, allowance of free forms modelling and specific design.

Benefits from composite materials are especially important where weight control is critical, for example in the aerospace industry. They are also utilized at a greater extent

L. Wang et al. (Eds.): AMP 2020, LNME, pp. 325–344, 2020.
https://doi.org/10.1007/978-3-030-46212-3_25

due to its beneficial properties in the wind power industry, automotive, electronics and medicine (implants realized using advanced composite materials.

In industry, there are many discussions regarding the use of composite materials. The study of VDI (Association of German Engineers) shows an increased use of these type of materials from 28000 to/year in 2008 to 42000 to/year in 2012. The study expects an increase up to 130000 to/year in 2020 [7].

The challenges for automotive engineering are the reduction of gas emissions prescribed by the EU. Lightweight construction is an important lever for reducing fuel consumption. For this reason, automotive manufacturers are providing their developers and suppliers with quantitative guidelines for the continuous weight reduction of the components.

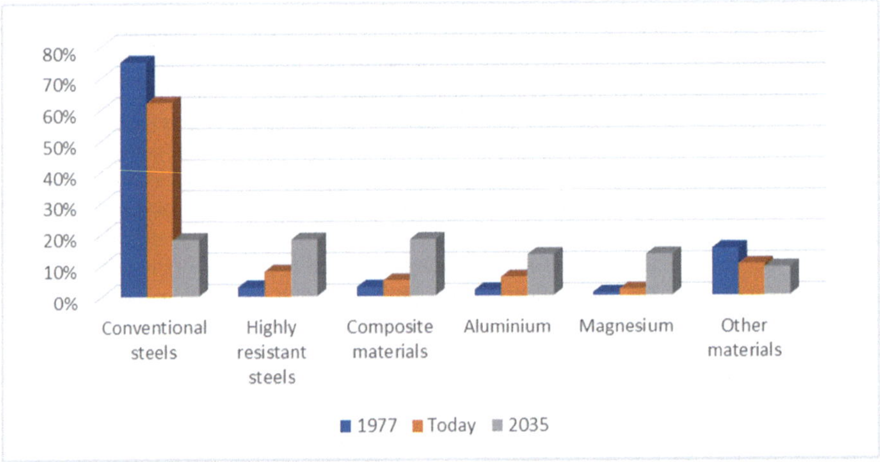

Fig. 1. Time evolution of types of materials used in a car (according to VDI)

Lightweight construction materials such as carbon fiber reinforced composites are increasingly replacing conventional steel, which thus loses its dominant role as a recent study of VDI also shows (see Fig. 1).

Nationally, for the moment at least, composite materials are not used in the manufacturing of the frames of mechanical machines, but internationally there are studies and researches regarding the replacement of classic materials of which mechanical parts of the machines are manufactured with components manufactured from carbon fiber reinforced composite materials. Analysing the published articles and the state of the art regarding the machining of composite materials, there are a few, but representative studies, in these areas [9, 15]. The researches presented in this chapter will be used in lecturers in higher education [16, 17].

The demand for high strength fiber composite components is set to increase across many industries until 2020, leading to solid market growth. According to VDMA (Verband Deutscher Maschinen und Anlagenbau, Mechanical Engineering Industry Association) the study of the demand for high-strength carbon-fiber composite components is rising by 17% a year. The results of this study present also the explicit demand of the companies to accelerate the research in this area. Another attractive aspect beside their performant mechanical properties is that the experts expect that the costs of fiber composite components will come down by around 30% by 2020. A conclusion of the studies indicates that there will be a strong growth in demand for these products, with the growing importance of lightweight construction across various industries. Challenge for engineering would be to drive down production costs through technological development, indicating a clear wish of the companies (sectors of industry, like automotive, aeronautics and wind energy), to intensify the researches in order to gain new knowledge in machining of carbon fiber reinforced composites [10].

2 Research Methodology

The methods of investigation involve solving the existing problems, such as: reduction of working times, increased productivity, cost reduction and improvement of the quality of the processed part.

First stage of this research has the role to identify and generate in working hypotheses. The existing researches were analysed and the relevant studies for the proposed theme were identified. After a detailed examination of the publications and of the knowledge existing in production, the work was focused on a step by step documentation of the information needed for identifying the machining parameters, which have influence on the quality of the machined carbon fiber reinforced composites.

Due to the variety of the machinability characteristics of different types of carbon fiber reinforced composites, an optimization of the process parameters is necessary for each type of material. Currently, the process parameters are experimental optimized by the users, which implies bigger costs.

The research was focused on three different tool paths for milling and drilling operations (Helical-Figs. 2 and 3, Zig-Zag Figs. 4 and 5 and iMachining, Figs. 6 and 7).

The three different tool paths were selected with the role of highlighting the degree of influence of the material removal method as well as of the tool trajectory on the quality of the material. The used strategies for this chapter are presented above.

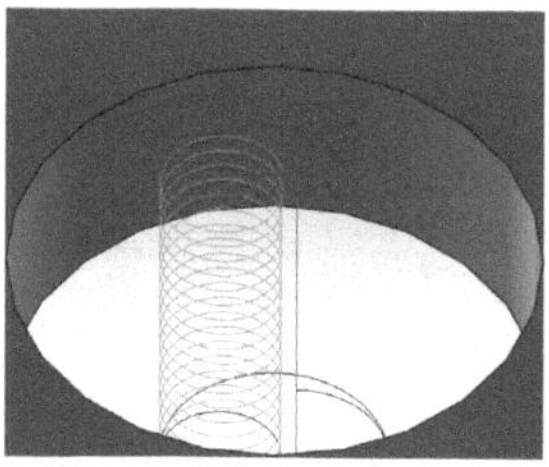

Fig. 2. Helical strategy for drilling

The Helical milling strategy uses a helical path for machining parts, while rotating around its own axis. The strategy combines the movement of the vertical z-axis with the horizontal x-y axis. The strategy has the advantage of generating low cutting forces while reducing the tool wear.

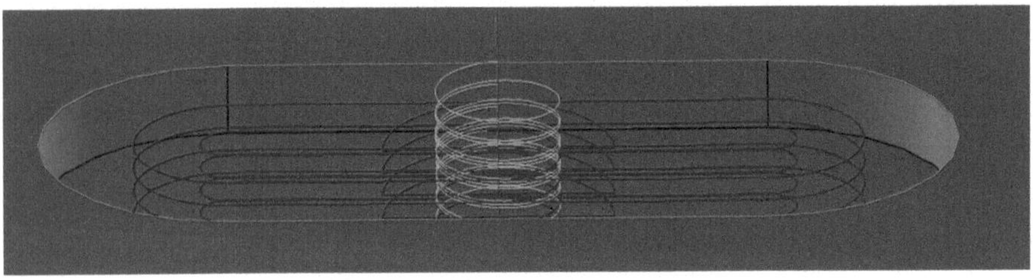

Fig. 3. Helical strategy for milling

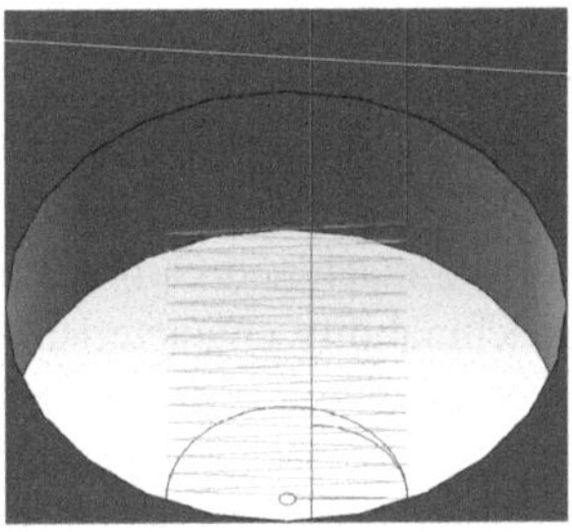

Fig. 4. Zig-Zag strategy for drilling

The Zig-Zag strategy uses parallel-linear paths for the material removal. The material is removed both in forward and backward movement, using a combination of two axis movement: the z axis with x or y axis.

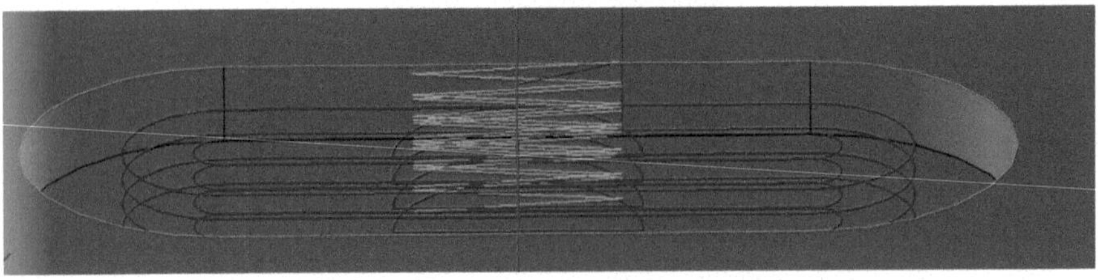

Fig. 5. Zig-Zag strategy for milling

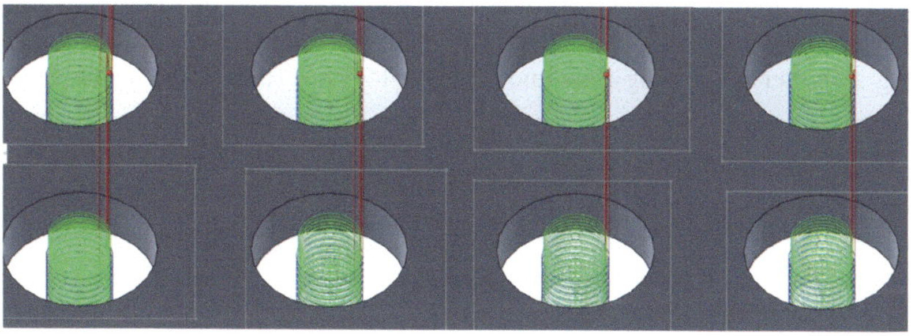

Fig. 6. iMachining strategy for drilling

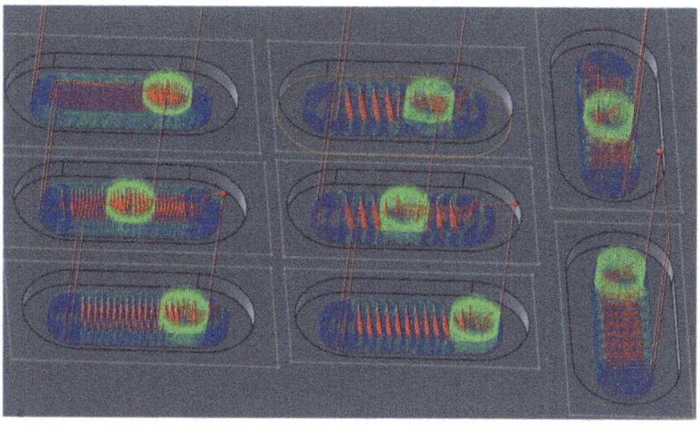

Fig. 7. iMachining strategy for milling

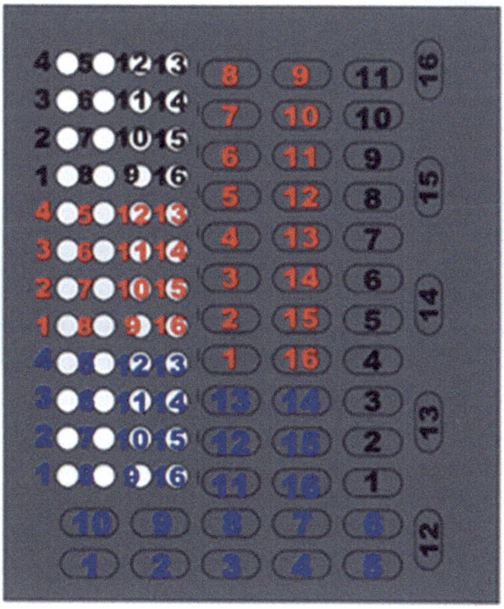

Fig. 8. Tool path strategy for each processed hole and slot (Blue-Helical, Red-Zig-Zag, Black-iMachining)

iMachining involves the establishment, by the program, of the optimal route necessary to be machined in order to remove the material with constant volume thus allowing a shorter processing time and a longer tool life (Fig. 8 and Table 1).

Table 1. Process parameters for each strategy

Operation	Helical		Zig-Zag		iMachining	
	Speed (rot/min)	Feed (mm/min)	Speed (rot/min)	Feed (mm/min)	Speed (rot/min)	Feed (mm/min)
1	11500	690	11500	690	11500	690
2	6500	390	6500	390	6500	390
3	7500	450	7500	450	7500	450
4	8500	510	8500	510	8500	510
5	7000	420	7000	420	7000	420
6	10000	600	10000	600	10000	600
7	9500	570	9500	570	9500	570
8	8000	480	8000	480	8000	480
9	10500	630	10500	630	10500	630
10	12000	720	12000	720	12000	720
11	5000	300	5000	300	5000	300
12	6000	360	6000	360	6000	360
13	5500	330	5500	330	5500	330
14	9000	540	9000	540	9000	540
15	11000	660	11000	660	11000	660
16	4500	270	4500	270	4500	270

3 Results

Surface roughness evaluation is very important for many fundamental problems such as friction, contact deformation, heat and electric current conduction, tightness of contact joints and positional accuracy. For this reason, surface roughness has been the subject of experimental and theoretical investigations for many decades. Also, surface roughness imposes one of the most critical constrains for the selection of machines and cutting parameters in process planning. Although many factors affect the surface condition of a machined part, parameters such as cutting speed, work piece condition, feed rate and depth of cut have more influences on the surface roughness for a given machine tool and work piece set-up [1].

Dimensional positioning, and geometrical tolerances such as cylindricity, parallelism and flatness [4], of each processed hole or slot was measured using the 3D measuring machine Axiom Aberlink twoo (Fig. 9), and the results were printed as shown in Fig. 10.

Fig. 9. 3D measurements

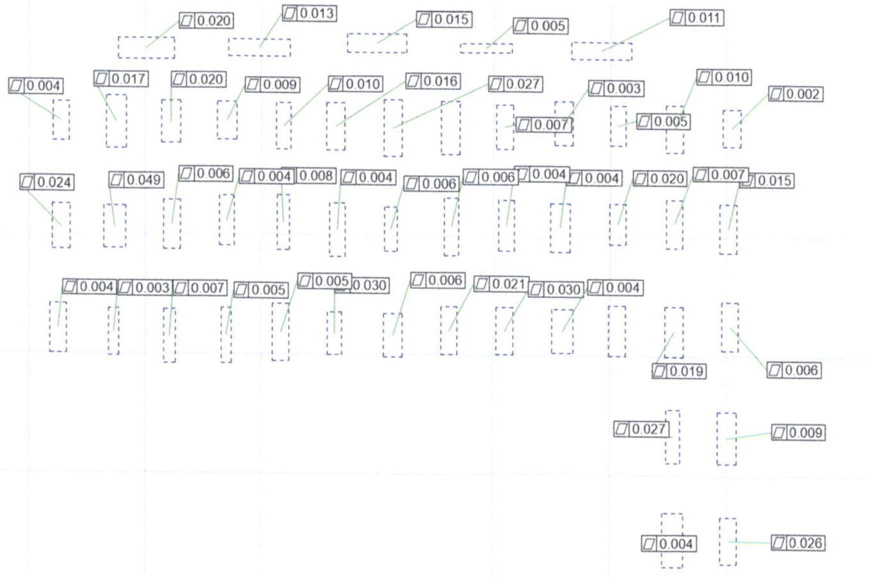

Fig. 10. Flatness deviation measurement report

Based on the results, using Design-Expert software v12 the following correlation charts were plotted in order to identify the influence of the process parameters on the obtained surface quality, for each tool path strategy (Figs. 11, 12 and 13).

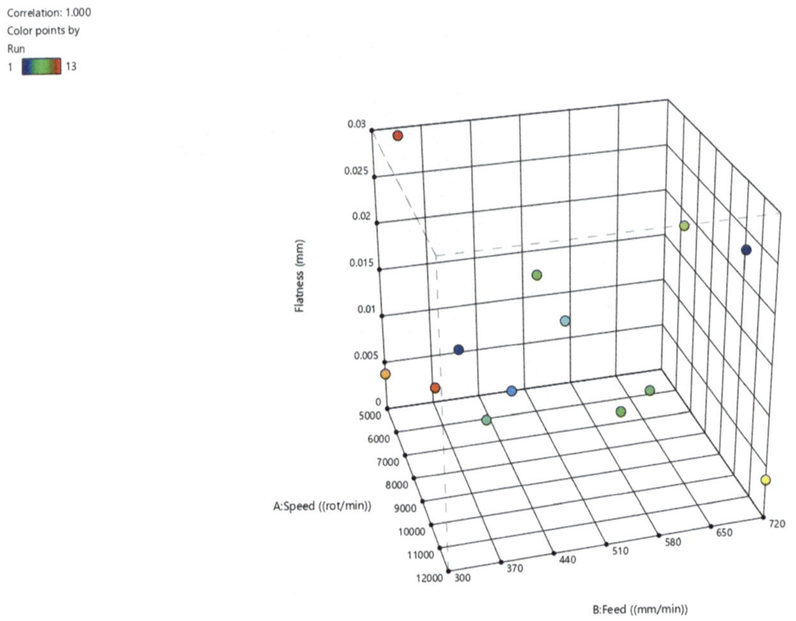

Fig. 11. Correlation chart for Helical strategy

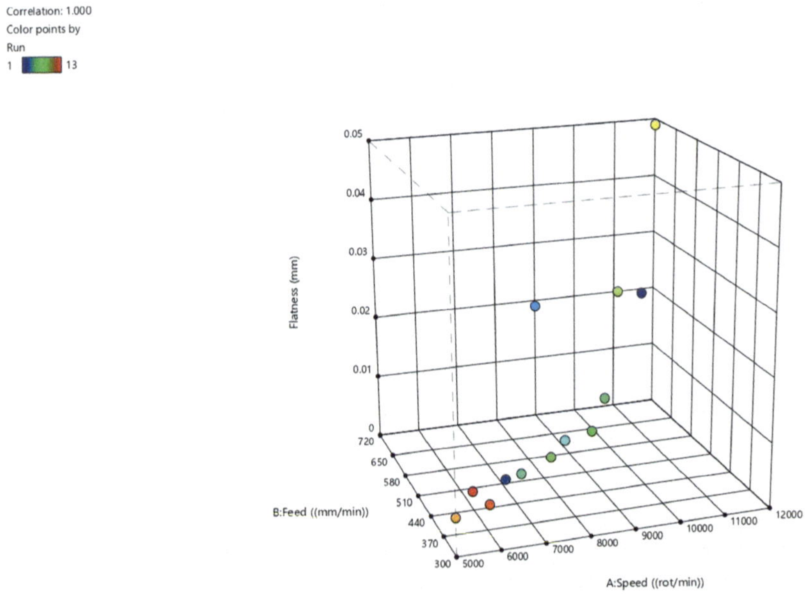

Fig. 12. Correlation chart for Zig-Zag strategy

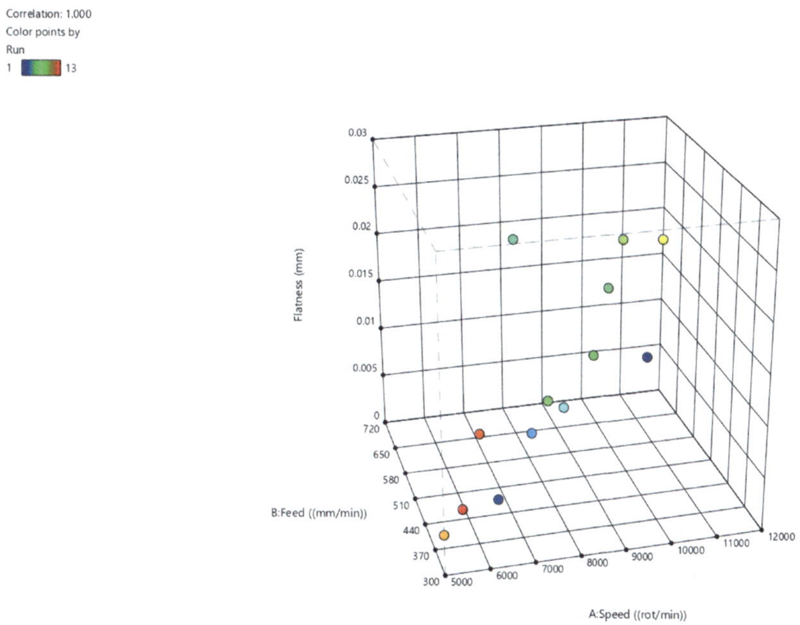

Fig. 13. Correlation chart for iMachining strategy

The best results were obtained for the iMachining method, where the values were constant for all the machining process parameters. The strategy that uses helicoidal displacements is the one for which the highest measured value of the flatness deviation was 0.027 mm. Removing material using only circular interpolation generates higher values for the flatness deviation.

Further the 3D measurement reports were plotted in order to determine the dimensional accuracy for each processed hole and the results are shown in Figs. 14 and 15.

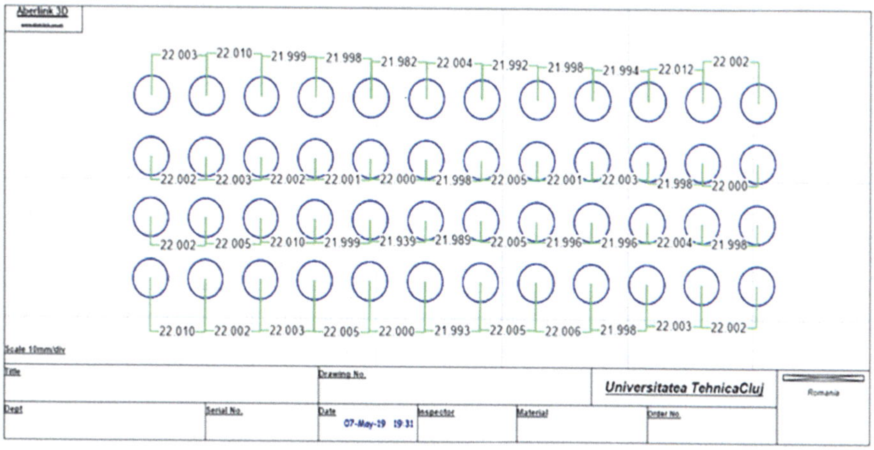

Fig. 14. Measurement report on x-axis

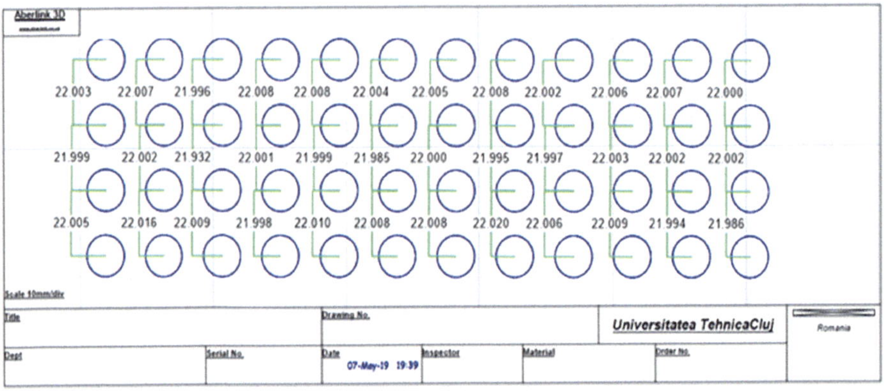

Fig. 15. Measurement report on y-axis

As for the milled slots the distances and parallelism between the flanks was also measured, the results were plotted in Fig. 16.

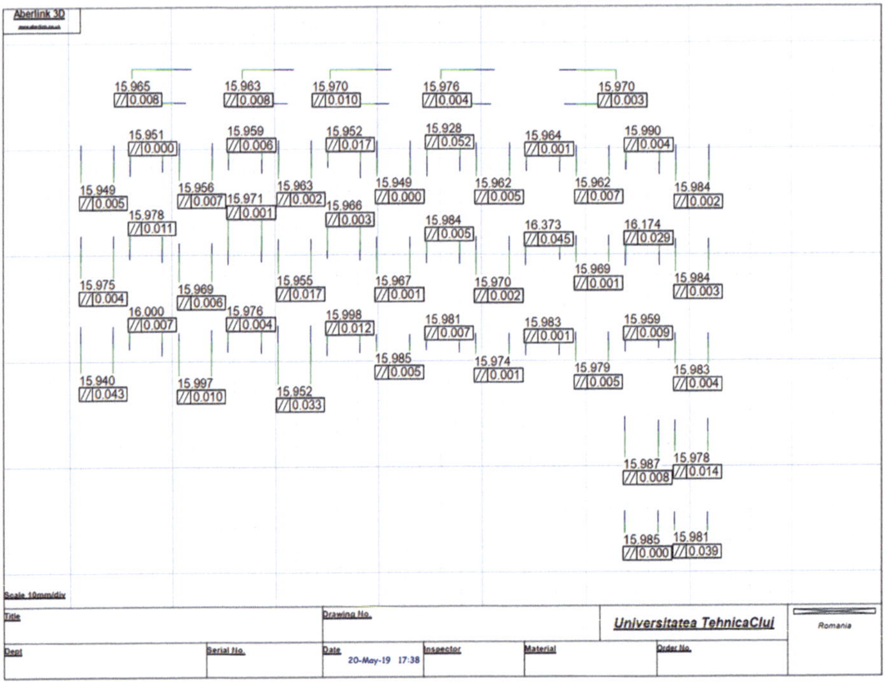

Fig. 16. Measurement report for parallelism deviation and thickness of the slots

For each processed hole the cylindricity deviation and the diameter was also measured, and the results were plotted into the correlation charts presented in Figs. 17, 18, 19 and 20.

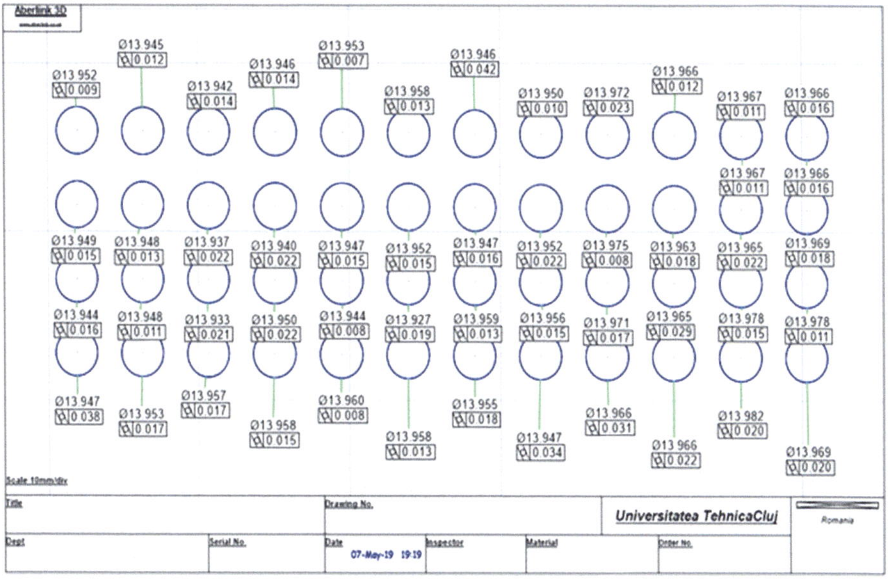

Fig. 17. Measurement report for cylindricity deviation and diameter

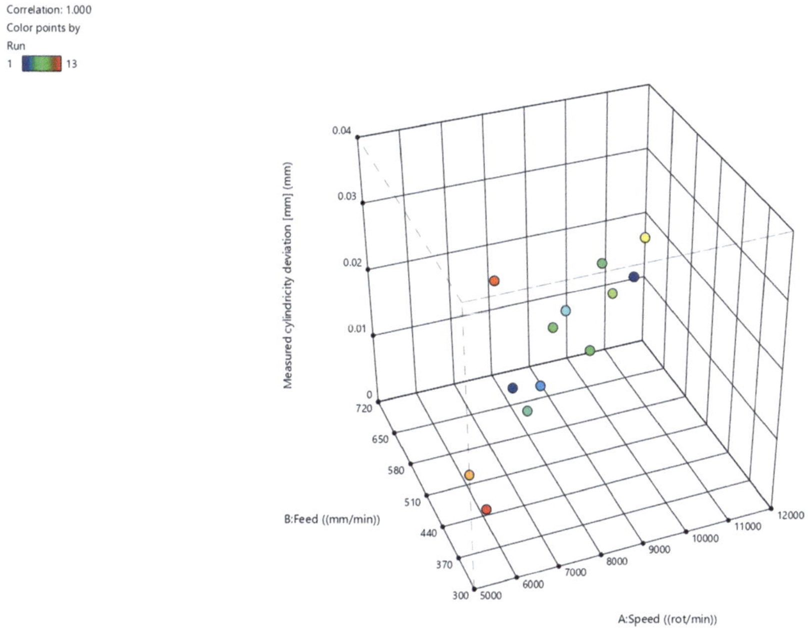

Fig. 18. Correlation chart for Helical strategy

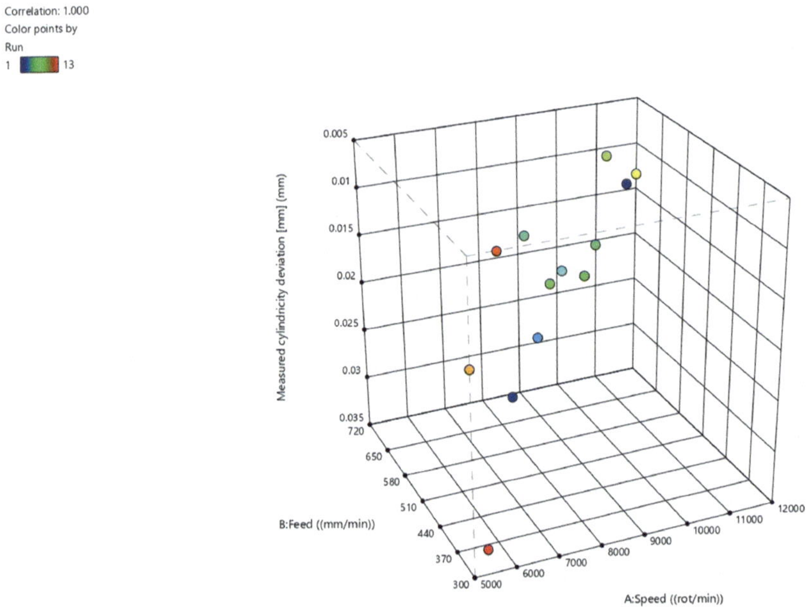

Fig. 19. Correlation chart for Zig-Zag strategy

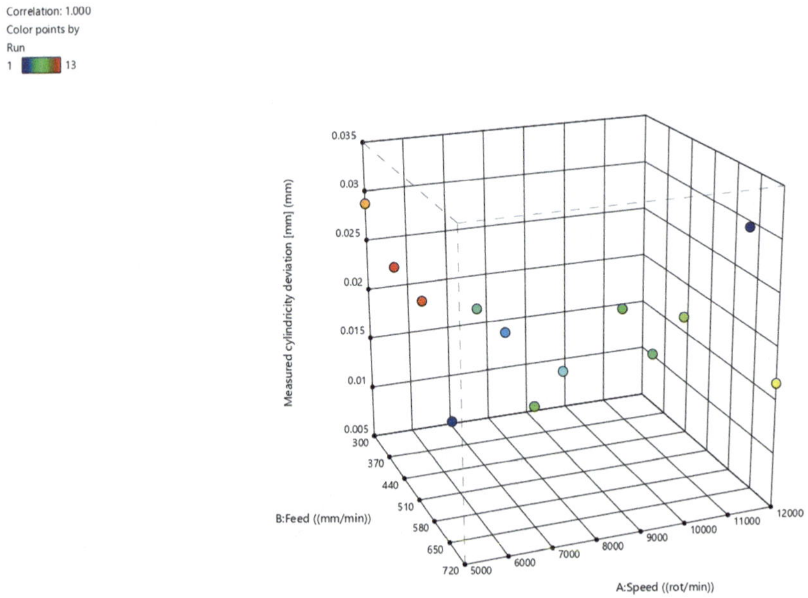

Fig. 20. Correlation chart for iMachining strategy

As in the case of the flatness analysis, the results obtained show that the iMachining strategy generates the most constant values. This strategy allows a constant machining. The biggest deviations were obtained for the Zig-Zag strategy, concluding that machining only on two axis does not lead to good result in terms of cylindricality.

Further the roughness parameters were measured using a roughness tester from Namicon, TR220. The evaluation length, ln, was set to 12.5 mm and the sampling length, lr, was set to 2.5 mm according to ISO 4288. The results are presented in Tables 2, 3 and 4.

Table 2. Measured roughness values for each processed slot

Nr op	Helical						
	Ra[μm]	Rq[μm]	Rz[μm]	Rt[μm]	Rp[μm]	Rv[μm]	Ry[μm]
1	0.67	0.902	3.667	5.722	2.101	1.566	5.722
2	1.096	1.329	5.031	6.953	2.484	2.546	6.953
3	0.682	0.873	4.273	5.82	1.789	2.484	5.82
4	1.144	1.461	6.207	10.05	2.699	3.507	10.05
5	0.985	1.235	5.21	6.894	2.402	2.808	6.894
6	0.991	1.229	5.421	7.48	2.972	2.449	7.48
7	0.594	0.79	3.316	5.722	1.265	2.05	5.722
8	1.327	1.6	6.421	8.808	2.875	3.546	8.808
9	0.953	1.203	4.835	7.91	2.531	2.304	7.91
10	0.797	1.012	4.984	8.906	2.082	2.902	8.906
11	0.871	1.13	4.976	7.089	2.292	2.683	7.089
12	0.835	1.021	4.48	6.289	1.878	2.601	6.289
13	1.195	1.459	6.269	8.261	3.074	3.195	8.261
14	1.164	1.42	5.132	7.363	2.64	2.492	7.363
15	0.833	1.08	5.171	7.148	2.488	2.683	7.148
16	0.809	1.04	4.703	7.148	2.339	2.363	7.148

For the first 13 operations, of each tool path strategy shown, correlation charts were plotted in order to establish the relation between the roughness values of Ra, Rz and the process parameters (see Figs. 21, 22, 23, 24, 25, 26, 27, 28 and 29).

The Helical strategy seems to be more constant regarding the Ra parameter, having 7 values between 0.5–1 μm. Comparing these strategies over the range of operations, iMachining optimization strategy manages to obtain the best Ra values.

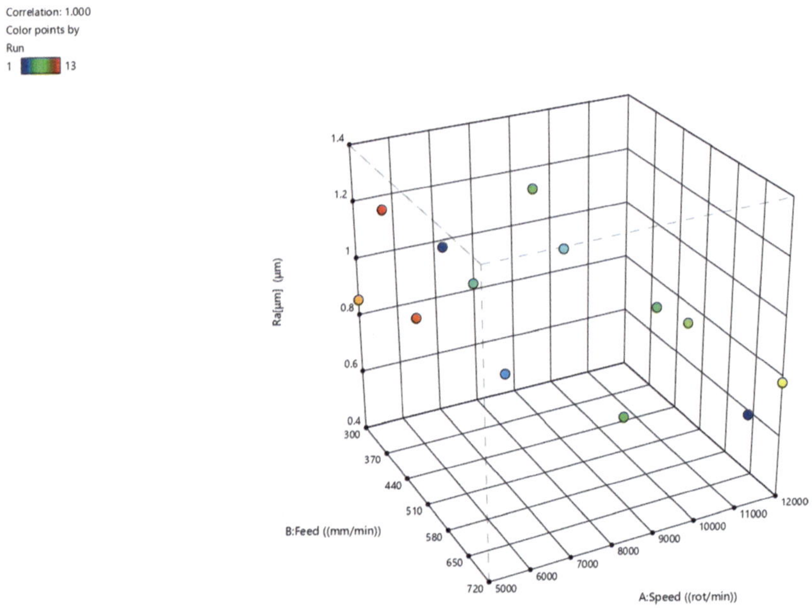

Fig. 21. Correlation chart for Helical strategy, Ra values

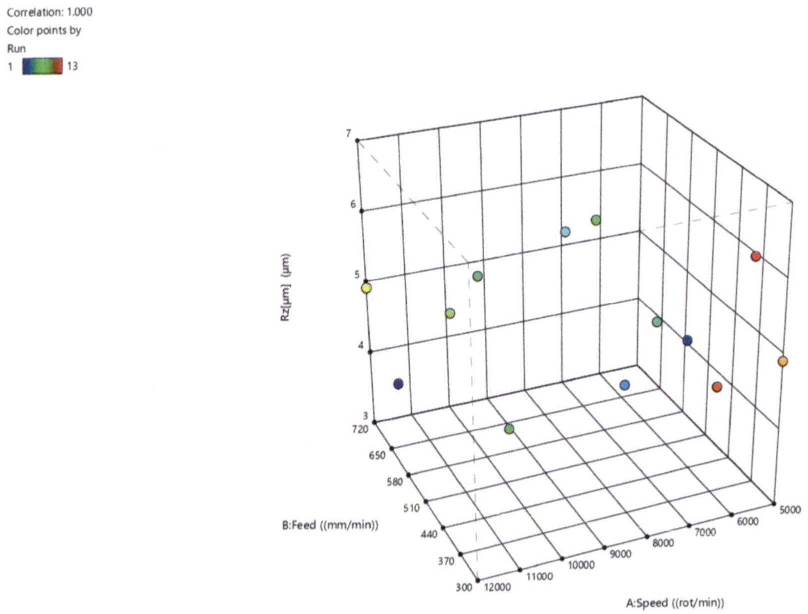

Fig. 22. Correlation chart for Helical strategy, Rz values

The Zig-Zag method involves surface machining using the maximum tool path length and maximum cycle time. For that reason, using the same tool and tool parameters, the Zig-Zag strategy generates constant values for Ra parameters, values that are higher than the one obtained for the other strategies, partly due to the large surface area to be milled in a single pass.

Table 3. Measured roughness values for each processed slot

Run	Zig-Zag						
	Ra[μm]	Rq[μm]	Rz[μm]	Rt[μm]	Rp[μm]	Rv[μm]	Ry[μm]
1	0.628	0.839	3.082	6.132	1.671	1.41	6.132
2	0.971	1.193	4.554	7.167	1.656	2.898	7.167
3	0.816	0.986	4.496	6.035	2.097	2.398	6.035
4	0.766	0.998	4.285	6.367	2.109	2.175	6.367
5	0.623	0.811	4.125	6.894	1.789	2.335	6.894
6	0.857	1.143	4.941	10.39	2.117	2.824	10.39
7	0.654	0.826	4.527	5.976	2.132	2.394	5.976
8	0.698	0.922	5.121	7.656	1.953	3.167	7.656
9	1.522	1.947	9.558	12.92	4.07	5.488	12.92
10	1.475	1.838	8.679	13.28	4.089	4.589	13.28
11	0.84	1.04	4.777	7.441	2.007	2.769	7.441
12	0.765	0.973	5.289	6.601	2.199	3.089	6.601
13	0.574	0.722	4.769	8.242	2.363	2.406	8.242
14	0.917	1.255	7.335	14.68	3.167	4.167	14.68
15	0.648	0.898	4.402	7.812	1.664	2.738	7.812
16	0.934	1.174	4.906	7.851	1.914	2.992	7.851

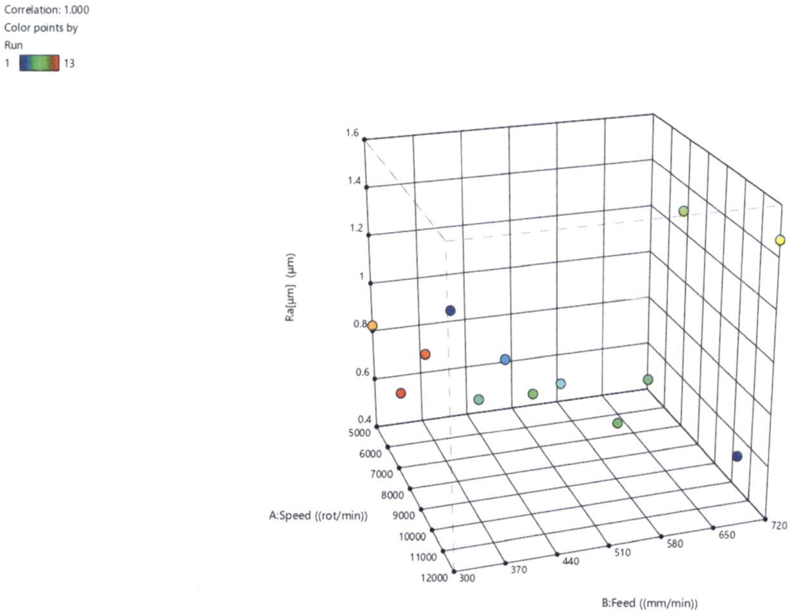

Fig. 23. Correlation chart for Zig Zag strategy, Ra values

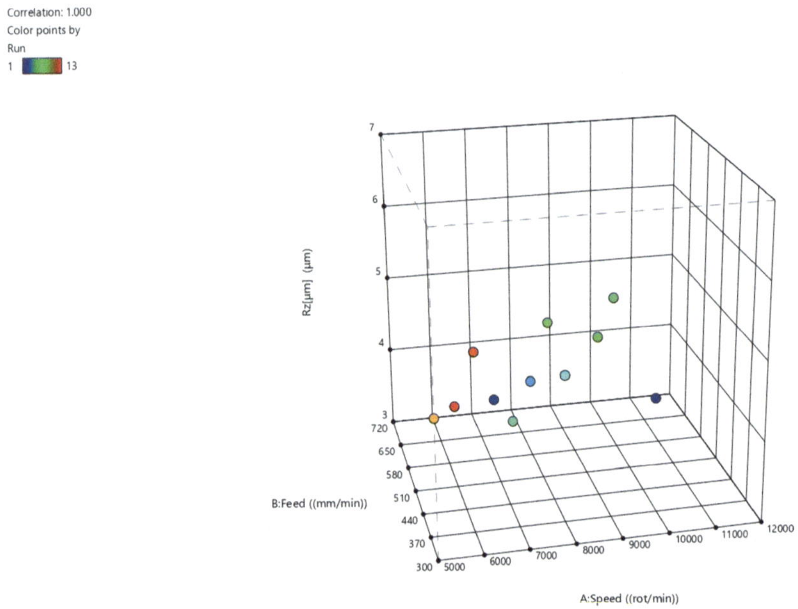

Fig. 24. Correlation chart for Zig Zag strategy, Rz values

The iMachining strategy uses constant volume removal of the material. Taking into consideration the amount of material that needs to be machined would be the reason for obtaining the smallest results. In this way, we obtain a shorter processing time and a longer tool life.

Table 4. Measured roughness values for each processed slot

Run	iMachining						
	Ra[µm]	Rq[µm]	Rz[µm]	Rt[µm]	Rp[µm]	Rv[µm]	Ry[µm]
1	0.472	0.619	2.734	4.199	1.058	1.675	4.199
2	0.635	0.787	3.382	5.253	1.507	1.875	5.253
3	0.722	0.938	3.781	6.621	1.73	2.05	6.621
4	0.541	0.683	2.945	4.707	1.21	1.734	4.707
5	0.513	0.703	3.781	6.171	1.359	2.421	6.171
6	0.664	0.868	5.425	9.863	2.605	2.82	9.863
7	0.593	0.754	2.902	5.839	1.687	1.214	5.839
8	0.636	0.792	3.945	6.132	1.925	2.019	6.132
9	2.239	2.837	13.07	28.78	6.039	7.035	28.78
10	1.773	2.214	10	15.15	4.015	5.988	15.15
11	0.458	0.665	3.425	7.421	1.562	1.863	7.421
12	1.06	1.434	8.398	10.66	2.82	5.578	10.66
13	1.216	1.598	8.304	11.52	3.039	5.265	11.52
14	1.22	1.549	7.363	9.765	3.679	3.683	9.765
15	1.026	1.29	5.703	8.769	2.867	2.835	8.769
16	1.169	1.527	7.128	12.69	4.105	3.023	12.69

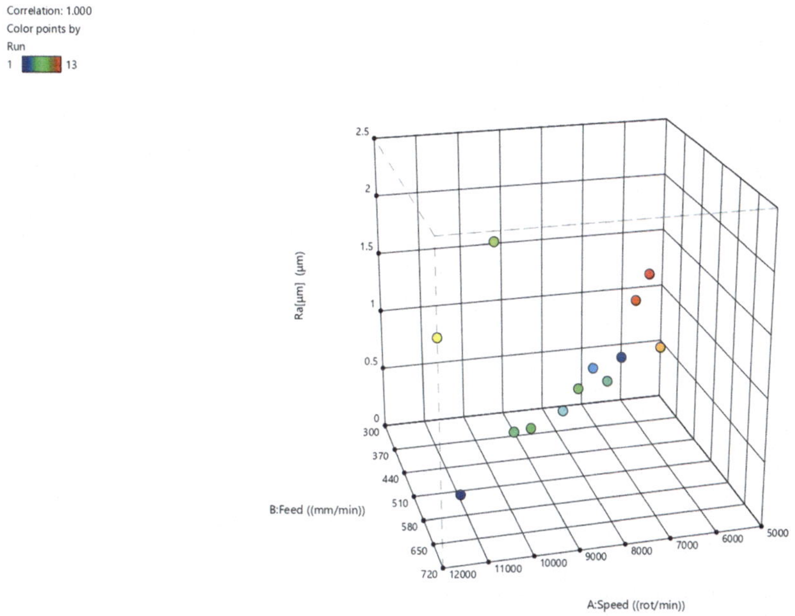

Fig. 25. Correlation chart for iMachining strategy, Ra values

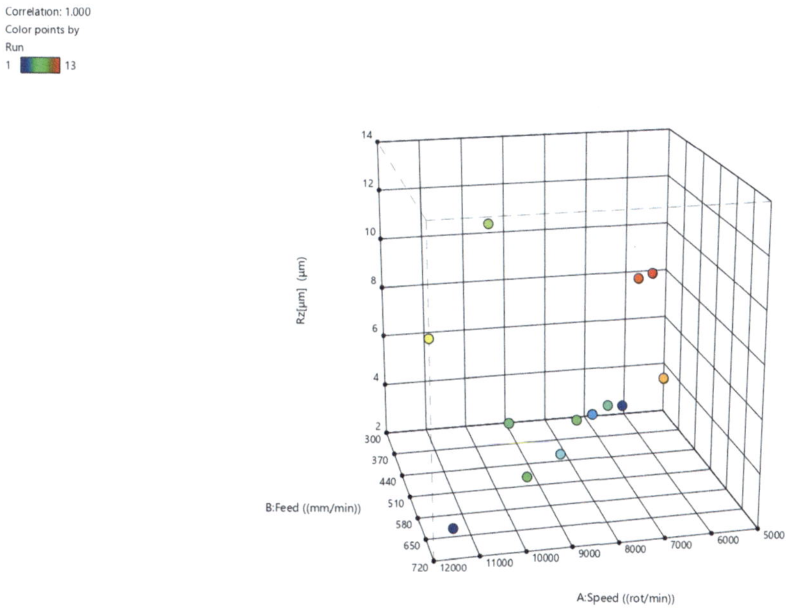

Fig. 26. Correlation chart for iMachining strategy, Rz values

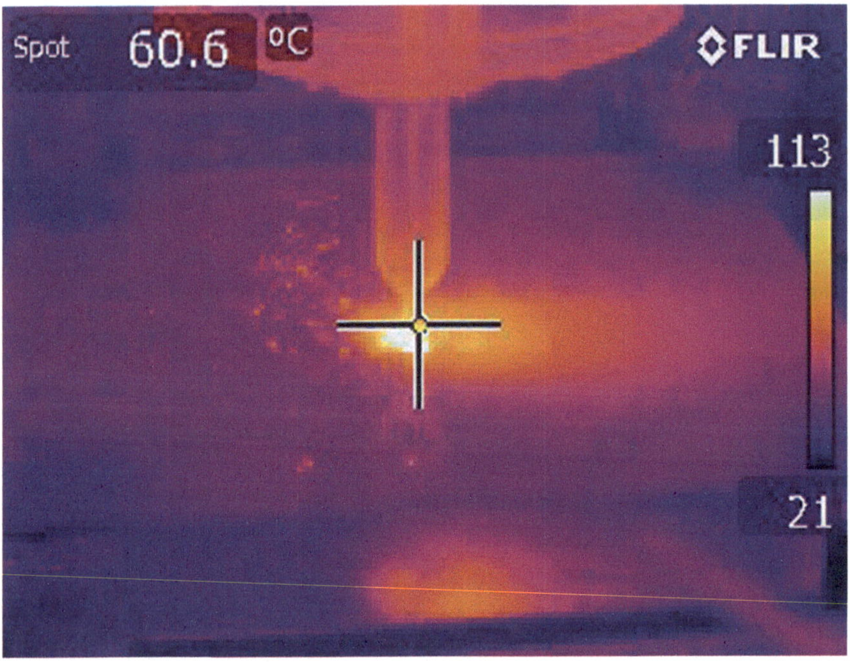

Fig. 27. Thermal image registered during the milling process

When analysing the feed and speed of the process, the surface roughness values increase whit the feed and speed rates. Therefore, all the parameter combinations and tool path strategies studied in this chapter influence the surface quality of the machined part.

Previous research show that the thermo-physical properties of the fibre reinforced polymers causes high temperatures at the tool tip when machined [13]. Another reason that must be considered is the fact that especially thermoplastic polymers have a glass transition temperature ranging from 150–250 °C, meaning that the heat formation during the milling process can lead to plasticizing of the polymer matrix [13]. The machining of fibre reinforced polymer materials presented in this chapter was performed without cooling lubricant. The coolant can imbue the plastic material and can induce chemical reactions with certain functional groups of the macromolecules [13]. Process temperatures have been recognized as an important factor influencing the tool wear rate and tool lifetime [14]. The process temperatures were also measured, for each machining operation, using a thermal camera, Flir ThermaCam E50. An example of a registered thermal image is presented in Fig. 27. During the experiments the temperatures did not exceed 113 °C.

4 Conclusions and Further Research Strategies

The researches presented in this chapter were focused on experiments to determine the process parameters that have influence on the quality of the finished product (form deviations, positional accuracy and roughness parameters).

Correlation charts were plotted using Design Expert v12 software in order to identify the best suited strategy and process parameters, to obtain the best quality of the finished products.

The measured values were linked to the process parameters used for machining and the aim was to determine the optimal process parameters and tool geometries in order to obtain the best quality of the processed parts.

The Zig-Zag strategy. generated, as expected, the lowest results of the measured parameters for all strategies.

iMachining strategy generates the best quality compared to Zig-Zag and Helical strategy, due to the constant machining of the tool.

The process temperatures were also be measured for each machining operation, using infrared thermography. The aim of these research is to further investigate if there is a relation between the process temperatures and the machined surface quality.

Testing and analysis of the proposed solutions and identifying the improved directions can give relevant results that can be used in further researches in this domain.

As for further research strategies based on the measurements presented in this chapter we can conclude:

- Mathematical models will be developed, equations that will correlate the important machining parameters with the values of the obtained roughness parameters and the measured dimensional and geometrical tolerances. These equations will be able to predict the values of the roughness parameters and the dimensional accuracy when machining these types of materials.
- Machining of different types of carbon fiber reinforced materials, having different thickness, containing different amounts of fibers, and different fiber orientation, using tools with different geometries. Afterwards, tool wear, delamination, dimensions, geometrical deviations and surface roughness parameters should be measured.
- Validating the generated equations by experimental researches. The values obtained with the developed equations will be compared with the values obtained by experimental measurements.

References

1. Palanikumar, K.: Modeling and analysis for surface roughness in machining glass fiber reinforced plastics using response surface methodology. Mater. Des. **28**(10), 2611–2618 (2007). www.elsevier.com/locate/matdes
2. Voss, R., et al.: Influence of fiber orientation, tool geometry and process parameters on surface quality in milling of CFRP. CIRP J. Manuf. Sci. Technol. (2016). https://doi.org/10.1016/j.cirpj.2016.10.002/

3. Abdul Nasir, A.A., Azmi, A.I., Khalil, A.N.M.: Measurement and optimisation of residual tensile strength and delamination damage of drilled flax fibre reinforced composites, Measurement **75**, 298–307 (2015). https://www.journals.elsevier.com/measurement
4. Dragomir, M., Popescu, S., Płowucha, W., Marxer, M.: Blended learning in the field of measurement uncertainty. experiences from the MUVOT project. In: Proceedings of the 2nd International Conference Quality and Innovation in Engineering and Management, Cluj-Napoca, Romania, 22–24 November. Published in a special issue of Quality - Access to Success, vol. 13, no. SUPPL. 5, pp. 127-130 (2012). ISSN 1582-2559
5. Uhlmann, E., Sammler, F., Richarz, S., Heitmüller, F., Bilz, M.: New production technologies in aerospace industry. In: 5th Machining Innovations Conference (MIC 2014) Machining of Carbon Fibre Reinforced Plastics, Procedia CIRP, vol. 24, pp. 19–24 (2014)
6. http://www.kfz-betrieb.vogel.de/cfk-raeder-gehen-2016-in-serie-a-505562/. Accessed 21 Dec 2016
7. http://www.ressourcedeutschland.de/fileadmin/user_upload/downloads/kurzanalysen/2014-Kurzanalyse-03-VDI-ZRE-CFK.pdf. Accessed 22 Dec 2016
8. https://www.welt.de/motor/article125550098/Auch-die-edlen-Carbon-Felgen-haben-ihre-Tuecken.html. Accessed 05 Jan 2017
9. http://www.io-journal.de/neu-auf-der-emo-hannover-steigerung-der-wirtschaftlichkeit-durch-cfk-traegerelemente-fuer-den-werkzeugmaschinenbau/. Accessed 05 Jan 2017
10. http://www.ingenieur.de/Themen/Werkstoffe/Serienfertigung-CFK-Alleine-es. Accessed 07 Jan 2017
11. http://www2.coromant.sandvik.com/coromant/eBook/Composite/eng/pdf/C-2940-155.pdf. Accessed 09 Jan 2017
12. http://www.sandvik.coromant.com/sitecollectiondocuments/downloads/global/technical%20guides/en-gb/c-2920-30.pdf. Accessed 09 Jan 2017
13. Wilfried, K., Peter, G.: Temperaturbestimung beim Bohren Faserverstaerkte Kunstoffe Hilfsmittel Thermographie. In: VDI-Z, pp. 132–138 (1990)
14. Wen-Chou, C.: Some experimental investigations in the drilling of carbon fibre-reinforced plastic (CFRP) composite laminates. Int. J. Mach. Tools Manuf **37**(8), 1097–1108 (1997)
15. http://www.cfk-maschinenbau.de/cfk-im-maschinenbau
16. Crişan, A.: Proiectarea cursurilor universitare, de la calitate la inovaţie, Editura Didactică şi Pedagogică, Bucureşti, p. 218 (2016). ISBN 978-606-31-0337-7
17. Crişan, A.: Strategii curriculare în învăţământul universitar, ed. Institutul European, Iaşi, p. 260 (2013). ISBN 978-606-24-0006-4

Industry 4.0 in Croatia – Perspective and Industrial Familiarity with the (New) Digital Concept

Maja Trstenjak(✉), Tihomir Opetuk, Danijel Pavković, and Davor Zorc

Faculty of Mechanical Engineering and Naval Architecture, University of Zagreb, Ivana Lucica 5, 10 000 Zagreb, Croatia
maja.trstenjak@fsb.hr

Abstract. The impact of Industry 4.0, almost ten years after its introduction to industry and science, has encouraged the manufacturers to change their working environment, offer customized products and change the mind-set on the organizational level. In Croatia, so far there are only few examples its partial implementation in the industrial practice. That is why in this paper the overview of the current research about Industry 4.0 in Croatia previously conducted and companies' readiness is given, as well as the results of the new research regarding the familiarity of the local industry with the new concept. The flaws of the current implementation procedure will be recognized and strategic guidance for the future steps will be given. Also, the importance of the local support from the various institutions, like government or universities, will be considered with the definition of their role during the transformational digitization process.

Keywords: Industry 4.0 · Digitization · Croatia · Smart manufacturing

1 Introduction

The idea of the Industry 4.0 concept has been present on the market for almost a decade now. Presented at the Hannover fair in 2011, it has given a new vision and perspective for the industrial and service production. Some argue whether we are in the middle of the 4th industrial revolution or evolution. The digitization of the processes and complete transformation of the working environment as we know it is the simple result of the technology development, supported by the internet, the increase of its speed and capacity. Yet, to keep in track with the competition on the market, to achieve the flexible and modular manufacturing of service systems that result in smart products, customized for a single user, the changes have to occur in relatively short time, to keep in track with the big players on the market, with intensive R&D departments. These kinds of changes also require very high investments, and even the optimal development of the strategic transformational plan, that might be very challenging for SMEs. Also, the support from the local government is very important in the transformational process, especially if the goal is to achieve sustainable Industry 4.0. Financial incentives might be very helpful while targeting at renewable energy sources or other parts of the

L. Wang et al. (Eds.): AMP 2020, LNME, pp. 345–355, 2020.
https://doi.org/10.1007/978-3-030-46212-3_26

physical environment, as well if certain permits are required. The administrative process of the local governance can be an unbearable barrier for the local companies, sometimes even the large multi-national representatives. In countries like Croatia companies often go through such obstacle, which is why the research will be focused on the current state of Industry 4.0 and possibilities of its implementation.

2 Industry 4.0 in Croatia – Previous Research and Findings

The research of Industry 4.0 readiness of Croatian companies was first conducted in 2015, and the result have shown that the level of readiness is 2.15, on the scale 1 to 4, based on the characteristics of 4 industrial revolutions. The research was conducted on the target group of Croatian manufacturing industry, in which every company representative had to evaluate the features of the current state in their working environment in nine dimensions. Dimensions examined are: Product Development, Technology, Production Management, Production monitoring, Materials inventory management, Management of stocks of finished products, Quality assurance, PLM and TPS/GALP. The companies have rated their Technology, Materials inventory management and Management of stocks of finished products the highest. The lowest rate on average got the dimension of Product development, Production monitoring and TPS/GALP.

In the later stages of the research, the individual approach was taking place, aimed at the unique cases in order to recognize their development state by advanced audits, where data was also collected to create the learning factory that would be some sort of competence centre, and the development of the new, specific Industry 4.0 technologies and organizational features, aiming at Croatian industry [1].

Roland Berger Industry 4.0 index has defined the Croatian industry in the group of the “Hesitators”. The dimensions evaluated in this research are: production process sophistication, degree of automation, workplace readiness and innovation intensity (industrial excellence) in combination with high value added, industry openness, innovation network and internet sophistication. The dimensions were rated on the scale 1 to 5, where 5 represents a country that is “excellently prepared” for Industry 4.0. The countries who are also part of the “Hesitators” group were Italy, Spain, Estonia, Portugal, Poland and Bulgaria. The highest level of readiness is shown for the countries in the group “Frontrunners” and that are Germany, Ireland, Sweden and Austria [2].

Croatian Agency for Small Business, Innovation and Investments has conducted a research in 2017 about the readiness of the Croatian entrepreneurs for Industry 4.0. The results were given by the data collected from 26 small and medium entrepreneurs. Most of them have shown to understand the use of smart manufacturing and are familiar with new trends in the industry, with tendency of their implementation. Only 12% have already started the digitization and implementation process. They have acknowledged the high cost of implementation, lack of information and difficulty of new technologies as the greatest challenges, as their goal is better product quality, better relation with customers, decrease of manufacturing cost and better compliance with customer specifications or regulatory requirements [3].

Croatian Chamber of Commerce have advised the organization of the local communities for technological development, education aimed at Industry 4.0 knowledge at local communities and formation of the expert group for single company readiness calculation. Also, they have suggested the definition of Croatian version of Industry 4.0 with the important question of financial help left open [4].

In Croatia, there is an initiative for the functioning Industry 4.0 National platform since 2017. Its goal is to create smart factories, digitize the business, service and manufacturing processes to reduce manufacturing and service cost. With the platform, seven main activities towards digitization in Croatia were identified: Networking and digital connection, Education of human labour for Industry 4.0 working environment, Optimal use of resources, Digitization of public administration, the existence of legal regulation, technology standardization, system safety and data protection [5].

In 2019 the research about the economic aspects of the Industry 4.0 transformation in Croatia was conducted. The focus was on the influence and predictions of key indicators of deindustrialization and Industry 4.0 and its projections to 2025. Those are GDP, trends of industrial production, labour productivity trends, employment trends, foreign direct investments, investments in research and development, etc. Eurostat and World Bank data are used to find the results and therefore aren't connected to the direct transformation of the specific industry or a single company, but give an overview on the local industry as a whole and compare it with the other European countries [6].

3 Research and Methodology

As an upgrade to the previously conducted research, the new research is conducted within the 33 Croatian companies about their familiarity with Industry 4.0 concept, motivation and future steps they intend to take in for the complete digital transformation towards the new concept. The research has been conducted by online questionnaire formed in Google Forms application.

The distribution of the participant's industry type are shown on the Fig. 1.

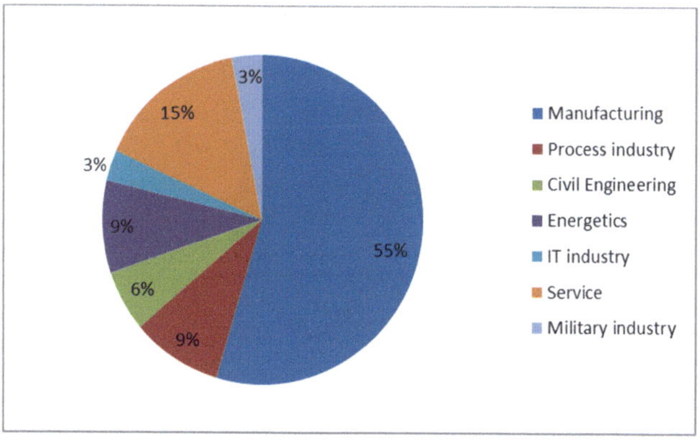

Fig. 1. Percentage of research participants by industry type

Most of the participants (55%) are from the manufacturing industry, while there are also participants from service, process, civil engineering, energetics, IT and Military industry. The results will be discussed by the industry type, company size, yearly revenue and the position of the participant within the company. Distribution of the participants by these criteria is shown below (Figs. 2 and 3).

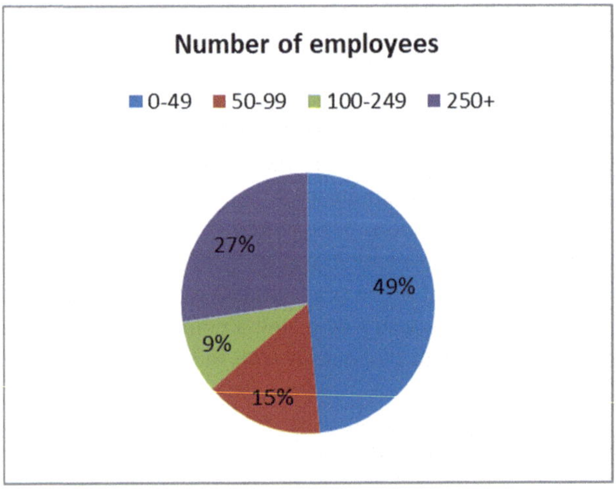

Fig. 2. Percentage of participants by company size

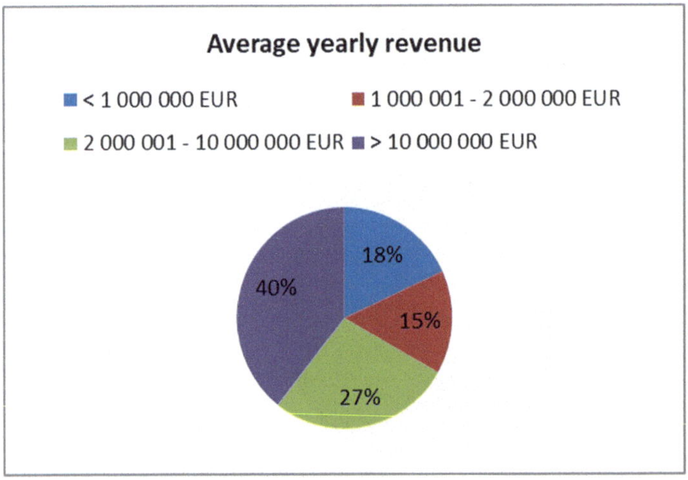

Fig. 3. Percentage of participants by company's average yearly revenue

By the Croatian and EU law, the classification of company size in micro, small, medium and large companies is shown in Table 1.

Table 1. Official classification of company size

Size	Number of employees	Average yearly revenue (EUR)
Micro	<10	<2 000 000
Small	<50	<10 000 000
Medium	<250	<50 000 000
Large	>250	>50 000 000

4 Results and Discussion

The participants have given answers regarding their innovation level, Industry 4.0 dimensions that they find most useful in their company, their opinion on the readiness of the certain dimensions in the value chain and where do they think they should seek support and knowledge about the new concept. Those variables will be discussed in the following chapters.

4.1 Innovation Level

Innovation level is one of the key factors in Industry 4.0 implementation. The general overview of the concept is given in the literature, but each company should create a unique strategic approach in digitization period. Innovation level also creates the motivational atmosphere in the working environment, which enables the successful horizontal and vertical integration, but also the interesting solutions for the digital smart factory. The participants were asked if they thing that their innovation level is average compared to other companies within the same industry type or they are very innovative, ahead the competition (Fig. 4).

Majority of the participants think that their innovativity level is similar to industry average (58%). Interesting fact are the small companies (0–49 employees) in which 63% of the participants consider their innovation level above average. Results like this can be explained by the smaller community and acceptance of the single person's ideas and its easier implementation in practice. Also, the large systems often are slow with the acceptance of the change and possibility of the innovation implementation. That is why those with over 10 000 000 EUR yearly revenue have in general said that their innovation level is similar to industry average (77%) (Fig. 5).

Another interesting result is that both CEO and participants from other managerial positions in the company both think in majority that their innovation level is similar to average. It would be expected for CEOs to give a more opportunistic view on the company's innovation, which is now a sign for the need of the change within the system. Participant from the military industry has acknowledged that their company has high innovation level, as well as the majority from the process industry. Participants from the manufacturing and service industry both in majority think that their company in similar to average, but the ratio is smaller in the manufacturing industry. The Industry 4.0

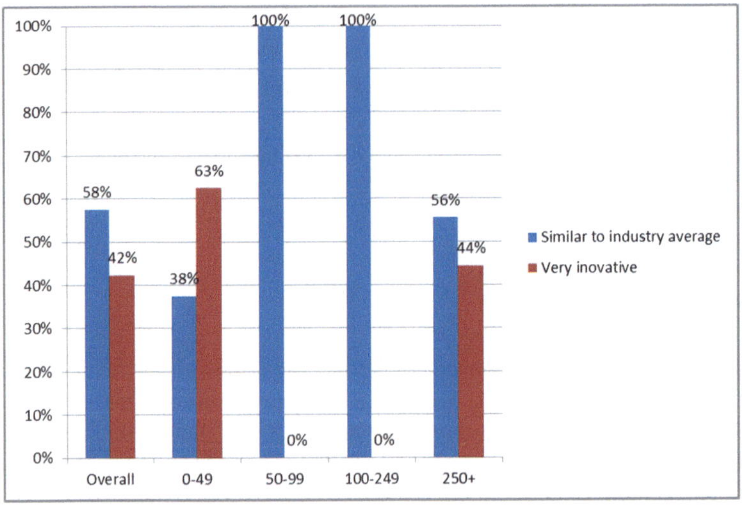

Fig. 4. Innovation level by company size

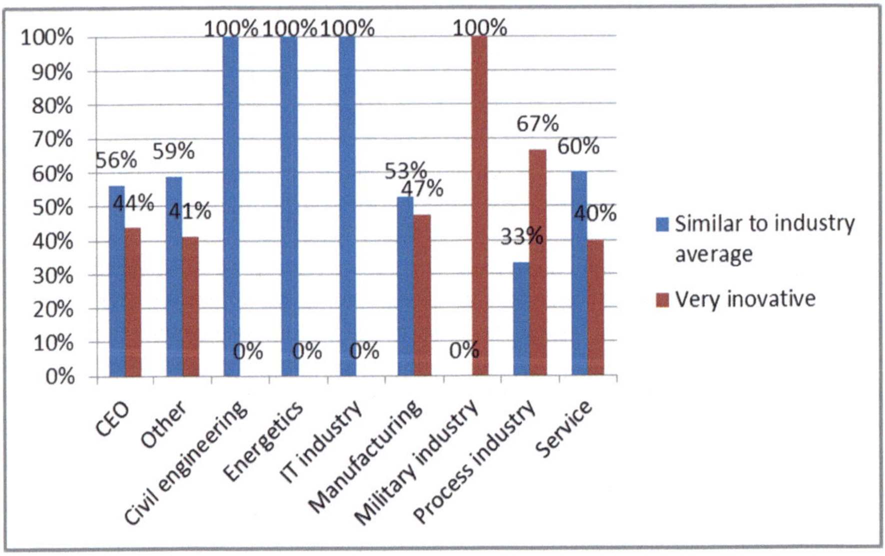

Fig. 5. Innovation level by industry type and participant's profession

concept has mostly been presented as the digitization of the manufacturing, there are much less evidences and possibilities yet available for the transformation of the service industry, which also has to be part of the transformational process.

4.2 Importance and Priority of Industry 4.0 Dimensions

Digitization, Internet of Things, Cloud Computing, Manufacturing process automation, CPS and Big Data manipulation are dimensions of Industry 4.0 for which, in this research, the participants had to choose the priority and its importance of implementation within their company (Figs. 6, 7 and 8).

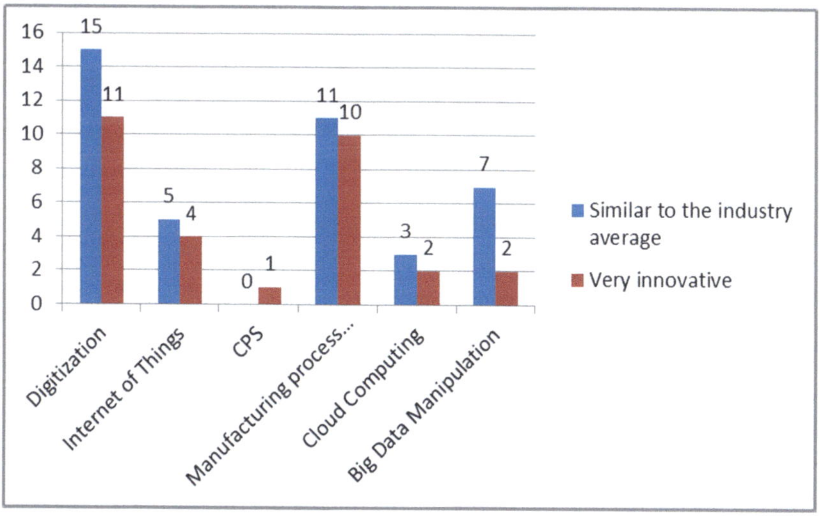

Fig. 6. Industry 4.0 dimensions by innovation level

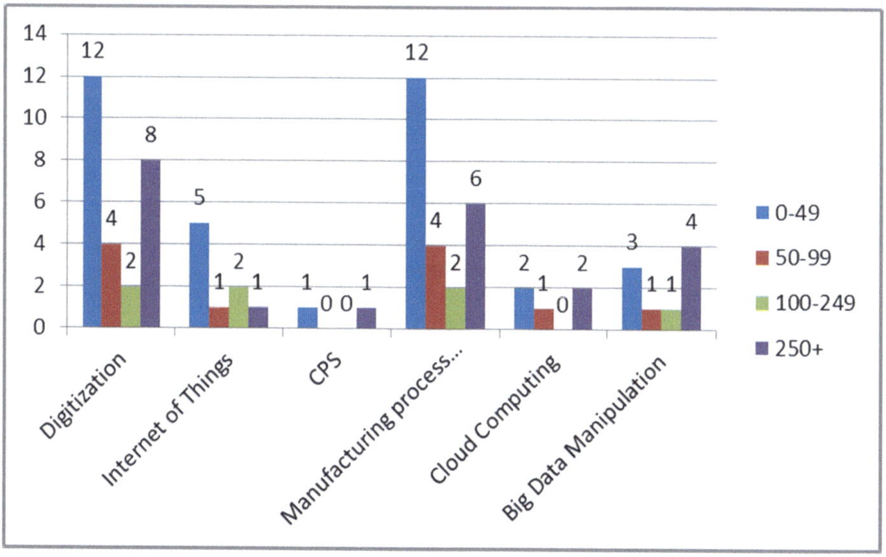

Fig. 7. Industry 4.0 dimensions by company size

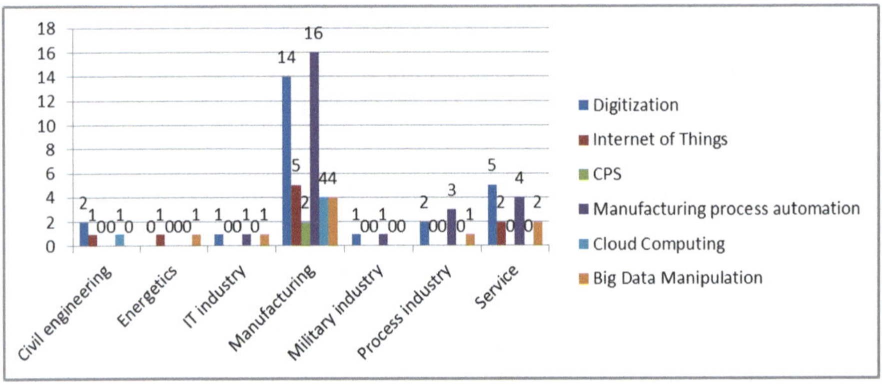

Fig. 8. Industry 4.0 dimensions by industry type

The results have shown that participants have found "Digitization" as most common and most important for the implementation in their working environment. Only two participants from the manufacturing industry have acknowledged "CPS (Cyber Physical Systems)" as important for their transformational process. Also, those who think that have average innovation level have interest in "Big Data Manipulation" implementation, while every industry type, size and innovation level have acknowledge the importance of manufacturing process automation. The participant's profession hasn't shown as an influential factor on the priority of Industry 4.0 dimensions implementation.

4.3 Maturity Level Estimation

The participants were asked to estimate the level of maturity for Industry 4.0. The dimensions "Strategy", "Technology", "Process and activities", "Organization and culture", "Customers and Clients" were evaluated on the scale less than 20%, 20–40%, 40–60%, 60–80% and more than 80% where the higher number means the highest maturity estimation for the Industry 4.0 (Fig. 9).

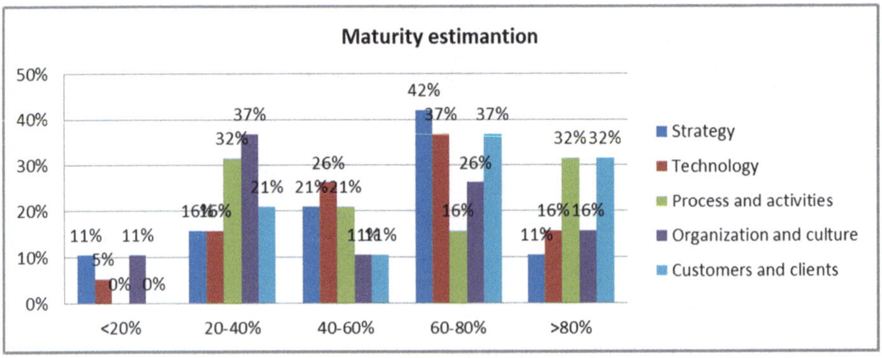

Fig. 9. Maturity estimation of the Industry 4.0 dimensions

Maturity of their company's strategy most of the participants have rated with 60–80% readiness, which can be said that is pretty high level of the maturity. It is also a good sign for the future steps that need to be taken, since the good strategy enables the optimal transition to digital concept. 37% of the participants have rated technology maturity with 60–80% of maturity and 26% with 40–60%. This is also higher than expected, compared to the innovation level results and shows the good potential. Process and activities were mostly rated with 20–40%, but the same number of participants have rated their process and activities with >80% maturity level, which shows the demand for the unique approach to digitization changes. Organization and culture, on te other hand, most participants have rated with 20–40% maturity. This might be in contradiction with the strategy dimension and is the key Industry 4.0 dimension to complete the digitization process. The relationship with customers and clients were also rated as very mature with 37% of answers in favour of 60–80% category and 32% in >80% category.

4.4 Local Support and Education

The participants were asked if they are aware of the National Platform for Industry 4.0 existance, is their company a member of the platform and who do they think should help them and support during the digitization period. The support of Government, Universities and Industry 4.0 users were rated on the scale 1 to 5 where 1 is top priority.

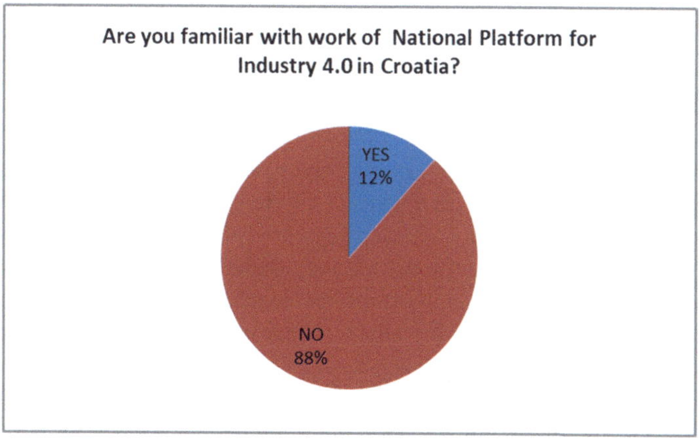

Fig. 10. Familiarity with national platform for Industry 4.0

Only 12% of the participants (Fig. 10) are aware of National Platform existance, and among those, not a single company is a member of it. This raises the awareness of the need for better education in the field of Industry 4.0 and its local findings and approaches.

Participants think that the government has the biggest role in providing education and support for the digital transformation of the local companies, as shown of Fig. 11, it has given an average rate of 2,81. The European practice of learning factory concept, placed in the local universities isn't recognized with the participants, because they claim that the Industry 4.0 users have bigger role in the education and support than the local universities and science institutions.

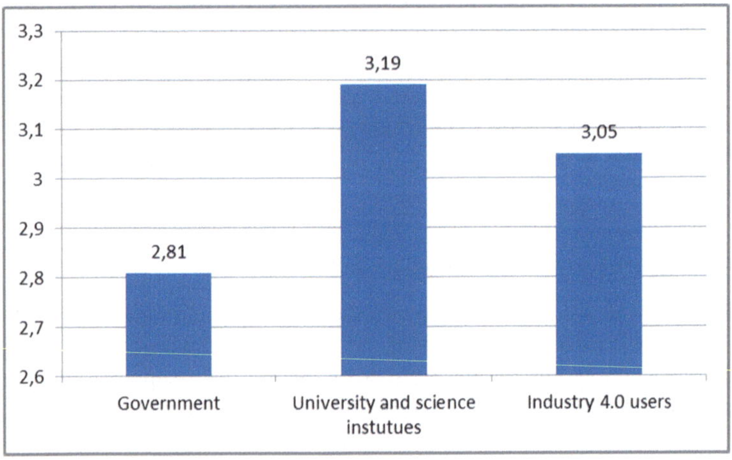

Fig. 11. The key Industry 4.0 supporters

5 Conclusion

As the decade of Industry 4.0 has almost passed, in the countries like Croatia, this concept is yet unknown to many companies. Only few have started to implement parts of the concept, while others are approaching very slowly, only accepting the new technology needed. The complete transformation demands high investments under the era of uncertainty and unclear support of the local government. Compared to the research conducted in past years, the slight change in awareness can be noticed, but none of the radical transformation has yet taken place. The role of Universities and other scientific institutions is also one of the key factors in industry digitization – firstly as the educators of the future human labor, and training enablers for the professions of the future but also as the competence centers and learning factory service providers. Companies have a concern about the need for digitization and automation of the manufacturing processes, but other Industry 4.0 dimensions like Cloud Computing, CPS or Internet of Things, still aren't familiar enough to them. The transformational basis is in high-speed internet infrastructure which still isn't available in many parts of Croatia. The work ethics also is a priority of a change, since the innovation level is not above the average. This includes the mindset of the workers and their functioning by innovative approaches in their everyday environment.

For the future research, the role of the National Platform for Industry 4.0 should be studied, their impact on the local companies and connection to the government who should provide the support in both financial and administrative way to provide optimal transformation of local companies to the complete Industry 4.0 concept implementation.

References

1. Veza, I., Mladineo, M., Peko, I.: Analysis of the current state of Croatian manufacturing industry with regard to Industry 4.0. In: Proceedings of 15th International Scientific Conference on Production Engineering, Vodice, Croatia (2015)
2. Blanchet, M., Rinn, T., Von Thanden, G., Thieulloy, G.: Industry 4.0 - the new industrial revolution - how Europe will succeed. Roland Berger Strategy Consultants GMBH (2014)
3. Hamag Bicro: Industrija 4.0 u Hrvatskoj. https://hamagbicro.hr/hamag-bicro-potice-pametnu-proizvodnju-u-hrvatskoj/.
4. Peric, E.: Industry 4.0, Croatian Chamber of Commerce. https://www.hgk.hr/documents/hgk-industrija-4058d8c59722f1e.pdf. Accessed 10 Mar 2020
5. Stefanic, N.: EU Digitalna agenda 2020 i Nacionalna platforma za digitalizaciju industrije RH. Liderova 4IR konferencija, Zagreb (2017)
6. Tomljanovic, M., Grubisic, Z., Kamenkovic, S.: Deindustrialization and implementation of Industry 4.0 - case of The Republic of Croatia. J. Central Banking Theor. Pract. **8**(3), 133–160 (2019)

Author Index

L. Wang et al. (Eds.): AMP 2020, LNME, pp. 357–358, 2020.
https://doi.org/10.1007/978-3-030-46212-3

The manufacturer's authorised representative in the EU is Springer Nature Customer Service Centre GmbH, Europaplatz 3, 69115 Heidelberg, Germany. If you have any concerns regarding our products, please contact ProductSafety@springernature.com

Printed and bound by CPI Group (UK) Ltd, Croydon, CR0 4YY
05/10/2026
02228124-0014